DYNAMIC OPTIMIZATION

DYNAMIC OPTIMIZATION

The Calculus of Variations and Optimal Control in Economics and Management

SECOND EDITION

Morton I. Kamien
Nancy L. Schwartz

DOVER PUBLICATIONS, INC.
Mineola, New York

Bibliographical Note

This Dover edition, first published in 2012, is an unabridged republication of the work originally published in 1991 by Elsevier Science B. V., Amsterdam.

Library of Congress Cataloging-in-Publication Data

Kamien, Morton, I.

Dynamic optimization : the calculus of variations and optimal control in economics and management / Morton I. Kamien and Nancy L. Schwartz. — 2nd ed.

p. cm.

Originally published: 2nd ed. Amsterdam : Elsevier Science, c1991.

Includes bibliographical references and index.

ISBN-13: 978-0-486-48856-1 (pbk.)

ISBN-10: 0-486-48856-X (pbk.)

1. Mathematical optimization. 2. Control theory. 3. Calculus of variations. I. Schwartz, Nancy Lou. II. Title.

QA402.5.K32 2012
519.6—dc23

2012013820

Manufactured in the United States by LSC Communications
48856X06 2018
www.doverpublications.com

Contents

PART II: OPTIMAL CONTROL

APPENDIX A. CALCULUS AND NONLINEAR PROGRAMMING

APPENDIX B. DIFFERENTIAL EQUATIONS

Preface to the Fourth Printing

I continue to be gratified by the enthusiasm of the current generation of students for this book. It is good to know that it aids their efforts to gain knowledge of the subjects covered.

As an application of the theory that "less is more", I have attempted to eliminate as many typographical errors as I could find in the previous printing. I have been aided in this task by Professor S. Devadoss of the University of Idaho.

Preface to the Second Edition

The enthusiastic reception by students all over the world to the first edition suggests that I follow the advice not to fix what is not broken. However, there have been a number of new theoretical developments and applications of dynamic optimization methods that should prove useful to students of this subject. New developments include advances in how to do comparative dynamics, how to optimally switch from one state equation to another during the planning period, and how to take into account the history of the system governing an optimization problem through the use of an integral state equation. Each one of these new developments has been included either in a previously existing section or in an entirely new section. An example of the application of these new developments is included in the text, or a reference to where it can be found is provided.

The most noticeable addition is a section on differential games. Since the publication of the first edition, game theoretic methods of analysis in general have become pervasive throughout economics and management science. In particular, interesting insights into economic and management science problems have been gained through their modeling as differential games. Optimal control and continuous time dynamic programming methods are the standard means for analyzing these types of games. Thus, the inclusion of this topic seemed warranted.

As far as exposition is concerned, I have sought to stick with the same style and level of presentation that proved so successful in the first edition. Thus, again the focus is on providing the student with the tricks of the trade on an informal intuitive level. This objective has been made easier to maintain because of the appearance of more formal expositions of optimal control methods in the texts by Feichtinger and Hartl, and Seierstad and Sydsaeter, and in differential games by Basar and Olsder, and by Mehlmann. Here and there throughout the text I have attempted to improve the exposition and provide some new insights. With the encouragement and aid of my favorite physicist I have included several classic applications of the calculus of variations to physics.

I want to express my appreciation to Richard Hartl for his many very thoughtful and useful comments on the first edition. Stuart Halliday provided excellent typing and very helpful editorial suggestions. Carolyn Settles urged me on when my enthusiasm for the task waned. Special thanks are also due my wife, Lenore, and my son, Randall, for their patience with me throughout this endeavor.

The most difficult and painful part of preparing this new edition was not having my co-author, Nancy Lou Schwartz, to share the success of the first edition and to help in writing this one. Were it not for her, there would never have been a first edition. Thus, this edition is dedicated to her memory.

Morton I. Kamien

Preface to the First Edition

We have long believed that the optimal action in the short run need not be optimal in the long run. Therefore, we have studied various economics and management science programs from a long run, dynamic perspective. In addition to our experience with dynamic optimization as a research tool, we have taught these methods for nearly ten years at Northwestern University. We have found a myriad of articles using these techniques, more every year, and some advanced mathematics and engineering texts, but a dearth of textbooks explaining the methods systematically at a relatively elementary level. This book has grown out of our research and classroom experience to meet the needs of economics and management students.

We have tried to explain the methods and illustrate their uses very simply; higher level books are available elsewhere. We have used a few running examples and a collection of illustrations culled from or stimulated by the literature. The result is, we believe, a highly readable introduction to the calculus of variations and optimal control and their uses in economics and management.

Our attention is restricted to continuous time problems. This reflects personal preference, our perception of the gap in the literature, and space limitation. Of course, there are many interesting problems that are as well or better handled through a discrete time formulation. We have elected to omit them, nonetheless recognizing their importance and utility.

This text borrows heavily from the literature, as will be apparent to the knowledgeable reader. We have not attempted to document each source, but hereby acknowledge our debt to the texts and papers upon which we have drawn heavily, cited in the references.

The text has been used for classes of first year doctoral students in economics and in management who have had some exposure to microeconomics and to nonlinear programming (the Kuhn-Tucker theorem). It is also appropriate for advanced undergraduates in mathematical social sciences or in economics and mathematics. Finally, it is quite suitable for self-study.

The calculus of variations is discussed in Part I, and optimal control is the subject of Part II. The two approaches are closely related, and the student is frequently asked to note these relationships in exercises throughout Part II. Part II has been written under the assumption that the reader is already familiar with the bulk of Part I.

We assume that the reader is comfortable with the calculus of several variables. Many of the results from calculus that we use are reviewed in Appendix A. This appendix may be used as an introduction to calculus optimization. Although some previous knowledge of differential equations is helpful, it is not needed; Appendix B is a self-contained introduction to the needed facts and techniques in differential equations.

Most sections contain exercises that are an integral part of a study program. Some are routine extensions of the theory developed in the section; these are intended to fix the ideas that have been used as well as to provide modest extensions. Others are routine numerical problems; again, we believe strongly that the working of routine problems is a useful complement to reading and following working examples of the text. Some are not routine. Finally, some exercises have been suggested by applications in the literature and not only provide practice but also indicate further uses of the methods. The table of contents indicates the main applications discussed in the text and exercises of each section.

Some sections contain a few remarks on further reading. Texts and papers where the theory may be found are cited, and some representative papers applying it are mentioned. Full citations and additional references are at the end of the text.

Equations are referred to by numbers in parentheses, using a hierarchical system. A single number (n) is the nth equation of the current section.; (s, n) is the nth equation of Section s in the same part of the book; and (Ps, n) is the nth equation of Section s in Part P or Appendix P of the text. Examples, sections, and exercises are referenced by an identical numbering system, by which Example 1 is the first example in this section, Example 5.1 is the first example of Section 5, and Example I5.1 is the first example of Section 5 in Part I.

Derivatives and partial derivatives are denoted by primes and subscripts, respectively. A prime labels the derivative of the function with respect to its argument. Thus, $f'(x) = df/dx$ and $x'(t) = dx/dt$. A double prime indicates the second derivative: $f''(x) = d^2f/dx^2$ and $x''(g) = d^2x/dt^2$. Partial derivatives are frequently indicated by subscripts; sometimes the name of the argument is given, sometimes its number or order: $f_x(x, y) = f_1(x, y) = \partial f/\partial x$. The argument t is frequently suppressed. Often x is a function of t, sometimes indicated $x(t)$.

We are indebted to ten years of students and many colleagues and reviewers for stimulation and for helping us clarify our thinking and presentation and reduce the errors. We especially note the help of R. Amit, R. Braeutigam, R. deBondt, M. Intriligator, S. Kawasaki, E. Muller, S. Ochiai, Y. Ohta, and H. Sonnenschein. Many typists helped us through the innumerable drafts of the manuscript, including Judy Hambourger, Laura Coker, and especially Ann Crost. We acknowledge gratefully the assistance of the J. L. Kellogg Graduate School of Management, the Center for Advanced Study in Managerial Economics and Decision Sciences, and IBM. Lenore and Randall Kamien exhibited the patience required while this book was being written.

Morton I. Kamien
and
Nancy L. Schwartz

PART I

CALCULUS OF VARIATIONS

Section 1

Introduction

We are taught early in life about the necessity to plan ahead. Present decisions affect future events by making certain opportunities available, by precluding others, and by altering the costs of still others. If present decisions do not affect future opportunities, the planning problem is trivial. One need then only make the best decision for the present.

This book deals with analytic methods for solving planning problems in continuous time, namely the dynamic optimization techniques of the calculus of variations and of optimal control theory. The solution to a continuous time dynamic problem is a continuous function (or a set of functions) indicating the optimal path to be followed by the variables through time (or space).

The origin of the calculus of variations is commonly traced to the posing of the brachistochrone problem by John Bernoulli in 1696 and its solution by him and independently by his brother James in 1697. (If a small object moves under the influence of gravity, which path between two fixed points enables it to make the trip in the shortest time?) Other specific problems were solved and a general mathematical theory was developed by Euler and Lagrange. The most fruitful applications of the calculus of variations have been to theoretical physics, especially in connection with Hamilton's principle or the Principle of Least Action. Early applications to economics appeared in the late 1920s and early 1930s by Roos, Evans, Hotelling, and Ramsey, with further applications published occasionally thereafter.

The modern era began in the early 1960s with a resurgence of interest by mathematicians and groups of economists and management scientists in certain dynamical problems. Optimal control theory, developed in Russia by Pontryagin and his co-workers in the late 1950s and published in English translation in 1962, generalizes the calculus of variations by extending its range of applicability. About the same time, economists were interested in models of optimal economic growth (initiated decades before by Ramsey) while management scientists were studying optimal inventory policies. These researchers were a "ready market" for the new tools of dynamical analysis, and application began quickly. Dynamic programming, developed by Richard Bellman in the 1950s, had also stimulated interest. Of course, these mathematical techniques have been used in many other fields, such as aeronautics and chemical engineering. The tools of calculus of variations and optimal control are now nearly standard in economics and management science. This text is devoted to expositing them and illustrating their range of application in these areas.

In a static problem, one seeks an optimal number or finite set of numbers. For instance, a firm may seek the production level x^* that maximizes the profit $F(x)$ generated by producing and selling x units:

$$\max_{x \geq 0} F(x). \tag{1}$$

The answer to this problem is a number, x^*. If $F(x)$ has a particular functional form, then the number x^* can be determined precisely. Otherwise, x^* is characterized in terms of the function F. If F is continuously differentiable and production is worthwhile, x^* typically satisfies the first order necessary condition $F'(x^*) = 0$.

In another case, several variables may be selected simultaneously:

$$\begin{aligned} &\max \quad F(x_1, \ldots, x_n) \\ &\text{subject to} \quad x_i \geq 0, \quad i = 1, 2, \ldots, n, \end{aligned}$$

where $F(x_1, \ldots, x_n)$ is the profit function,[1] and x_i is the amount of the ith good, $i = 1, \ldots, n$. The solution is a set of n numbers, $x_1^*, \ldots, x_n^*$, representing the amounts of each good to produce and sell for maximum profit.

A multiperiod discrete time generalization of (1) involves choosing the amount x_t of the good to be produced and sold in each period t:

$$\begin{aligned} &\max \quad \sum_{t=1}^{T} F(t, x_t) \\ &\text{subject to} \quad x_t \geq 0, \quad t = 1, \ldots, T \end{aligned} \tag{2}$$

The optimal solution is a set of T numbers, $x_1^*, \ldots, x_T^*$. Since the output in any period affects the profit in that period only, problem (2) is reduced to a sequence of static problems, namely, to choose a production level in each period to maximize current profit. Thus the T first order conditions satisfied by the T variables separate into T separate conditions, each in a single variable. The extension to n goods should be apparent.

The problem becomes truly dynamic if the current production level affects not only present profit but also profit in a later period. For example, current profit

[1] This is a different function than in (1). We use the name F for profit in several places, but the particular form of the function may be different in each case.

may depend on both current and past output due to costs of changing the production rate (e.g., costs of hiring or firing):

$$\begin{aligned} \max \quad & \sum_{t=1}^{T} F(t, x_t, x_{t-1}) \\ \text{subject to} \quad & x_i \geq 0, \quad t = 1, \ldots, T, \end{aligned} \tag{3}$$

with a production level x_0 at the moment of planning, $t = 0$. Note x_0 must be specified since it affects the profit available in period 1. The T first order conditions satisfied by an optimal solution do not separate; they must be solved simultaneously.

The continuous time analog to (2) is

$$\begin{aligned} \max \quad & \int_0^T F(t, x(t))\, dt \\ \text{subject to} \quad & x(t) \geq 0. \end{aligned} \tag{4}$$

The solution will be a function $x^*(t)$, $0 \leq t \leq T$, that gives the firm's optimal output rate at each point in time over its planning period. As in (2), this is not really a dynamic problem since the output at any t affects only current profit. The optimal solution is to choose $x(t)$ to maximize profit $F(t, x(t))$ for each t.

The continuous time analog of (3) is less immediate. Since time is continuous, the meaning of "previous period" is not clear. It relates to the idea that current profit depends on both current output and the rate at which output changes with respect to time. The rate output changes over time is $x'(t)$. Thus, the problem may be written

$$\begin{aligned} \max \quad & \int_0^T F(t, x(t), x'(t))\, dt \\ \text{subject to} \quad & x(t) \geq 0, \quad x(0) = x_0. \end{aligned}$$

Necessary conditions for solution will be deferred until a few more examples of continuous dynamic optimization problems have been given.

Example 1. A firm has received an order for B units of product to be delivered by time T. It seeks a production schedule for filling this order at the specified delivery date at minimum cost, bearing in mind that unit production cost rises linearly with the production rate and that the unit cost of holding inventory per unit time is constant. Let $x(t)$ denote the inventory accumulated by time t. Then we have $x(0) = 0$ and must achieve $x(T) = B$. The inventory level, at any moment, is the cumulated past production; the rate of change of inventory is the production rate $dx/dt = x'(t)$. Thus the firm's total cost at any moment t is

$$[c_1 x'(t)]x'(t) + c_2 x(t) = c_1[x'(t)]^2 + c_2 x(t),$$

where the first term is the total production cost, the product of the unit cost of production and the level of production; the second term is the total cost of holding inventory; and c_1 and c_2 are positive constants. The firm's objective is to determine a production rate $x'(t)$ and inventory accumulation $x(t)$ for $0 \le t \le T$ to

$$\min \int_0^T \left[c_1(x'(t))^2 + c_2 x(t)\right] dt$$

$$\text{subject to} \quad x(0) = 0, \quad x(T) = B, \quad x'(t) \ge 0.$$

One possible production plan is to produce at a uniform rate, namely $x'(t) = B/T$. Then $x(t) = \int_0^t (B/T)\, ds = Bt/T$ and the total cost incurred will be

$$\int_0^T \left[c_1(B/T)^2 + c_2 Bt/T\right] dt = c_1 B^2/T + c_2 BT/2.$$

While this is a feasible plan, it may not minimize cost.

Example 2. From a stock of capital equal at time t to $K(t)$ an output can be produced at rate $F(K)$. The production function F is assumed to be twice continuously differentiable, increasing, and concave. This output can be consumed, yielding immediate satisfaction, or it can be invested to augment the capital stock and hence future productive capacity. Output $F(K)$ is therefore the sum of consumption $C(t)$ and investment $K' = dK/dt$ (the change in capital stock). The problem is to maximize utility from consumption over some specified period by choosing the portion of output to be invested at each moment t. This amounts to selecting the rate of growth K' of the capital stock to

$$\max \int_0^T U(C(t))\, dt, \quad \text{or} \quad \max \int_0^T U[F(K(t)) - K'(t)]\, dt$$

$$\text{subject to} \quad K(0) = K_0, \quad K(T) \ge 0,$$

where the utility function U of consumption is twice continuously differentiable, increasing, and concave.

If capital is not perfectly durable, but decays at a constant proportionate rate b, then reinvestment at rate $bK(t)$ is required to keep its stock intact (see the appendix to this section) and therefore $F(K) = C + K' + bK$. Hence the amount available for consumption is output less both the amount devoted to replenishing the capital stock and the net change in capital. If, in addition, future satisfactions are to be discounted at rate r (see the appendix to this section), the problem is

$$\max \int_0^T e^{-rt} U[F(K(t)) - K'(t) - bK(t)]\, dt$$

$$\text{subject to} \quad K(0) = K_0, \quad K(T) \ge 0,$$

Example 3. Let $P(K)$ be the profit rate, exclusive of capital costs, that can be earned with a stock of productive capital K. If the firm has little control over price, then $P(K) = pF(K)$, where p is the market price and $F(K)$ is the output that capital stock K yields. The functions P or F are typically assumed to be twice differentiable, increasing (at least for small and moderate values of their arguments), and concave. The capital stock decays at a constant proportionate rate b, so that $K' = I - bK$, where I is gross investment, that is, gross additions of capital. The cost of gross additions to capital at rate I is $C(I)$, where C is an increasing convex function. If investment goods have a constant price c, then $C(I) = cI$.

We seek to maximize the present value of the net profit stream over the fixed planning period $0 \leq t \leq T$:

$$\max \int_0^T e^{-rt}[P(K) - C(K' + bK)]\, dt$$

$$\text{subject to} \qquad K(0) = K_0, \quad K(T) \geq 0,$$

where $K = K(T)$ and $K' = K'(t)$. Recall that $I = K' + bK$.

This example can be reinterpreted as a problem in which K is "human capital," $P(K)$ is the earnings of an individual with human capital K, $C(I)$ is the cost of education or training, and b is the rate of forgetting. Alternatively, K may be a firm's stock of goodwill; $P(K)$ is then the firm's maximum earnings rate if its goodwill is K, and $C(I)$ is expenditure on advertising and promotion to enhance goodwill. In still another context, K is the stock of a durable good that is leased to others, $P(K)$ is the rentals collected, and $C(I)$ is the cost of producing or acquiring additional units. Finally, K may be the state of healthiness or health capital, $P(K)$ the earnings or well-being associated with health K, and $C(I)$ the expenditure on preventative health care.

The range of applications is far wider than suggested by these three examples. Also, the same mathematical techniques apply to problems over space as well as time and have been used, for example, in the optimal design of a city. Some classic geometry problems are noted next.

Example 4. Find the shortest distance in the plane between the points, (a, A) and (b, B). To state the problem more formally, recall that the square of the length of the hypotenuse of a right triangle is equal to the sum of the squares of lengths of the other two sides. Thus a small distance ds can be expressed as $ds = [(dt)^2 + (dx)^2]^{1/2} = [1 + x'(t)^2)]^{1/2}\, dt$ (see Figure 1.1). Hence the length of the path to be minimized is

$$\int_a^b \left[1 + x'(t)^2)\right]^{1/2} dt \qquad \text{subject to} \quad x(a) = A, \quad x(b) = B.$$

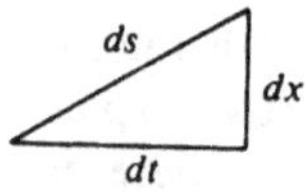

Figure 1.1

Example 5. Find the maximum area in the plane that can be enclosed by a curve of length L and a straight line. This is a generalization of a familiar algebra exercise for finding the rectangle of perimeter L that encloses the largest area. The algebra problem involves selection of two parameters only (length and width) since a rectangle is required. In the present problem, the shape of the curve is to be found as well.

Specifically, we seek the maximum area that can be enclosed by a string of length L and a straight line, with the string emanating from the point (0, 0) and terminating at $(T, 0)$ on the straight line (Figure 1.2):

$$\max \quad \int_0^T x(t)\, dt$$

$$\text{subject to} \quad \int_0^T \left[1 + x'(t)^2\right]^{1/2} dt = L, \quad x(0) = 0, \quad x(T) = 0.$$

Example 6. Find the path $y(x)$ in the x, y plane such that a particle of mass, m, propelled by the force of gravity travels from an initial point (x_0, y_0) to a final point (x_1, y_1), in the shortest time. The particle may be thought of as a bead and the path as the configuration of the wire along which it slides. The total time of the particle's journey is:

$$T = \int dt.$$

Now, $dt = ds/(ds/dt) = ds/v$, where v is the particles velocity ds/dt, and ds is a short distance along its path. But $(ds)^2 = (dx)^2 + (dy)^2$ so $ds = (1 + y'^2)^{1/2}\, dx$ where $y' = dy/dx$. Also, it is supposed that the particle neither gains nor loses energy along its journey. This means that its kinetic energy, $mv^2/2 = mgy$, its potential energy everywhere along the path, where mg is the particle's weight and g refers to the acceleration of gravity. (We suppose the particle's initial velocity is zero.) It then follows that $v = (2gy)^{1/2}$. Upon

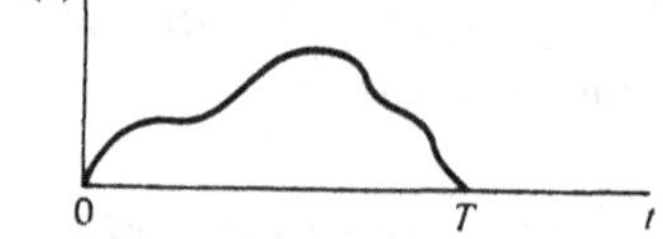

Figure 1.2

making all the right substitutions, the problem becomes

$$\min \int_{x_0}^{x_1} [(1 + y'^2)/y]^{1/2}\, dx/(2g)^{1/2}.$$

This is the brachistochrone problem.

As noted above, the solution of a dynamic optimization problem is a function, through time (or space). In the course of the discussion, analogy will be made, wherever possible, to concepts familiar from static optimization, where calculus can be employed. Three methods of dynamic optimization will be discussed—first the calculus of variations, second, optimal control, and third, dynamic programming. The calculus of variations is analogous to classical calculus in its arena of applicability. That is, it can most easily be employed when all the functions describing the problem are smooth and the optimum is strictly inside the feasible region. Optimal control is applicable in these instances but can also accommodate boundary solutions. It may be viewed as an analog of Kuhn-Tucker theory; see Section A6. Dynamic programming is a generalization of both methods. It is especially useful in dealing with problems involving uncertainty and in differential games.

The three questions of the existence of an optimum, the necessary conditions for optimality, and the sufficiency of the necessary conditions for optimality, arising in calculus (see Section A4) have their counterparts in dynamic optimization. Existence is the most difficult question to answer and will not be treated here. Some optimization problems appear reasonable yet have no optimal solution. For example, even though a curve of shortest length connecting any two given points in a plane exists, there is no connecting curve of greatest length. There is no continuous function $x(t)$ connecting $(t, x) = (0, 1)$ and $(1, 0)$ with minimum area under the graph of the function. The problem (see Figure 1.3)

$$\min \int_0^1 x(t)\, dt$$

$$\text{subject to} \quad x(0) = 1, \quad x(1) = 0,$$

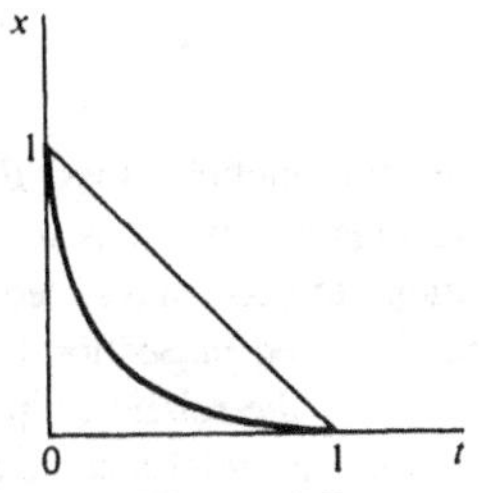

Figure 1.3

has no solution in the class of continuous functions. (The area approaches zero as the path approaches the L-shaped broken straight line joining (0, 1), (0, 0) and (1, 0). But that broken line is not the graph of a function; a unique value of x is not associated with each t.)

Necessary conditions for optimality will be emphasized with some attention to sufficiency. Examples from economics and management science are provided throughout.

APPENDIX TO SECTION 1

Discounting

If A dollars were invested at $100r$ percent per year, the amount would grow to $A + rA = (1 + r)A$ after 1 year, to $(1 + r)A + r(1 + r)A = (1 + r)^2A$ after 2 years, and to $(1 + r)^tA$ after t years. If the interest were compounded, not annually, but twice a year, then the interest rate per 6 month period would be $r/2$ and the initial amount A would grow to $(1 + r/2)^2A$ after 1 year and to $(1 + r/2)^{2t}A$ after t years. More generally, if interest were compounded m times per year, then the rate per period is r/m. The amount A would grow to $(1 + r/m)^mA$ after 1 year and to $(1 + r/m)^{mt}A$ after t years. Continuous compounding amounts to letting $m \to \infty$. Since

$$\lim_{m \to \infty}(1 + r/m)^{mt} = e^{rt},$$

it follows that A dollars invested at annual rate r, continuously compounded, grow to Ae^{rt} dollars in t years.

What amount X now would grow to B dollars in t years, if interest is compounded continuously at rate r? The unknown sum X will be worth Xe^{rt} in t years, so $Xe^{rt} = B$. Thus, $X = e^{-rt}B$. That is, the *present value of B dollars available* t *years in the future, if the interest rate or discount rate is r*, is $e^{-rt}B$.

Decay

If a stock K decays at a constant proportionate rate $b > 0$ and if it is not replenished, then $K'(t)/K(t) = -b$; that is $K'(t) = -bK(t)$. Since the solution to this differential equation is $K(t) = K(0)e^{-bt}$, we sometimes say that the stock K decays *exponentially at rate b*.

FURTHER READING

References on the calculus of variation include Bliss; Elsgolc, Gelfand and Fomin; and D. R. Smith. A history of the calculus of variations is provided by Goldstine. For early applications of calculus of variations to economics, see Evans (1924, 1930), Hotelling, and Ramsey. Many of the references cited in Section II1 on optimal control discuss the calculus of variations as well and show the relationships between the two approaches to dynamic optimization problems. Certain references cited in Section II20 on dynamic programming likewise show relationships among the techniques. Modern applications of the calculus of variations are now too numerous to list; a brief selected bibliography appears at the end of this book.

Solution to discrete time problems like (1)–(3) is discussed in Sections A4–A6, where some references are also mentioned. Concavity and convexity of functions are discussed in Section A3.

Example 1 will be pursued throughout this book. Example 2 is known as the neoclassical growth model; see also Section 10. Generalizations of this basic growth model comprise, perhaps, the largest set of papers on any single application area of these techniques in economics. Takayama; Intriligator (1971); Hadley and Kemp; Tu, Miller, Seierstad, and Sydsaeter (1987); Sethi and Thompson; and Feichtinger and Hartl all review some of this literature. Example 3 and variants of it will be discussed in a variety of contexts.

Section 2

Example Solved

Suppose in Example 1.1 that the cost of holding inventory is zero: $c_2 = 0$. Although now $x(t)$ does not enter into the integrand, the problem is dynamic since total production over the time span of length T must be B. Since c_1 is a positive constant, an equivalent problem is obtained by setting $c_1 = 1$.

$$\min \int_0^T [x'(t)]^2\, dt$$
$$\text{subject to} \qquad x(0) = 0, \quad x(T) = B, \quad x'(t) \geq 0. \tag{1}$$

The solution to a discrete time approximation to this problem will suggest the form of the solution to (1), whose optimality will then be verified.

Divide the interval $[0, T]$ into T/k segments of equal length k. The function $x(t)$ can be approximated by the polygonal line with vertices y at the end points of each segment: $(0, 0), (k, y_1), (2k, y_2), \ldots, (T, B)$. The decision variables are the inventory levels $y_1, y_2, \ldots, y_{T/k-1}$ (Figure 2.1).

The rate of change $x'(t)$ is approximated by $\Delta x/\Delta t = (y_i - y_{i-1})/k$. Thus the approximating problem is to find y_i, $i = 1, \ldots, (T/k) - 1$ so as to

$$\min \sum_{i=1}^{T/k} [(y_i - y_{i-1})/k]^2 k, \qquad \text{where} \quad y_0 = 0 \quad \text{and} \quad y_{T/k} = B, \tag{2}$$

recalling that dt is approximated by $\Delta t = k$. Set the partial derivative of (2) with respect to each y_i equal to zero

$$(y_i - y_{i-1})/k - (y_{i+1} - y_i)/k = 0, \qquad i = 1, \ldots, T/k - 1.$$

Therefore

$$y_i - y_{i-1} = y_{i+1} - y_i, \qquad i = 1, \ldots, T/k - 1.$$

Thus successive differences are equal. The change in inventory, or the production rate, is the same during each time interval of length k. Inventory grows linearly with the y_i lying on a straight line from $(0, 0)$ to (T, B).

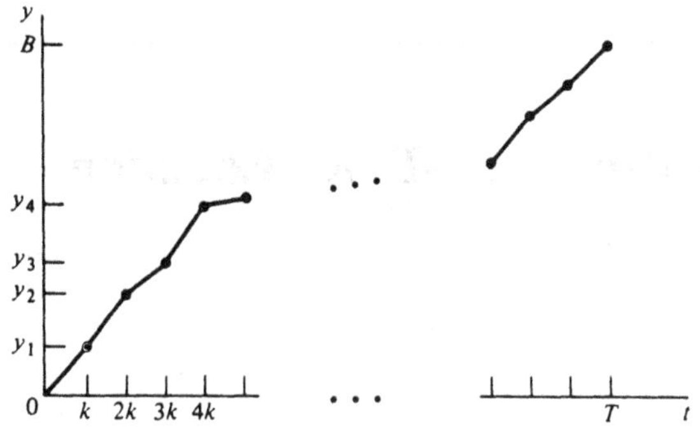

Figure 2.1

The continuous time optimization problem (1) is obtained as $k \to 0$ in (2). The preceding calculation suggests that the solution to the continuous time problem may also involve production at a constant rate with inventory building linearly: $x(t) = tB/T$. Since $x'(t) = B/T \geq 0$, this path is feasible. This conjecture will be verified by showing that no feasible path for inventory accumulation can give lower cost.

All feasible paths obey the same initial and terminal conditions. Let $z(t)$ be some continuously differentiable comparison path satisfying $z(0) = 0$, $z(T) = B$. Define $h(t) = z(t) - x(t)$, the deviation between the comparison path $z(t)$ and the candidate path $x(t) = tB/T$ at time t. Then $h(0) = 0$ and $h(t) = 0$, since the paths x and z coincide at the initial and terminal time. Since $z(t) = tB/T + h(t)$, it follows that $z'(t) = B/T + h'(t)$ and the difference in cost between using plans z and x is

$$\int_0^T \{[z'(t)]^2 - [x'(t)]^2\}\, dt = \int_0^T \{[B/T + h'(t)]^2 - (B/T)^2\}\, dt$$

$$= 2(B/T)\int_0^T h'(t)\, dt + \int_0^T [h'(t)]^2\, dt$$

$$= 2(B/T)[h(T) - h(0)] + \int_0^T [h'(t)]^2\, dt$$

$$= \int_0^T [h'(t)]^2\, dt \geq 0$$

since $h(T) = h(0) = 0$. Thus, the cost of using any feasible production plan (comparison path) can be no less than the cost of following the candidate plan (path) $x(t) = tB/T$, $0 \leq t \leq T$. Hence, the candidate path is optimal.

Typically problems will not be solved in this manner, although both discrete time approximation and verification of optimality are useful. However, the notion of comparison path is important since it plays a central role in developing the method of solution.

Section 3

Simplest Problem—Euler Equation

We seek properties of the solution to the problem

$$\max_{x(t)} \int_{t_0}^{t_1} F(t, x(t), x'(t))\, dt \tag{1}$$

$$\text{subject to} \qquad x(t_0) = x_0, \quad x(t_1) = x_1. \tag{2}$$

The function F is assumed to be continuous in its three arguments t, x, x′ and to have continuous partial derivatives with respect to the second and third, x and x'. Note that while the third argument of F is the time derivative of the second, F is to be viewed as a function of three *independent* arguments. Thus if $F(a, b, c) = a^2 + bc - c^2$, then $F(t, x, x') = t^2 + xx' - (x')^2$. The *admissible* class of functions $x(t)$, among which the maximum is sought, consists of all continuously differentiable functions defined on the interval $[t_0, t_1]$ satisfying the fixed endpoint conditions (2).

Suppose that the function $x^*(t)$, $t_0 \le t \le t_1$ provides the maximum to (1). Let $x(t)$ be some other admissible function. Define the function $h(t)$ to be the deviation between the optimal path $x^*(t)$ and the comparison path $x(t)$ at each t:

$$h(t) = x(t) - x^*(t).$$

Since both x^* and x must obey (2), we have

$$h(t_0) = 0, \quad h(t_1) = 0. \tag{3}$$

We say the deviation h is *admissible* if the function $x = x^* + h$ is admissible.

For any constant a, the function $y(t) = x^*(t) + ah(t)$ will also be admissible since it is continuously differentiable and obeys (2) (since x^* does and h is

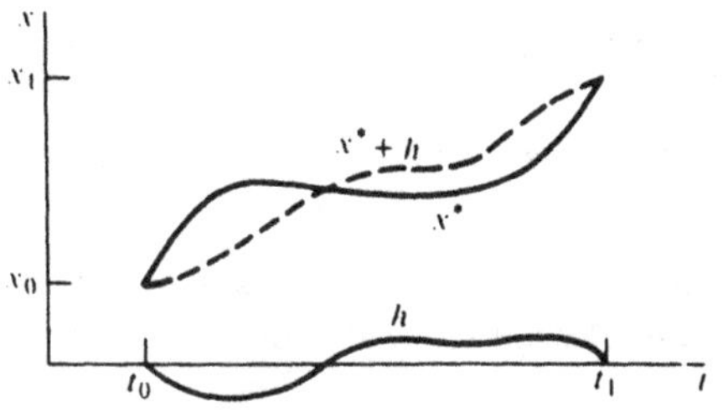

Figure 3.1

zero at the endpoints). With the functions x^* and h both held *fixed*, compute the value of the integral in (1) for $y(t)$ as a function of the parameter a (see Figure 3.1). The result is a function of a, say $g(a)$:

$$g(a) = \int_{t_0}^{t_1} F(t, y(t), y'(t))\, dt$$

$$= \int_{t_0}^{t_1} F(t, x^*(t) + ah(t), x^{*\prime}(t) + ah'(t))\, dt. \qquad (4)$$

Since x^* maximizes (1), the function g must assume its maximum at $a = 0$. But this implies that $g'(0) = 0$ by the first order necessary condition for a maximum of a function of a single variable; see (A4.2). To compute $g'(a)$, first apply the chain rule (A1.5) to the integrand of (4);

$$dF(t, x^*(t) + ah(t), x^{*\prime} + ah'(t))/da = F_x h(t) + F_{x'} h'(t),$$

where F_x and $F_{x'}$ denote partial derivatives of F with respect to its second and third arguments, respectively, and are evaluated at $(t, x^*(t) + ah(t), x^{*\prime}(t) + ah'(t))$. Second, Leibnitz's rule for differentiating under an integral (AI.10) is applied to compute $g'(a)$, and the result is evaluated at the point that maximizes $g(a)$, namely, $a = 0$, yielding

$$g'(0) = \int_{t_0}^{t_1} [F_x(t, x^*(t), x^{*\prime}(t))h(t) + F_{x'}(t, x^*(t), x^{*\prime}(t))h'(t)]\, dt$$

$$= 0 \qquad (5)$$

The condition that $g'(0)$ be zero is necessary since x^* is assumed optimal. Recall that the function h, held fixed to this point, was chosen arbitrarily, restricted only to being continuously differentiable and satisfying endpoint conditions (2). The right side of (5) must be zero for any choice of h satisfying these two restrictions.

Expression (5) can be put into a more convenient form by integrating the second term by parts (A1.8). In $\int u\, dv = uv - \int v\, du$, we let $F_{x'}$ play the role of u and $h'(t)\,dt$ play dv, obtaining

$$\int_{t_0}^{t_1} F_{x'} h'\,dt = F_{x'} h \,|_{t_0}^{t_1} - \int_{t_0}^{t_1} h(t)(dF_{x'}/dt)\,dt. \tag{6}$$

(We have assumed that $dF_{x'}/dt$ exists. This assumption is examined in the appendix to this section.) Recall (3) and substitute it into (5):

$$\int_{t_0}^{t_1} [F_x(t, x^*(t), x^{*\prime}(t)) - dF_{x'}(t, x^*(t), x^{*\prime}(t))/dt] h(t)\,dt = 0. \tag{7}$$

Equation (7) must hold if x^* maximizes (1), and it must hold for every continuously differentiable function h that is zero at the endpoints. It will certainly hold if the coefficient of $h(t)$ is zero for every t, i.e.

$$F_x(t, x^*(t), x^{*\prime}(t)) = dF_{x'}(t, x^*(t), x^{*\prime}(t))/dt, \qquad t_0 \le t \le t_1. \tag{8}$$

Equation (8) is called the *Euler equation* corresponding to problem (1)—(2). It may be viewed as a generalization of the standard calculus first order necessary conditions $f'(x^*) = 0$ for a number x^* to maximize the function $f(x)$. Indeed if $dF_{x'}/dt = 0$, then (8) reduces to that standard calculus condition.

In fact, (7) holds for the entire class of functions h only if (8) holds:

Lemma 1. Suppose that $g(t)$ is a given, continuous function defined on $[t_0, t_1]$. If

$$\int_{t_0}^{t_1} g(t)h(t)\,dt = 0 \tag{9}$$

for every continuous function $h(t)$ defined on $[t_0, t_1]$ and satisfying (3), then $g(t) = 0$ for $t_0 \le t \le t_1$.

PROOF. Suppose the conclusion is not true, so $g(t)$ is nonzero, say positive, for some t. Then, since g is continuous, $g(t) > 0$ on some interval $[a, b]$ in $[t_0, t_1]$. We construct a particular $h(t)$ satisfying the conditions of Lemma 1, namely (Figure 3.2),

$$h(t) = \begin{cases} (t-a)(b-t), & a \le t \le b, \\ 0, & \text{elsewhere}. \end{cases}$$

Compute

$$\int_{t_0}^{t_1} g(t)h(t)\,dt = \int_a^b g(t)(t-a)(b-t)\,dt > 0,$$

since the integrand is positive. This contradicts the hypothesis that (9) holds for every function h in the specified class. A similar argument shows that

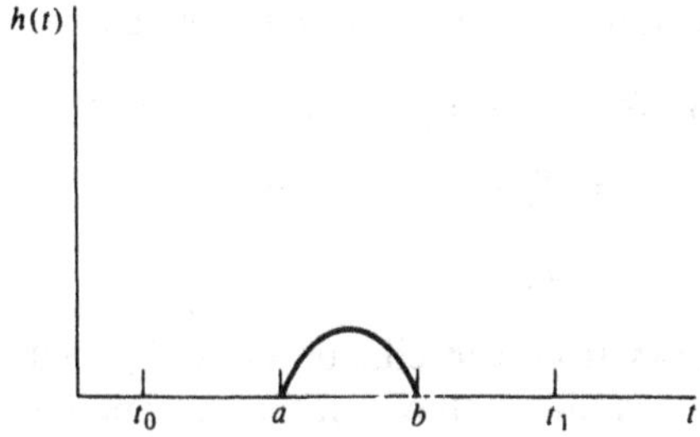

Figure 3.2

"$g(t) < 0$ for some t" leads to a contradiction. Supposition that the conclusion of the lemma was false while the hypotheses were true led to an inconsistency, verifying the lemma. □

Applying Lemma 1 to (7) indicates that (8) must hold. Equation (8), the Euler equation, is a fundamental necessary condition for optimality of the function $x^*(t)$ in the problem (1)–(2).

It is important to note the Euler equation must hold for each t in the interval $[t_0, t_1]$. Equally important is the fact that $dF_{x'}/dt$ is the *total* derivative with respect to t. In interpreting this notation, it must be remembered that the partial derivative $F_{x'}(t, x, x')$ is itself a function of three variables. This function is to be differentiated with respect to t. The total rate of change of the value of the function $F_{x'}$ with advance in t is due, not only to change in t itself, but also the concomitant changes in x and x'. Apply the chain rule to compute

$$dF_{x'}/dt = F_{x't} + F_{x'x}x' + F_{x'x'}x'',$$

noting that each of F's arguments depends on t. Subscripts indicate partial differentiation. Thus the Euler equation can be written

$$F_x = F_{x't} + F_{x'x}x' + F_{x'x'}x'', \qquad t_0 \le t \le t_1, \tag{10}$$

where the partial derivatives are all evaluated at $(t, x^*(t), x^{*\prime}(t))$ and $x' = x'(t)$, $x'' = x''(t)$. The Euler equation is a second order differential equation for $x(t)$, to be solved with the two boundary conditions (2). Another useful form of the Euler equation is

$$F_{x'}(t, x^*(t), x^{*\prime}(t)) = \int_{t_0}^{t} F_x(s, x^*(s), x^{*\prime}(s))\, ds + c, \tag{11}$$

where c is a constant. This form is called the ***duBois-Reymond equation*** and applies as well when the admissible functions $x(t)$ are allowed to have corners. If F_x is continuous, differentiation of (11) with respect to t yields (8).

Still another useful form of the Euler equation can be found by evaluating

$$\begin{aligned} d(F - x'F_{x'})/dt &= F_t + F_x x' + F_{x'} x'' - x'' F_{x'} - x' dF_{x'}/dt \\ &= F_t + x'(F_x - dF_x/dt) \\ &= F_t \end{aligned} \tag{12}$$

The second line follows from cancellation of $x'' F_{x'}$ and collection of terms involving x'. The third line follows from recognition that the bracketed expression is just the Euler equation (8) and must be satisfied by any function $x(t)$ that solves (1). This form, (12), of the Euler equation is especially useful when the function F does not depend on t explicitly, i.e., $F = F(x(t), x'(t))$.

In general, the coefficients in (10) (the partial derivatives of F) are not constant and the differential equation is quite difficult to solve. Indeed, there may be no closed form analytic solution to this differential equation for a given specification of F. In such cases, however, it may be possible to gain qualitative insight into the behavior of an optimal function x^* even without finding an analytic expression for it. If the problem is stated with implicit functions only (as Examples 1.2 and 1.3), then usually one seeks qualitative characterization of the solution.

Dealing with the Euler equation in its form as a second order differential (10) can be avoided by replacing it with two first order differential equations. This can be done by letting

$$p(t) = F_{x'}(t, x, x'). \tag{13}$$

Then, if $F_{x'x'} \neq 0$, x' can be expressed as a function of t, x, and p. Now a new function, called the Hamiltonian, is defined as

$$H(t, x, p) = -F(t, x, x') + px' \tag{14}$$

where p is referred to as a generalized momenta in physics. In economics it turns out to be a shadow price. The total differential of the Hamiltonian is

$$dH = -F_t\, dt - F_x\, dx - F_{x'}\, dx' + p\, dx' + x'\, dp = -F_t\, dt - F_x\, dx + x'\, dp,$$

since $F_{x'}\, dx' = p\, dx'$ by the definition of p. Thus, it follows that

$$\partial H/\partial x = -F_x \quad \text{and} \quad \partial H/\partial p = x'.$$

Now, if the function $x(t)$ satisfies the Euler equation (8), then $-F_x = -dF_{x'}/dt = -dp/dt = -p'$. Finally, then,

$$p' = -\partial H/\partial x \quad \text{and} \quad x' = \partial H/\partial p \tag{15}$$

These two first order differential equations are referred to as the *canonical form of the Euler equation*. The Hamiltonian plays an important role in optimal control theory.

The Euler equation is also a necessary condition for a *minimum* of (1) subject to (2). Reviewing the argument, the idea of maximization entered at Equation (4), where we noted that maximization of g at $a = 0$ implies $g'(0) = 0$. But, if x^* were a minimizing function for (1)–(2), then $g(a)$ would assume its minimum at $a = 0$, so that $g'(0) = 0$.

Solutions of an Euler equation are called *extremals*. They are the analog of solutions to the equation $f'(x) = 0$, called stationary points. If a point is to maximize the function $f(x)$, it must be a stationary point. On the other hand, a stationary point need not provide a maximum; it could be a minimizing point, or neither maximizing nor minimizing. We do know that the maximizing point is a stationary point, so we can restrict our search for the maximum to stationary points. In similar fashion, we know that if a maximizing function for the problem of (1) and (2) exists, it must be an extremal. Thus we find solutions to the Euler equation, i.e., we find extremals and search among extremals for the optimizing path.

APPENDIX TO SECTION 3

The integration by parts to get from (5) to (7) rests on the supposition that the derivative $dF_{x'}/dt$ exists. We show, through a pair of lemmas, that the supposition is justified.

Lemma 2. *If $g(t)$ is a given continuous function on $[t_0, t_1]$ with the property that*

$$\int_{t_0}^{t_1} g(t)h'(t)\,dt = 0 \tag{16}$$

for every continuously differentiable function h defined on $[t_0, t_1]$ satisfying $h(t_0) = h(t_1) = 0$, then the function $g(t)$ must be constant on $[t_0, t_1]$.

PROOF. Let c be the average value of g so by the Mean-Value theorem (A2.1)

$$\int_{t_0}^{t_1} [g(t) - c]\,dt = 0.$$

For any h satisfying the hypothesis, compute

$$\int_{t_0}^{t_1} [g(t) - c]h'(t)\,dt = \int_{t_0}^{t_1} g(t)h'(t)\,dt - c[h(t_1) - h(t_0)] = 0 \tag{17}$$

in view of (16) and (3). In particular, consider

$$h(t) = \int_{t_0}^{t} [g(s) - c]\,ds,$$

that satisfies the hypotheses. Then, by Leibnitz' rule, $h'(t) = g(t) - c$, so

$$\int_{t_0}^{t_1} [g(t) - c]h'(t)\,dt = \int_{t_0}^{t_1} [g(t) - c]^2\,dt \geq 0 \tag{18}$$

The integral must be nonnegative since the integrand is, but the expressions on the left in (17) and (18) are identical; since (17) is zero, (18) must also be, and therefore

$$g(t) = c, \qquad t_0 \leq t \leq t_1 \qquad \square$$

Lemma 3. *If $g(t)$ and $f(t)$ are continuous on $[t_0, t_1]$ and if*

$$\int_{t_0}^{t_1} [g(t)h(t) + f(t)h'(t)]\, dt = 0 \tag{19}$$

for every continuously differentiable function $h(t)$ defined on $[t_0, t_1]$ with $h(t_0) = h(t_1) = 0$, then the function $f(t)$ must be differentiable and $f'(t) = g(t)$ for $t_0 \geq t \geq t_1$.

PROOF. Since h is differentiable, we may integrate by parts:

$$\int_{t_0}^{t_1} g(t)h(t)\, dt = -\int_{t_0}^{t_1} G(t)h'(t)\, dt, \qquad \text{where} \quad G(t) = \int_{t_0}^{t} g(s)\, ds,$$

since $h(t_0) = h(t_1) = 0$. Substituting into (19) gives

$$\int_{t_0}^{t_1} [f(t) - G(t)]\, h'(t)\, dt = 0.$$

Applying Lemma 2 indicates that $f(t) - G(t)$ is constant on $[t_0, t_1]$. Thus $f(t) = \int_{t_0}^{t} g(s)\, ds + c$ for some constant c, demonstrating that f is a differentiable function with $f'(t) = g(t)$ as claimed. $\square$

Applying Lemma 3 to (5), we note that since $x(t)$ and $x'(t)$ are specified functions, the coefficients of the $h(t)$ and $h'(t)$ in (5) are functions of t alone. Identify those coefficients with $g(t)$ and $f(t)$ in Lemma 3 to conclude that $dF'_{x'}/dt$ does exist and that

$$dF_{x'}/dt = F_x.$$

This is the Euler equation. It can be shown further that since $dF_{x'}/dt$ exists, the second derivative x'' of the extremal exists for all values of t for which $F_{x'x'}(t, x(t), x'(t)) \neq 0$. The Euler equation can be derived from Lemmas 2 and 3 alone without Lemma 1. Its derivation by means of Lemma 1 is less rigorous but simpler.

Section 4

Examples and Interpretations

Example 1. Find a function $x(t)$ to solve the problem of Section 2 above:

$$\min \int_0^T x'^2(t)\,dt$$

$$\text{subject to} \qquad x(0) = 0, \quad x(T) = B.$$

The integrand is $F(t, x, x') = x'^2$. Thus, $F_{x'} = 2x'$. Since F is independent of x in this case, $F_x = 0$. Hence Euler's equation is $0 = 2x''(t)$, that is equivalent to $x'' = 0$, integrating once yields $x'(t) = c_1$, where c_1 is a constant of integration. Integrating again gives $x(t) = c_1 t + c_2$, where c_2 is another constant of integration. The two constants are determined from the boundary conditions of the problem; evaluating the function x at $t = 0$ and $t = T$ gives conditions that c_1 and c_2 must satisfy, namely

$$x(0) = 0 = c_2, \qquad x(T) = B = c_1 T + c_2.$$

Solving, one obtains $c_1 = B/T$, $c_2 = 0$; therefore the solution to the Euler equation with the given boundary conditions is

$$x(t) = Bt/T, \qquad 0 \le t \le T.$$

If there is a solution to the problem posed, this must be it. Of course, this is the solution found in Section 2.

Example 2. Find extremals for

$$\int_0^1 \{[x'(t)]^2 + 10tx(t)\}\,dt \qquad \text{subject to} \quad x(0) = 1, \quad x(1) = 2.$$

Since $F(t, x, x') = x'^2 + 10tx$, we have $F_x = 10t$, $F_{x'} = 2x'$ and $dF_{x'}/dt = 2x''$; thus the Euler equation is $10t = 2x''$ or, equivalently,

$$x''(t) = 5t.$$

The variables x and t are separated. Integrate, introducing constants of integration c_1 and c_2:

$$x'(t) = 5t^2/2 + c_1$$
$$x(t) = 5t^3/6 + c_1 t + c_2.$$

The constants are found by using the boundary conditions; they obey

$$x(0) = 1 = c_2, \qquad x(1) = 2 = 5/6 + c_1 + c_2.$$

Solving, $c_1 = 1/6$, $c_2 = 1$, and the extremal sought is

$$x(t) = 5t^3/6 + t/6 + 1.$$

Example 3. Find extremals for

$$\int_{t_0}^{t_1} \left[tx'(t) + (x'(t))^2 \right] dt \qquad \text{subject to} \quad x(t_0) = x_0, \quad x(t_1) = x_1,$$

where t_0, t_1, x_0 and x_1 are given parameters. Write $F(t, x, x') = tx' + x'^2$ and compute $F_x = 0$ and $F_{x'} = t + 2x'$. Therefore, the Euler equation is

$$dF_{x'}/dt = d(t + 2x')/dt = 0.$$

Since the right side is zero, there is no need to carry out the differentiation; a function whose derivative is zero must itself be constant. Hence

$$t + 2x'(t) = c_1$$

for some constant c_1. Separate the variables, integrate again, and rearrange the result slightly to obtain

$$x(t) = c_2 + c_1 t/2 - t^2/4.$$

The constants of integration must satisfy the pair of equations

$$x(t_0) = x_0 = c_2 + c_1 t_0/2 - t_0^2/4,$$
$$x(t_1) = x_1 = c_2 + c_1 t_1/2 - t_1^2/4.$$

Example 4. Returning to Example 1.1, we seek a production and inventory accumulation plan to minimize the sum of production and storage costs:

$$\min \int_0^T \left\{ c_1 [x'(t)]^2 + c_2 x(t) \right\} dt$$

$$\text{subject to} \qquad x(0) = 0, \quad x(T) = B, \quad x'(t) \geq 0.$$

where c_1 and c_2 are given nonnegative constants. Suppose the optimal solution satisfies the nonnegativity condition $x'(t) \geq 0$ with strict inequality, so that this constraint is never binding. Since $F_x = c_2$, $F_{x'} = 2c_1 x'$, the Euler equation is $2c_1 x'' = c_2$ or

$$x''(t) = c_2/2c_1. \tag{1}$$

Integration twice yields

$$x(t) = c_2 t^2/4c_1 + k_1 t + k_2,$$

where k_1 and k_2 are constants of integration determined by the boundary conditions:

$$x(0) = 0 = k_2, \qquad x(T) = B = c_2 T^2/4c_1 + k_1 T + k_2.$$

Thus

$$k_1 = B/T - c_2 T/4c_1, \qquad k_2 = 0$$

so

$$x(t) = c_2 t(t - T)/4c_1 + Bt/T, \qquad 0 \leq t \leq T \tag{2}$$

is the extremal sought.

We check whether (2) obeys $x' \geq 0$. From the Euler equation (1), it follows immediately that $x'' > 0$ so that x' is an increasing function of t. Therefore $x'(t) \geq 0$ for all t if and only if it holds initially; $x'(0) = k_1 \geq 0$. This means the constraint $x'(t) \geq 0$ will be satisfied by (2) provided that

$$B \geq c_2 T^2/4c_1. \tag{3}$$

Therefore (2) is the solution to the problem if required total production B is sufficiently large relative to the time period T available, and the storage cost c_2 is sufficiently small relative to the unit production cost c_1. If (3) does not hold, then start of production is postponed in the optimal plan. This will be demonstrated later (Section II.10) when we study the case where $x' \geq 0$ is a tight constraint.

The Euler equation $2c_1 x'' = c_2$ has an interpretation. Recall that c_2 is the cost of holding one additional unit of inventory for one time period. Also $c_1[x'(t)]^2$ is the total production cost at t, so $2c_1 x'$ is the instantaneous marginal cost of production and $2c_1 x''$ is its time rate of change. Therefore, the Euler equation calls for balancing the rate of change of the marginal production cost against the marginal inventory holding cost to minimize the cost of delivering B units of product at time T.

This interpretation may be clearer after integrating the Euler equation over a very small segment of time, say Δ. Since the equation must hold for all t along the path, we have

$$\int_t^{t+\Delta} 2c_1 x''(t)\, ds = \int_t^{t+\Delta} c_2\, ds,$$

that is, using (A1.1)

$$2c_1[x'(t+\Delta) - x'(t)] = c_2\Delta$$

or, rearranging,

$$2c_1 x'(t) + c_2\Delta = 2c_1 x'(t+\Delta).$$

Thus, the marginal cost of producing a unit at t and holding it for an increment of time Δ must be the same as the marginal cost of producing it at $t + \Delta$. That is, we are indifferent between producing a marginal unit at t or postponing it a very little while. Indeed, all along the optimal path, no shift in the production schedule can reduce cost.

Example 5. In Example 4, suppose expenditures are discounted at a continuous rate r (review the appendix to Section 1).

$$\min \int_0^T e^{-rt}[c_1 x'^2 + c_2 x]\, dt$$

$$\text{subject to} \qquad x(0) = 0, \quad x(T) = B.$$

Again $x'(t) \geq 0$ is needed for economic sense, and again this requirement is temporarily set aside. Compute $F_x = e^{-rt}c_2$ and $F_{x'} = 2e^{-rt}c_1 x'(t)$. The Euler equation is

$$e^{-rt}c_2 = d(2e^{-rt}c_1 x'(t))/dt.$$

It calls for balancing the present value of the marginal inventory cost at t with the rate of change of the corresponding marginal production cost. Integrating this equation over a small increment of time, we get

$$\int_t^{t+\Delta} e^{-rt}c_2\, ds = \int_t^{t+\Delta} [d(2e^{-rt}c_1 x'(s))/ds]\, ds;$$

that is,

$$2e^{-rt}c_1 x'(t) + \int_t^{t+\Delta} e^{-rt}c_2\, ds = 2e^{-r(t+\Delta)}c_1 x'(t+\Delta).$$

The marginal cost of producing a unit at t and holding it over the next little increment of time Δ equals the marginal cost of producing a unit at $t + \Delta$. Consequently, we are indifferent between producing a marginal unit at t or a little later. No change in this production schedule can reduce total discounted cost.

Expanding

$$dF_{x'}/dt = -2re^{-rt}c_1x'(t) + 2e^{-rt}c_1x''(t)$$

and simplifying also gives the Euler equation:

$$x''(t) = rx'(t) + c_2/2c_1. \tag{4}$$

Note that since x' must be nonnegative, the right side of (4) must be positive. Hence, in an optimal plan $x'' > 0$; an optimal plan involves a strictly increasing production rate over time. Even without holding costs, a time preference for money ($r > 0$) leads the firm to produce at an increasing rate over time to maintain a constant present value of the marginal production cost. The larger unit holding cost c_2 and the larger the interest rate r, the more advantageous it is to postpone production (to save holding costs, to postpone expenditures) and thus the steeper will be the path of production with respect to time.

To solve the Euler equation (4), note that neither x nor t appears in the differential equation. Make the change of variable $x' = u$, so that $x'' = u'$. This gives a first order linear differential equation with constant coefficients (see Section B2):

$$u' = ru + c_2/2c_1$$

with solution

$$x' = u = k_1e^{rt} - c_2/2rc_1,$$

where k_1 is a constant of integration. Integration yields

$$x(t) = k_1e^{rt}/r - c_2t/2rc_1 + k_2.$$

Boundary conditions

$$x(0) = 0 = k_1/r + k_2, \qquad x(T) = B = k_1e^{rT}/r - c_2T/2rc_1 + k_2$$

give values for the constants of integration

$$k_2 = -k_1/r, \qquad k_1 = r(B + c_2T/2rc_1)/(e^{rT} - 1).$$

Hence the solution is

$$x(t) = (B + c_2T/2rc_1)(e^{rt} - 1)/(e^{rT} - 1) - c_2t/2rc_1, \qquad 0 \le t \le T, \tag{5}$$

provided $x'(t) \ge 0$ is obeyed. This may be checked as before. Since $x'' > 0$, we know that $x' \ge 0$ throughout provided $x'(0) \ge 0$. In turn, it can be shown that $x'(0) \ge 0$ if and only if

$$B \ge (e^{rT} - 1 - rT)c_2/2r^2c_1. \tag{6}$$

If (6) holds, then (5) satisfies the nonnegativity condition $x' \geq 0$ and it will be the optimal solution. Production optimally begins at $t = 0$ in qualitatively similar circumstances as when $r = 0$, although the functional form of (6) is somewhat different from (3). If (6) is violated, the start of production is postponed.

Example 6. Continuing, suppose production cost is a monotone increasing, convex function $g(x')$ of the production rate x':

$$g(0) = 0, \qquad g' \geq 0, \qquad g'' > 0 \qquad \text{for } x \geq 0.$$

The quadratic production cost is clearly a special case. For the problem

$$\min \int_0^T e^{-rt}[g(x') + c_2 x]\,dt$$
$$\text{subject to} \qquad x(0) = 0, \quad x(T) = B$$

compute

$$F_x = e^{-rt}c_2, \qquad F_{x'} = e^{-rt}g'(x'),$$
$$dF_{x'}/dt = -re^{-rt}g'(x') + e^{-rt}g''(x')x''.$$

The Euler equation is therefore

$$g''(x'(t))x''(t) = rg'(x'(t)) + c_2. \tag{7}$$

All terms in (7) except for x'' are known to be nonnegative. It follows that $x'' > 0$, so the optimal production rate (where positive) will be strictly increasing with respect to time until T, when a total of B is accumulated. As in Example 5, holding costs and time preferences both lead the firm to postpone production to save inventory cost and postpone spending. From (7), for a given production rate x', the larger the interest rate r and the larger the holding cost c_2, the greater the rate of change of production x'' must be. Note that the general qualitative behavior of the solution has been determined without exact specification of the production cost function g. Of course, an explicit analytic solution generally rests on this specification (as in the quadratic case discussed above).

Example 7. An individual seeks the consumption rate at each moment of time that will maximize his discounted utility stream over a period of known length T. The utility of consumption $U(C(t))$ at each moment t is a known increasing concave function (diminishing marginal utility of consumption): $U' > 0$ and $U'' < 0$. Future utility is discounted at rate r. The objective is

$$\max \int_0^T e^{-rt}U(C(t))\,dt \tag{8}$$

subject to a cash flow constraint. The individual derives current income from exogenously determined wages $w(t)$ and from interest earnings iK on his holdings of capital assets $K(t)$. For simplicity, the individual may borrow capital ($K < 0$) as well as lend it at interest rate i. Capital can be bought or sold at a price of unity. Thus income from interest and wages is allotted to consumption and investment:

$$iK(t) + w(t) = C(t) + K'(t). \tag{9}$$

Both the initial and terminal capital stocks are specified:

$$K(0) = K_0, \qquad K(T) = K_T. \tag{10}$$

Using (9) to eliminate C from (8) leads to a calculus of variations problem in one function $K(t)$. Denoting the integrand of (8) by F, and taking account of (9), we compute (using the chain rule) $F_K = e^{-rt}U(C)i$ and $F_{K'} = -e^{-rt}U'(C)$. The Euler equation is

$$d(-e^{-rt}U'(C))/dt = e^{-rt}U'(C)i. \tag{11}$$

To facilitate interpretation, integrate (11) over a small interval of time and rearrange to

$$e^{-rt}U'(C(t)) = \int_t^{t+\Delta} e^{-rs}U'(C(s))i\,ds + e^{-r(t+\Delta)}U'(C(t+\Delta)). \tag{12}$$

Along an optimal consumption plan, the individual cannot increase utility by shifting the time of consumption of a marginal dollar. The marginal discounted utility from consumption at t (the left side of (12)) must equal the marginal discounted utility achieved by postponing that consumption to $t + \Delta$ (the right side of (12)). To explain further, note that if consumption is postponed, a dollar earns income at rate i that may be consumed as earned. Since a marginal dollar consumed at time s contributes incremental utility of $U'(C(s))$, a fraction i of a dollar consumed at s contributes $iU'(C(s))$. Thus the first term on the right in (12) is the discounted incremental utility achieved during the period of postponement. Finally, at the end of the period of postponement, the dollar itself is consumed, yielding incremental utility $U'(C(t + \Delta))$. The second term on the right of (12) is this discounted marginal utility.

Performing the indicated differentiation in (11) and collecting terms gives

$$-U''C'/U' = i - r. \tag{13}$$

The proportionate rate of change of marginal utility equals the difference between the earnings rate on invested capital and the impatience rate. Since $-U''/U' > 0$ by hypothesis, the optimal solution is characterized by $dC/dt > 0$ if and only if $i > r$. The optimal consumption path rises if the rate of earnings on capital i exceeds the individual's rate of impatience r. (A relatively patient individual foregoes some current consumption to let capital grow so that a higher rate of consumption may be enjoyed later).

If the functional form of U is specified, more may be said. Let $U(C) = \ln C$, $w(t) = 0$ for $0 \le t \le T$, and let $K_T = 0$. In this case (13) becomes

$$C'/C = i - r.$$

Integrate and substitute into (9):

$$K' - iK = -C = -C(0)e^{(i-r)t}.$$

Multiply through by e^{-rt}, integrate, and use boundary conditions $K(0) = K_0$ and $K(T) = 0$ to find the constants of integration, yielding

$$K(t) = e^{it}K_0\left[1 - (1 - e^{-rt})/(1 - e^{-rT})\right]. \tag{14}$$

Then

$$C(t) = rK_0 e^{(i-r)t}/(1 - e^{-rT}).$$

EXERCISES

1. Find the Euler equation and its solution for

$$\int_1^2 \left[x + tx' - (x')^2\right] dt \qquad \text{subject to} \quad x(1) = 3, \quad x(2) = 4.$$

2. Find the Euler equation and its solution for

$$\int_a^b F(t, x, x')\, dt \qquad \text{subject to} \quad x(a) = A, \quad x(b) = B.$$

but do not evaluate constants of integration where
a. $F(t, x, x') = (x')^2/t^3$,
b. $F(t, x, x') = (x')^2 - 8xt + t$.

3. Show that in an optimal solution to Example 6, one is indifferent to producing a marginal unit at t or producing it at $t + \Delta$, since the sum of the marginal discounted cost of production at t and discounted holding cost from t to $t + \Delta$ just about equals the marginal discounted production cost at $t + \Delta$.

4. Solve Example 6 for the case $r = 0$. Explain the economic reason for your answer.

5. Find the consumption plan $C(t)$, $0 \le t \le T$, over a fixed period to maximize the discounted utility stream

$$\int_0^T e^{-rt}C^a(t)\, dt$$

subject to

$$C(t) = iK(t) - K'(t), \qquad K(0) = K_0, \qquad K(T) = 0$$

where $0 < a < 1$ and K, as in Example 7, represents the capital stock.

FURTHER READING

Solution of differential equations is discussed in Appendix B; see especially Sections B1–B2. Example 4 will be discussed throughout this book; the nonnegativity condition is handled in Section II10. Example 7 is drawn from Yaari (see also Levhari and Sheshinksi for an application of this model to taxation).

Section 5

Solving the Euler Equation in Special Cases

The Euler equation can be difficult to solve. If any of the three arguments (t, x, x') do not appear or if the integrand has a special structure, then hints for solutions are available. Some of these instances are mentioned below, both for specific guidance and for practice. Note that sometimes the "hints" are of little use and a direct solution to the Euler equation may be the easier route.

Case 1. F depends on t, x' only: $F = F(t, x')$.

Since F does not depend on x, the Euler equation (3.11) reduces to

$$F_{x'} = \text{const}.$$

This is a first order differential equation in (t, x') only and is referred to as a first integral of the Euler equation. Examples 4.1 and 4.3 fit this case.

Example. The Euler equation for

$$\int_{t_0}^{t_1} (3x' - tx'^2)\, dt \qquad \text{subject to} \quad x(t_0) = x_0, \quad x(t_1) = x_1,$$

is

$$F_{x'} = 3 - 2tx' = c_0;$$

therefore

$$tx' = (c_0 - 3)/(-2) \equiv c_1.$$

Separate variables:

$$x' = c_1/t;$$

and integrate:

$$x = c_1 \ln t + c_2.$$

The constants of integration c_1 and c_2 satisfy the pair of equations

$$x_0 = c_1 \ln t_0 + c_2, \qquad x_1 = c_1 \ln t_1 + c_2.$$

Case 2. F depends on x, x' only: $F = F(x, x')$.
The Euler equation (3.12) reduces to the *first integral*

$$F - x' F_{x'} = \text{const}, \qquad t_0 \le t \le t_1,$$

which is the first order differential equation to be solved.

Example. Among the curves joining (t_0, x_0) and (t_1, x_1), find one generating a surface of minimum area when rotated about the t axis. That is,

$$\min \int_{t_0}^{t_1} 2\pi x\left[1 + (x')^2\right]^{1/2} dt$$

$$\text{subject to} \qquad x(t_0) = x_0, \quad x(t_1) = x_1.$$

Since (ignoring the constant 2π)

$$F_{x'} = xx'/\left[1 + (x')^2\right]^{1/2},$$

we solve

$$F - x'F_{x'} = x\left[1 + (x')^2\right]^{1/2} - x(x')^2/\left[1 + (x')^2\right]^{1/2} = c.$$

Manipulating algebraically,

$$x = c(1 + x'^2)^{1/2}, \qquad \text{or} \qquad x^2 = c^2 + c^2 x'^2.$$

Rearranging, providing $c \neq 0$,

$$x'^2 = (x^2 - c^2)/c^2, \qquad \text{or} \qquad x' = \pm\left[(x^2 - c^2)/c^2\right]^{1/2}.$$

We can deal with the positive root only because of the symmetry of the problem. Separate variables:

$$dx/(x^2 - c^2)^{1/2} = dt/c,$$

provided $x \neq c$. Integrate (using an integral table):

$$\ln\left[\left(x + (x^2 - c^2)^{1/2}\right)/c\right] = (t + k)/c,$$

where k is the constant of integration. Note that the derivative of $\ln c = 0$ so that when we differentiate both sides we get back the original differential equation. Taking antilogs gives

$$x + (x^2 - c^2)^{1/2} = ce^{(t+k)/c}.$$

Now

$$x - (x^2 - c^2)^{1/2} = \left[x - (x^2 - c^2)^{1/2}\right]\left[x + (x^2 - c^2)^{1/2}\right]/\left[x + (x^2 - c^2)^{1/2}\right]$$
$$= \left[x^2 - (x^2 - c^2)\right]/\left[x + (x^2 - c^2)^{1/2}\right]$$
$$= c^2/ce^{(t-k)/c} = ce^{-(t+k)/c}.$$

That last step follows by substitution for $x + (x^2 - c^2)^{1/2}$. Finally, by addition we get

$$x = c\left[e^{(t+k)/c} + e^{-(t+k)/c}\right]/2.$$

This is the equation of a figure called a catenary; c and k can be found using the conditions $x(t_0) = x_0$ and $x(t_1) = x_1$.

Example. The brachistochrone problem.

$$\min \int_{x_0}^{x_1} \left[(1 + y'^2)/y\right]^{1/2} dx,$$

where the constant $(2g)^{-1/2}$ is ignored. As the integrand does not involve x explicitly (i.e., here the general form of the integrand is $F(x, y(x), y'(x))$ rather than $F(t, x(t), x'(t))$ as in our previous examples), we have

$$F - y'F_{y'} = \left[(1 + y'^2)/y\right]^{1/2} - y'^2\left[y(1 + y'^2)\right]^{-1/2}$$
$$= \left[y(1 + y'^2)\right]^{-1/2}\left[1 + y'^2 - y'^2\right] = \text{a constant}.$$

This implies that

$$\left[y(1 + y'^2)\right]^{-1/2} = \text{a constant}$$

which, in turn, means that

$$y(1 + y'^2) = 2k,$$

where k is a constant, so

$$y' = \left[(2k - y)/y\right]^{1/2}.$$

Separating variables gives

$$\left[y/(2k - y)\right]^{1/2} dy = dx.$$

Multiply the numerator and denominator in the bracketed expression by y to get

$$y\,dy/(2ky - y^2)^{1/2} = dx.$$

Both sides can now be integrated to get

$$x = -(2ky - y^2)^{1/2} + k \arccos(1 - y/k) + c,$$

where c is a constant. This is the equation of a cycloid.

Example. Newton's Second Law of Motion is $F = ma = md^2x/dt^2$, a second order differential equation. The problem is to find the integral for which this differential equation is the Euler equation. It turns out to be the integral

$$\int_{t_0}^{t_1} (mx'^2/2 - V(x)] \, dt.$$

Since the integrand does not involve t,

$$mx'^2/2 - V(x) - mx'^2 = c,$$

a constant. Thus,

$$-V(x) - mx'^2/2 = c.$$

Differentiation with respect to t yields

$$-V'x' - mx'x'' = 0$$

or

$$-V'(x) = mx''.$$

Identifying $-V'(x) = F$ yields the desired result.

The term $mx'^2/2$ in the integrand is a particle's kinetic energy, while $V(x)$ is defined as its potential energy. Physicists let $T = mx'^2/2$ and call

$$\int_{t_0}^{t_1} (T - V) \, dt = \int_{t_0}^{t_1} L \, dt$$

the *action*, or Hamilton's integral, where $L = T - V$ is called the Lagrangian. The description of the motion of a particle through space in terms of the Euler equation of this integral is referred to as the Principle of Least Action or Hamilton's Principle of Stationary Action. This principle plays a unifying role in theoretical physics in that laws of physics that are described by differential equations have associated with them an appropriate action integral. Indeed, discovering the action integral whose Euler equation is the desired physical law is a major achievement.

The "special methods" are *not* always the easiest way to solve a problem. Sometimes an applicable, special form of the Euler equation is easier to work with than the ordinary form and sometimes it is more difficult. The easiest route is determined by trial and error. For example, consider finding extremals for

$$\int_{t_0}^{t_1} 2x^2 + 3xx' - 4(x')^2] \, dt \qquad \text{subject to} \quad x(t_0) = x_0, \quad x(t_1) = x_1.$$

Since the integrand does not depend on t, we could write the Euler equation in the form $F - x'F_{x'} = c$; that is

$$2x^2 + 4(x')^2 = c. \tag{1}$$

This nonlinear differential equation is not readily solved.

On the other hand, the standard form of the Euler equation ($F_x = dF_{x'}/dt$) for this problem is the second order linear differential equation

$$2x'' + x = 0, \tag{2}$$

whose solution is easily found (see Section B3). The characteristic equation associated with this differential equation is $2r^2 + 1 = 0$, with roots $r = \pm i/2^{1/2}$. Hence extremals are of the form

$$x(t) = c_1 \sin t/2^{1/2} + c_2 \cos t/2^{1/2}.$$

The constants c_1 and c_2 are found using the given boundary conditions. (Differentiating the Euler equation found first (1) totally with respect to t leads to the second form (2).) Note that several of the exercises of this section are of the form of this illustration and are more readily solved by the standard Euler equation.

Case 3. F depends on x' only: $F = F(x')$.

The Euler equation is $F_{x'x'}x'' = 0$. Thus along the extremal at each t, either $F_{x'x'}(x') = 0$ or $x''(t) = 0$. In the latter case, integration twice indicates that the extremal is of the linear form $x(t) = c_1 t + c_2$. In the former case, either $F_{x'x'}(x') \equiv 0$ or else x' is constant, i.e., $F_{x'} = 0$. The case of x' constant was just considered. If F is linear in x', $F(x') = a + bx'$, the Euler equation is an identity and any x satisfies it trivially (see also Case 5 to come).

We conclude that if the integrand F depends solely on x' but is not linear, then graphs of extremals are straight lines. Even if the functional form of F or $F_{x'x'}$ appears complicated, we know that extremals must be linear. Boundary conditions determine the constants.

This result may be applied immediately to Example 1.4 to conclude that the *shortest distance between two points* in a plane is the straight line connecting them.

Example. Extremals of

$$\int_{t_0}^{t_1} (x')^2 \exp(-x')\, dt \qquad \text{subject to} \quad x(t_0) = x_0, \quad x(t_1) = x_1$$

must be of the form $x(t) = c_1 t + c_2$.

Case 4. F depends on t, x only: $F = F(t, x)$.

The Euler equation is $F_x(t, x) = 0$, which is not a differential equation. It calls for optimizing the integrand at each t. The dynamic problem is degenerate. (Review (1.4), for example).

Case 5. F is linear in x': $F = A(t, x) + B(t, x)x'$.

The Euler equation is $A_x + B_x x' = B_t + B_x x'$; that is, $A_x(t, x) = B_t(t, x)$, which is not a differential equation. This may be viewed as an implicit equation for x in terms of t. If the solution $x(t)$ of this equation satisfies the boundary conditions, it may be the optimal solution.

Alternatively, the Euler equation $A_x = B_t$ may be an identity, $A_x(t, x) \equiv B_t(t, x)$, satisfied by any function x. Then, according to the integrability theorem for exact differential equations (see Appendix B), there is a function $P(t, x)$ such that $P_t \equiv A$, $P_x \equiv B$ (so $P_{tx} \equiv A_x \equiv B_t$) and

$$F(t, x, x') = A + Bx' = P_t + P_x x' = dP/dt.$$

Thus, the integrand is the total derivative of a function P and

$$\int_{t_0}^{t_1} (A + Bx')\, dt = \int_{t_0}^{t_1} (dP/dt)\, dt = P(t_1, x(t_1)) - P(t_0, x(t_0)).$$

The value of the integral depends only on the endpoints; the path joining them is irrelevant in this case. Any feasible path is optimal. This is analogous to the problem of maximizing a constant function; any feasible point would yield the same value.

Case 5 is the only instance in which an Euler equation is an identity. To understand this, suppose that (3.10) is an identity, satisfied for any set of four values t, x, x', x''. The coefficient of x'' must be zero if (3.10) is to hold for every possible value of x''. Thus, $F_{x'x'} \equiv 0$. Then $F_x - F_{x't} - x'F_{xx'} \equiv 0$ for any t, x, x'. The first identity implies that F must be linear in x', so F has the form $A(t, x) + B(t, x)x'$. Then the second identity becomes $A_x \equiv B_t$, as was to be shown.

Two integrands that appear quite different can lead to the same Euler equation and thus have the same extremals. This happens if the integrands differ by an exact differential. For example, let $P(t, x)$ be a twice differentiable function and define

$$Q(t, x, x') \equiv dP/dt = P_t(t, x) + P_x(t, x)x'(t).$$

Then for any twice differentiable function $F(t, x, x')$, the two integrals

$$\int_{t_0}^{t_1} F(t, x, x')\, dt \qquad \text{and} \qquad \int_{t_0}^{t_1} [F(t, x, x') + Q(t, x, x')]\, dt$$

subject to $\quad x(t_0) = x_0, \quad x(t_1) = x_1$

differ by a constant (namely, $P(t_1, x_1) - P(t_0, x_0)$) and the Euler equations associated with the respective integrals are identical.

Example 1. $\int_{t_0}^{t_1} x'(t)\, dt$ subject to $x(t_0) = x_0$, $x(t_1) = x_1$. Since $F = x'$, $F_x = 0$ and $F_{x'} = 1$, so that the Euler equation is $0 = 0$, satisfied always. The integrand is an exact differential, so

$$\int_{t_0}^{t_1} x'(t)\, dt = x(t_1) - x(t_0) = x_1 - x_0$$

for any differentiable function. The value of the integral depends only on the endpoint conditions and not on the path connecting the endpoints.

Example 2. Suppose the production cost in Example 1.1 is linear:

$$\min \int_0^T \left[c_1 x'(t) + c_2 x(t)\right] dt$$

$$\text{subject to} \qquad x(0) = 0, \quad x(T) = B.$$

Then $F_x = c_2$ and $F_{x'} = c_1$, so that the Euler equation is $c_2 = 0$. This means that there is no optimal production plan if there is a positive holding cost ($c_2 > 0$) and that any feasible plan is optimal if the holding cost is zero. The underlying economics is as follows. If the unit production cost is constant, then the total cost of manufacturing B units is $c_1 B$ irrespective of the time schedule:

$$\int_0^T c_1 x'(t)\, dt = c_1\left[x(T) - x(0)\right] = c_1 B.$$

If the cost of holding inventory is zero, then all feasible production plans are equally good. If the inventory holding cost is positive, then postponing production to the last moment reduces the total storage cost. The limiting answer is $x(t) = 0$, $0 < t < T$, $x(T) = B$, which is a plan that can be approached but is not itself a continuous function.

Example 3. To find extremals for

$$\int_0^T txx'\, dt \qquad \text{subject to} \quad x(0) = 0, \quad x(T) = B,$$

compute $F_x = tx'$, $F_{x'} = tx$, $dF_{x'}/dt = x + tx'$. The Euler equation is $tx' = x + tx'$ or $0 = x$, which can be satisfied only if $B = 0$.

To verify that there is no solution to the problem, integrate by parts by letting $xx'\, dt = dv$ and $t = u$, sc $x^2/2 = v$ and $dt = du$. Then

$$\int_0^T txx'\, dt = \left(B^2 T - \int_0^T x^2\, dt\right)\Big/ 2 \le B^2 T/2.$$

The upper bound of $B^2T/2$ can be realized only by setting $x(t) = 0$, $0 \le t \le T$. This function satisfies the Euler equation but not the boundary conditions (except if $B = 0$). Evidently, there is no minimum; the integral can be made arbitrarily small.

Example 4. For

$$\int_{t_0}^{t_1} e^{-rt}(x' - ax)\, dt \qquad \text{subject to} \quad x(t_0) = x_0, \quad x(t_1) = x_1,$$

we compute

$$F_x = -ae^{-rt}, \qquad F_{x'} = e^{-rt}, \qquad dF_{x'}/dt = -re^{-rt}.$$

The Euler equation is $a = r$. If, in fact $a = r$, then the Euler equation is an identity and the integrand is an exact differential, namely $d(e^{-rt}x(t))/dt$. The value of the integral is $e^{-rt_1}x_1 - e^{-rt_0}x_0$, independent of the path between the given endpoints. On the other hand, if $a \ne r$, then the Euler equation cannot be satisfied; therefore there is no optimum. To verify this, add $rx - rx$ to the integrand and then use the observation just made:

$$\begin{aligned} &\int_{t_0}^{t_1} e^{-rt}(x' - rx + rx - ax)\, dt \\ &\quad = \int_{t_0}^{t_1} e^{-rt}(x' - rx)\, dt + \int_{t_0}^{t_1} e^{-rt}(rx - ax)\, dt \\ &\quad = x_1 e^{-rt_1} - x_0 e^{-rt_0} + (r - a)\int_{t_0}^{t_1} e^{-rt} x\, dt. \end{aligned}$$

If $r = a$, all feasible paths give the same value. If $r \ne a$, the value of the integral may be made arbitrarily large or small by choosing a path with a very high peak or low trough.

Example 5. Consider the discounted profit maximization problem (compare Example 1.3),

$$\begin{aligned} &\max \int_0^T e^{-rt}[p(t)f(K(t)) - c(t)(K' + bK)]\, dt \\ &\text{subject to} \quad K(0) = K_0, \quad K(T) = K_T, \end{aligned} \tag{3}$$

where $c(t)$ is the cost per unit of gross investment, $p(t)$ the price per unit of output (given functions of time), $K(t)$ the stock of productive capital, $f(K)$ output, and $I = K' + bK$ gross investment (net investment plus depreciation). Compute

$$F_K = e^{-rt}[pf'(K) - cb] \qquad \text{and} \qquad F_{K'} = -e^{-rt}c.$$

The Euler equation is

$$d[-e^{-rt}c(t)]/dt = e^{-rt}[p(t)f'(K(t)) - c(t)b].$$

To interpret this, integrate over a small interval of time:

$$e^{-rt}c(t) - e^{-r(t+\Delta)}c(t+\Delta) = \int_t^{t+\Delta} e^{-rs}[p(s)f'(K(s)) - c(s)b]\,ds.$$

The cost difference between purchasing a marginal unit of capital at t rather than at $t + \Delta$ is just offset by the marginal profit earned by that capital over the period $[t, t + \Delta]$.

Performing the indicated differentiation in the Euler equation yields the equivalent requirement

$$e^{-rt}[pf'(K) - cb] = e^{-rt}[rc - c'],$$

so the rule for choosing the optimal level of capital $K^*(t)$ is

$$p(t)f'(K^*(t)) = (r + b)c(t) - c'(t). \tag{4}$$

This is a static equation for $K^*(t)$, not a differential equation. It is feasible only if $K^*(0) = K_0$ and $K^*(T) = K_T$. It says that, if possible, capital stock should be chosen so that the value of the marginal product of capital at each moment equals the cost of employing it. The "user cost" of capital $(r + b)c - c'$ includes not only foregone interest on the money invested in capital and decline in money value because of its physical deterioration but also capital gains (or losses). Capital gains may occur, for example, if the unit price of capital rises, thereby increasing the value of the capital stock held by the firm. On the other hand, capital loss is possible, for instance, through the invention of a new productive method that makes the firm's capital stock outmoded, diminishing its value.

Example 6. The Euler equation for

$$\int_0^1 (x'^2 - 2xx' + 10tx)\,dt \qquad \text{subject to} \quad x(0) = 1, \quad x(1) = 2$$

is $x'' = 5t$ with solution $x(t) = 5t^3/6 + t/6 + 1$. This problem, its Euler equation, and its solution should be compared with Example 4.2. The Euler equations and solutions are identical. The integrands differ by the term $-2xx' = d(-x^2)/dt$, an exact differential. The value of the two integrals evaluated along the extremal will differ by

$$-x^2(1) + x^2(0) = -4 + 1 = -3.$$

EXERCISES

1. Find candidates to maximize or minimize

$$\int_{t_0}^{t_1} [t(1 + (x')^2)]^{1/2}\, dt \qquad \text{subject to} \quad x(t_0) = x_0, \quad x(t_1) = x_1.$$

You need not find the constants of integration.

2. Find candidates to maximize or minimize

$$\int_{t_0}^{t_1} F(t, x, x')\, dt \qquad \text{subject to} \quad x(t_0) = x_0, \quad x(t_1) = x_1$$

(but do not find the constants of integration) where
a. $F(t, x, x') = x^2 + 4xx' + 2(x')^2$,
b. $F(t, x, x') = x^2 - 3xx' - 2(x')^2$,
c. $F(t, x, x') = x'(\ln x')^2$,
d. $F(t, x, x') = -x^2 + 3xx' + 2(x')^2$,
e. $F(t, x, x') = te^{x'}$

3. Find candidates to maximize or minimize

$$\int_{t_0}^{t_1} \left[x^2 + axx' + b(x')^2\right] dt \qquad \text{subject to} \quad x(t_0) = x_0, \quad x(t_1) = x_1.$$

Consider the cases $b = 0$, $b > 0$, and $b < 0$. How does the parameter a affect the solution? Why?

4. A monopolist believes that the number of units $x(t)$ he can sell depends not only on the price $p(t)$ he sets, but also on the rate of change of price, $p'(t)$:

$$x = a_0 p + b_0 + c_0 p'. \tag{5}$$

His cost of producing at rate x is

$$C(x) = a_1 x^2 + b_1 x + c_1. \tag{6}$$

Given the initial price $p(0) = p_0$ and required final price $p(T) = p_1$, find the price policy over $o \le t \le T$ to maximize profits

$$\int_0^T [px - C(x)]\, dt$$

given (5), (6), and the boundary conditions above. (Caution: this problem involves much messy algebra; it has been included for its historic interest. See the suggestions for further reading that follow.)

5. Suppose a mine contains an amount B of a mineral resource (such as coal, copper, or oil). The profit rate that can be earned from selling the resource at rate x is $\ln x$. Find the rate at which the resource should be sold over the fixed period $[0, T]$ to maximize the present value of profits from the mine. Assume the discount

[Hint: Let $y(t)$ be the cumulative amount sold by time t. Then $y'(t)$ is the sales rate at t. Find $y(t)$ to

$$\max \int_0^T e^{-rt} \ln y'(t)\, dt$$

$$\text{subject to} \qquad y(0) = 0, \quad y(T) = B.]$$

6. Reconsider the problem in Exercise 5 above, but suppose that the profit rate is $P(x)$ when the resource is sold at rate x, where $P'(0) > 0$ and $P'' < 0$.

a. Show that the present value of the marginal profit from extraction is constant over the planning period. (Otherwise it would be worthwhile to shift the time of sale of a unit of the resource from a less profitable moment to a more profitable one.) Marginal profit $P'(x)$ therefore grows exponentially at the discount rate r.

b. Show that the optimal extraction rate declines through time.

FURTHER READING

See Arrow (1964) and Jorgenson for discussions of optimal firm investment. Samuelson (1965) provides an important application of the catenary to economics. Very thorough analyses of the brachistochrone problem and the surface of revolution of minimum area problem are presented by Bliss, and D. R. Smith. An elegant presentation of the Principle of Least Action is provided by Feynman. Also see Goldstein. An interesting biographical sketch of Hamilton and his contribution to physics can be found in Boorse, Motz and Weaver. Exercise 4 is due to Evans and is discussed in Allen. It is the starting point for Roos' differential game discussed in II.23. Exercises 5 and 6 are discussed in a more general context by Hotelling, as well as in Section 7 and 9.

See Section B3 regarding second order linear differential equations. Integrands that are linear in x' are discussed further in Section 16 for the case of infinite horizon autonomous problems.

Section 6

Second Order Conditions

In optimizing a twice continuously differentiable function $f(x)$ of a single variable x on an open interval, we know that if the number x^* maximizes $f(x)$, it is necessary that $f'(x^*) = 0$ and $f''(x^*) \leq 0$. If x^* satisfies $f'(x^*) = 0$ and $f''(x^*) < 0$, then x^* gives a local maximum to f. That is, if the function is stationary at x^* and locally concave in the neighborhood of x^*, then x^* must provide a local maximum.

Somewhat analogous conditions can be developed for the problem of finding a continuously differentiable function $x(t)$ that maximizes

$$\int_{t_0}^{t_1} F(t, x, x')\, dt \qquad \text{subject to} \quad x(t_0) = x_0, \quad x(t_1) = x_1, \qquad (1)$$

where F is twice continuously differentiable in its three arguments. We have seen that if the function $x^*(t)$ maximizes (1), then, if for any fixed admissible function $h(t)$ we define

$$g(a) = \int_{t_0}^{t_1} F(t, x^* + ah, x^{*\prime} + ah')\, dt,$$

we must have

$$g'(0) = \int_{t_0}^{t_1} (F_x h + F_{x'} h')\, dt = 0. \qquad (2)$$

The expression in the middle is called the *first variation*. The requirement that it be zero when evaluated along the optimal function $x^*(t)$ leads to the Euler equation, as discussed earlier.

The analog of the second derivative of a function is the *second variation*. It is

$$g''(0) = \int_{t_0}^{t_1} \left[F_{xx} h^2 + 2F_{xx'} hh' + F_{x'x'} (h')^2 \right] dt. \tag{3}$$

If $x^*(t)$ maximizes (1), it is *necessary* that the second variation (3) be nonpositive for all admissible functions $h(t)$. All second partial derivatives in the integrand of (3) are to be evaluated along $(t, x^*(t), x^{*\prime}(t))$. This second order condition for $x^*(t)$ to be maximizing leads to the Legendre condition to be derived shortly.

The second variation will surely be nonpositive if F is concave in (x, x'). To see this, note that the integrand of (3) is a quadratic form in h and h' and recall that such a quadratic form will be nonpositive if the coefficients are second partial derivatives of a concave function (see Section A3).

Indeed a stronger statement can be made: If the integrand $F(t, x, x')$ is jointly concave in its second and third arguments, and if $x^*(t)$ satisfies the Euler equation $F_x = dF_{x'}/dt$, then $x^*(t)$ maximizes (1). Thus, if F is concave in x and x', the Euler equation is *sufficient* for optimality. This statement is rather easily verified. (It is analogous to the situation where a stationary point of concave function is maximizing; see Section A4.)

Suppose $x^*(t)$ satisfies the Euler equation and F is concave in (x, x'). Abbreviate

$$F = F(t, x, x'), \qquad F^* = F(t, x^*, x^{*\prime}), \tag{4}$$

and let $h(t) = x(t) - x^*(t)$, so that $h'(t) = x'(t) - x^{*\prime}(t)$. Then since F is a concave function it follows from (A3.5) that

$$\begin{aligned} \int_{t_0}^{t_1} (F - F^*)\, dt &\le \int_{t_0}^{t_1} \left[(x - x^*) F_x^* + (x' - x^{*\prime}) F_{x'}^* \right] dt \\ &= \int_{t_0}^{t_1} \left(hF_x^* + h' F_{x'}^* \right) dt \\ &= \int_{t_0}^{t_1} h\left(F_x^* - dF_{x'}^*/dt \right) dt \\ &= 0. \end{aligned} \tag{5}$$

The next to last equality follows from an integration by parts; the last expression is zero because x^* satisfies the Euler equation by hypothesis. Relation (5) shows that no feasible path x provides a larger value than does a path x^* satisfying the Euler equation, so that x^* is maximizing as claimed.

For many problems of interest, F will not be concave in (x, x'). The *Legendre condition* requires that the integrand be locally concave in x' along the optimal path. It is obtained from the requirement that (3) be nonpositive by some manipulation. To see what nonpositivity of (3) requires, integrate the middle term by parts. Let $u = F_{xx'}$ and $dv = 2hh'\,dt$, so that $du = (dF_{xx'}/dt)\,dt$ and $v = h^2$. Then

$$\int_{t_0}^{t_1} 2F_{xx'} hh'\,dt = -\int_{t_0}^{t_1} h^2 (dF_{xx'}/dt)\,dt \tag{6}$$

if we recall that $h(t_1) = h(t_0) = 0$. Substitute from (6) into (3):

$$g''(0) = \int_{t_0}^{t_1} \left[(F_{xx} - dF_{xx'}/dt) h^2 + F_{x'x'} h'^2 \right] dt. \tag{7}$$

Now we need a lemma.

Lemma. *Let $P(t)$ and $Q(t)$ be given continuous functions on $[t_0, t_1]$ and let the quadratic functional*

$$\int_{t_0}^{t_1} \left\{ P(t)[h'(t)]^2 + Q(t)[h(t)]^2 \right\} dt \tag{8}$$

be defined for all continuously differentiable functions $h(t)$ on $[t_0, t_1]$ such that $h(t_0) = h(t_1) = 0$. A necessary condition for (8) to be nonpositive for all such h is that $P(t) \le 0$, $t_0 \le t \le t_1$.

PROOF. The proof appears in the appendix of this section.

Once the function x^* is fixed, the partial derivatives in (7) are functions of t alone. Identifying

$$P(t) = F_{x'x'}(t, x^*(t), x^{*\prime}(t)), \qquad Q(t) = F^*_{xx} - dF^*_{xx'}/dt,$$

the lemma says that nonpositivity of (7) requires

$$F_{x'x'}(t, x^*(t), x^{*\prime}(t)) \le 0. \tag{9}$$

This is the *Legendre condition*: A maximizing path x^* must satisfy not only the Euler equation but also the Legendre condition (9). As is to be shown in Exercise 3 below, for a minimizing path, the sign in (9) is reversed.

For Example 4.6, $F = e^{-rt}[g(x') + c_2 x]$, the Legendre condition requires that

$$F_{x'x'} = e^{-rt} g''(x') \ge 0$$

along a minimizing path. This is satisfied along any path since $g'' > 0$ was assumed; therefore the Legendre condition is surely satisfied. Moreover, F is convex in x, x' since not only $F_{x'x'} > 0$, but also

$$F_{xx} F_{x'x'} - F^2_{xx'} = 0$$

as the reader should verify. Thus the solution to the Euler equation is the minimizing path in Example 4.6.

In sum, *a maximizing path must satisfy the Euler equation and the Legendre condition* (9). These are *necessary* conditions. A minimizing path necessarily satisfies the Euler equation and the Legendre condition $F_{x'x'}(t, x^*, x^{*\prime}) \geq 0$, as is to be shown in Exercise 3. The strengthened Legendre condition, with a strong inequality (e.g., $F_{x'x'} < 0$) is *not* sufficient for an optimum, as is to be shown in Exercise 5. However, if the integrand F is concave in (x, x'), then a path that satisfies the Euler equation is maximizing; if the integrand is convex in (x, x'), a path satisfying the Euler equation is minimizing. Thus, *the Euler equation together with concavity* (convexity) *of the integrand in* (x, x') *are sufficient for a maximum* (*minimum*).

APPENDIX TO SECTION 6 (OPTIONAL)

Proof of Lemma

The lemma is proven by contradiction by showing that the assumptions that (8) is always nonpositive and that $P(t)$ is positive anywhere on the interval $[t_0, t_1]$ are inconsistent. Suppose for some time s and some $b > 0$, $P(s) = 2b > 0$. Since P is continuous, there will be a short interval of time around s during which P exceeds b. Thus, there is a $c > 0$ such that (Figure 6.1)

$$t_0 \leq s - c < s < s + c \leq t_1$$

and

$$P(t) > b, \qquad s - c \leq t \leq s + c.$$

Corresponding to this function P, we construct a particular continuously differentiable function $h(t)$ such that (8) will be positive. This will establish the necessity of $P(t) \leq 0$ for (8) to be nonpositive. In particular, let

$$h(t) = \begin{cases} \sin^2 \pi(t - s)/c & \text{for } s - c \leq t \leq s + c, \\ 0, & \text{elsewhere}. \end{cases} \tag{10}$$

Then

$$h'(t) = (\pi/c) 2 \sin \Theta \cos \Theta = (\pi/c) \sin 2\Theta$$

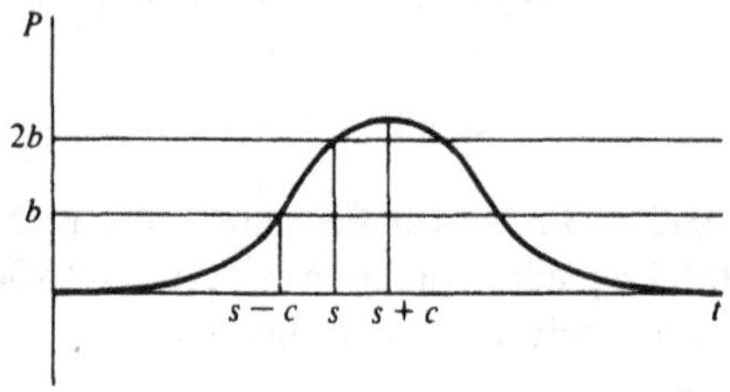

Figure 6.1

where $\Theta = \pi(t - s)/c$. This h satisfies all the requirements. Then

$$\int_{t_0}^{t_1} \left[P(h')^2 + Qh^2 \right] dt = \int_{s-c}^{s+c} P(\pi/c)^2 \sin^2 2\Theta \, dt + \int_{s-c}^{s+c} Q \sin^4 \Theta \, dt. \quad (11)$$

Since $P(t) > b$ for $s - c \leq t \leq s + c$, we have

$$\int_{s-c}^{s+c} P(\pi/c)^2 \sin^2 2\Theta \, dt \geq (b\pi/2c) \int_{-2\pi}^{2\pi} \sin^2 u \, du = b\pi^2/c \quad (12)$$

using a change of variable integration ($u = 2\Theta = 2\pi(t - s)/c$) and a table of integrals.

Since $Q(t)$ is continuous on the closed interval $[s - c, s + c]$, there is some positive M such that $-M \leq Q(t) \leq M$ for $s - c \leq t \leq s + c$. Then

$$\int_{s-c}^{s+c} Q \sin^4 \Theta \, dt \geq \int_{s-c}^{s+c} -M \, dt = -2cM. \quad (13)$$

Employing the lower bounds (12) and (13) in (11) yields

$$\int_{t_0}^{t_1} \left[P(h')^2 + Qh^2 \right] dt \geq b\pi^2/c - 2cM. \quad (14)$$

But $b\pi^2/c - 2cM > 0$ if $b\pi^2/2M > c^2$. Hence c can be chosen so small that the right side of (14) will be positive, and therefore the left side of (14) will also be positive. Hence (8) will be nonpositive for *any* admissible h only if $P(t) \leq 0$, $0 \leq t \leq t_1$. This completes the proof of the lemma. □

EXERCISES

1. Show that the solution to the Euler equation for Example 4.7 is a maximizing path.

2. Show that if $F(t, x, x')$ is jointly convex in (x, x'), then any function $x^*(t)$ satisfying the Euler equation $F_x = dF_{x'}/dt$ and boundary conditions of (1) minimizes the functional (1).

3. Show that a necessary condition for $x^*(t)$ to minimize (1) is that the inequality $F_{x'x'}(t, x^*(t), x^{*\prime}(t)) \geq 0$ be satisfied at every $t_0 \leq t \leq t_1$.

4. In Exercise 5.3 candidates were found for maxima and minima of

$$\int_{t_0}^{t_1} \left[x^2 + axx' + b(x')^2 \right] dt \qquad \text{subject to} \quad x(t_0) = x_0, \quad x(t_1) = x_1 \quad (15)$$

where a and b are known constants. The following questions ask what can be said about whether the externals are minimizing, maximizing, or neither for each possible sign of b.

   a. Show that if $b = 0$, then $x = 0$ is the only minimizing candidate. (Then a minimum exists only if $x_0 = x_1 = 0$.) Show that there is no maximizing path if $b = 0$.

b. Show that the extremal is minimizing if $b > (a/2)^2$.

c. Suppose that $(a/2)^2 > b > 0$. Find a path that minimizes

$$\int_{t_0}^{t_1} (x^2 + b(x')^2)\, dt \qquad \text{subject to} \quad x(t_0) = x_0, \quad x(t_1) = x_1.$$

How is this path related to the extremal for (15)? How do the resulting solution values differ? Can you now argue that the extremal found for (15), in the present case, is minimizing? (This shows that although the Legendre condition is necessary, convexity of the integrand is not necessary.) Now also prove that if $b = x_0 = x_1 = 0$, then $x(t) = 0$, $t_0 \le t \le t_1$, does provide the minimum to (15).

5. Consider

$$\max \int_0^{2\pi} [x^2 - (x')^2]\, dt \qquad \text{subject to} \quad x(0) = 0, \quad x(2\pi) = 0.$$

a. Show that extremals are of the form $x(t) = c \sin t$, yielding a value of zero to the integral. Is the Legendre condition satisfied? Is the integrand concave in (x, x')?

b. Show that $y(t) = t - t^2/2\pi$ is a feasible solution that yields a positive value to the integral. What conclusion can you draw about the sufficiency of the Legendre condition? What can be said about the existence of a solution to the stated problem?

FURTHER READING

The lemma and its proof are taken from Gelfand and Fomin. Additional second order conditions are given names of Weierstrass and Jacobi and can be found in the calculus of variations texts cited.

Section 7

Isoperimetric Problem

An optimization problem may be subject to an integral constraint:

$$\max \quad \int_{t_0}^{t_1} F(t, x, x')\, dt \tag{1}$$

$$\text{subject to} \quad \int_{t_0}^{t_1} G(t, x, x')\, dt = B, \quad x(t_0) = x_0, \quad x(t_1) = x_1, \tag{2}$$

where F and G are twice continuously differentiable functions and B is a given number. For example, the problem of maximizing the area enclosed by a straight line and a string of length B can be posed in this form. Let the straight line extend from $(t_0, x_0) = (0, 0)$ to $(t_1, x_1) = (t_1, 0)$. Then the area under the curve will be given by (1) with $F(t, x, x') = x$. The constraint on string length is given by (2) with $G(t, x, x') = [1 + (x')^2]^{1/2}$. (Recall Example 1.5). In this problem, the perimeter is constant, specified by (2)—hence the name "isoperimetric." Such an example has provided the name for a whole class of problems given by (1) and (2). Another example of the form of (1) and (2) was given in Exercises 5.5 and 5.6.

$$\max \quad \int_0^T e^{-rt} P(x)\, dt \tag{3}$$

$$\text{subject to} \quad \int_0^T x\, dt = B, \tag{4}$$

where $x(t)$ is the rate of extraction of a resource, B the initial endowment of the resource, and $P(x)$ the profit rate at t if the resource is extracted and sold at rate $x(t)$. Because of its special structure, the isoperimetric constraint (4) can be converted into a fixed endpoint constraint, by defining:

$$y(t) = \int_0^t x(s)\, ds \tag{5}$$

as the amount of resource extracted by time t. Then $y'(t) = x(t)$ and (3) and (4) are equivalently stated as

$$\max \quad \int_0^T e^{-rt} P(y')\, dt \tag{6}$$

$$\text{subject to} \qquad y(0) = 0, \quad y(T) = B. \tag{7}$$

Typically, there is no simple transformation to eliminate an isoperimetric constraint. However, recall that in a constrained calculus optimization problem, one may either use the constraint to eliminate a variable (yielding an equivalent unconstrained problem), or the constraint may be appended to the objective with a Lagrange multiplier and equivalent necessary conditions developed (see Section A5). A Lagrange multiplier technique works here also. For instance, appending (4) to (3) with a Lagrange multiplier gives

$$\begin{aligned} L &= \int_0^T e^{-rt} P(x)\, dt - \lambda\left(\int_0^T x\, dt - B\right) \\ &= \int_0^T \left[e^{-rt} P(x) - \lambda x\right] dt + \lambda B. \end{aligned} \tag{8}$$

A necessary condition for x to maximize the augmented integrand (8) is that it satisfy the Euler equation

$$e^{-rt} P'(x) = \lambda. \tag{9}$$

In agreement with the findings of Exercise 5.6, the present value of marginal profits is constant over the planning period.

In the general case, (1) and (2), we append constraint (2) to (1) by an undetermined multiplier λ. Any admissible function x satisfies (2), so for such an x,

$$\int_{t_0}^{t_1} F(t, x, x')\, dt = \int_{t_0}^{t_1} \left[F(t, x, x') - \lambda G(t, x, x')\right] dt + \lambda B. \tag{10}$$

The integral on the left attains its extreme values with respect to x just where the integral on the right does; λ then is chosen so that (2) is satisfied. The Euler equation for the integral on the right is

$$F_x - \lambda G_x = d(F_{x'} - \lambda G_{x'})/dt. \tag{11}$$

From (A5.11) the Lagrange multiplier method rests on the supposition that the optimal point is not a stationary point of the constraining relation; this prevents division by zero in the proof. An analogous proviso pertains here for a similar reason. Thus, a necessary condition for solution to (1) and (2) may be stated as follows: *If the function x^* is an optimum solution to* (1) *and* (2) *and if x^* is not an extremal for the constraining integral* (2), *then there is a number λ such that $x^*(t)$, λ satisfy* (1) *and* (2).

Example 1.

$$\min \quad \int_0^1 [x'(t)]^2 \, dt$$

$$\text{subject to} \quad \int_0^1 x(t)\, dt = B, \quad x(0) = 0, \quad x(1) = 2.$$

The augmented integrand is $(x')^2 - \lambda x$. Its Euler equation $\lambda + 2x'' = 0$ has the solution

$$x(t) = -\lambda t^2/4 + c_1 t + c_2.$$

Three constants are to be determined—λ, c_1, c_2—using the integral constraint and boundary conditions:

$$\int_0^1 x\, dt = \int_0^1 (-\lambda t^2/4 + c_1 t + c_2)\, dt = B,$$

$$x(0) = c_2 = 0, \qquad x(1) = -\lambda/4 + c_1 + c_2 = 2.$$

Hence

$$c_1 = 6B - 4, \qquad c_2 = 0, \qquad \lambda = 24(B - 1).$$

Example 2. For

$$\max \quad \int_0^T x\, dt$$

$$\text{subject to} \quad \int_0^T [1 + (x')^2]^{1/2}\, dt = B, \quad x(0) = 0, \quad x(T) = 0.$$

The augmented integrand $x - \lambda[1 + (x')^2]^{1/2}$ has Euler equation

$$1 = -d\left(\lambda x'/[1 + (x')^2]^{1/2}\right)/dt.$$

Separate the variables and integrate:

$$t = -\lambda x'/[1 + (x')^2]^{1/2} + k.$$

Solve for x' algebraically:

$$x' = (t - k)/[\lambda^2 - (t - k)^2]^{1/2}.$$

Let $u = \lambda^2 - (t-k)^2$, so $du = -2(t-k)\,dt$. Then

$$x(t) = \int x'(t)\,dt = -\int du/2u^{1/2} = -u^{1/2} + c,$$

so

$$(x-c)^2 + (t-k)^2 = \lambda^2.$$

The solution traces out part of a circle. The constraints k, c, λ are found to satisfy the two endpoint conditions and the integral constraint.

The Lagrange multiplier associated with (1) and (2) has a useful interpretation as the marginal value of the parameter B; that is, the rate at which the optimum changes with an increase in B. For instance, in the resource extraction problem (3) and (4), λ represents the profit contributed by a marginal unit of the resource. In Example 2, λ represents the rate at which area increases with string length.

To verify the claim, note that the optimal path $x^* = x^*(t; B)$ depends on the parameter B. Assume x^* is continuously differentiable in B. Define $V(B)$ as the optimal value in (1) and (2). Then

$$\begin{aligned} V(B) &= \int_{t_0}^{t_1} F(t, x^*, x^{*\prime})\,dt \\ &= \int_{t_0}^{t_1} [F(t, x^*, x^{*\prime}) - \lambda G(t, x^*, x^{*\prime})]\,dt + \lambda B \end{aligned} \tag{12}$$

since (2) is satisfied, where

$$x^* = x^*(t; B), \qquad x^{*\prime} = \partial x^*/\partial t. \tag{13}$$

Differentiating (12) totally with respect to B and taking (13) into account gives

$$V'(B) = \int_{t_0}^{t_1} [(F_x^* - \lambda G_x^*)h + (F_{x'}^* - \lambda G_{x'}^*)h']\,dt + \lambda. \tag{14}$$

where

$$h = \partial x^*/\partial B, \qquad h' = \partial x^{*\prime}/\partial B = \partial^2 x^*/\partial t\,\partial B. \tag{15}$$

But since the augmented integrand in (12) satisfies the Euler equation (11), the integral in (14), after integrating the last term by parts, is zero for any continuously differentiable function h satisfying the endpoint conditions. It follows that

$$V'(B) = \lambda \tag{16}$$

as claimed. (To see that the function h defined in (15) is admissible, one need only observe that the optimal path corresponding to any modified B must be feasible.)

EXERCISES

1. Find extremals for

$$\int_0^1 (x')^2\,dt \qquad \text{subject to} \qquad \int_0^1 x^2 dt = 2, \quad x(0) = 0, \quad x(1) = 0.$$

2. Minimize

$$\int_0^T e^{-rt} x\,dt \qquad \text{subject to} \qquad \int_0^T x^{1/2} dt = A.$$

   a. Solve using a multiplier.
   b. Solve by eliminating the isoperimetric constraint. [Hint: Define $y(t) = \int_0^t x^{1/2}(s)\,ds$.]

3. Minimize

$$\int_0^b (1 + x^2)^{1/2}\,dt \qquad \text{subject to} \qquad \int_0^b x\,dt = c,$$

given $b > 0$, $c > 0$.

4. Maximize $\int_0^1 (2x - x^2)\,dt$ subject to $\int_0^1 tx\,dt = 1$.

FURTHER READING

Cullingford and Prideaux use an isoperimetric formulation for project planning.

Section 8

Free End Value

Thus far both the initial and the terminal values of the function have been prespecified. Suppose that only the initial value is given with all subsequent values to be chosen optimally:

$$\begin{aligned} &\max \quad \int_{t_0}^{t_1} F(t, x(t), x'(t))\, dt \\ &\text{subject to} \qquad x(t_0) = x_0, \end{aligned} \tag{1}$$

given x_0, t_0, t_1. The terminal value $x(t_1)$ is free. Rather than joining two given points in the plane, we now seek the best differentiable curve joining a given point and a given vertical line (see Figure 8.1).

To find conditions necessarily obeyed by a function $x(t)$, $t_0 \le t \le t_1$, that is optimizing, we use the procedures already developed. For ease in writing, asterisks on optimal functions will sometimes be omitted where the meaning is clear from the context.

Suppose the function $x(t)$ is optimal and let $x(t) + h(t)$ be an admissible function. Thus, $x(t) + h(t)$ is defined on $[t_0, t_1]$, continuously differentiable, and satisfies the initial condition. This implies that $h(t_0) = 0$, but no restriction is placed on $h(t_1)$. The comparison curve may terminate at a point above or below (or at) $x(t_1)$ where the candidate ends, and therefore $h(t_1)$ may be positive, negative, or zero.

We consider the family of admissible curves $x(t) + ah(t)$, where $x(t)$ and $h(t)$ are *fixed*. The value of the integral (1) then depends on the parameter a, as

$$g(a) = \int_{t_0}^{t_1} F(t, x(t) + ah(t), x'(t) + ah'(t))\, dt, \tag{2}$$

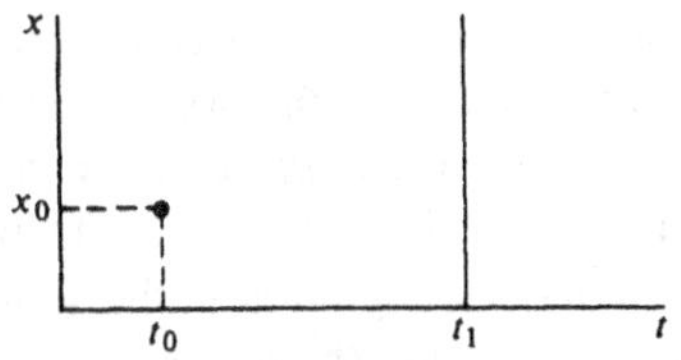

Figure 8.1

and assumes its maximum at $a = 0$ since x is optimal. Hence, as before,

$$g'(0) = \int_{t_0}^{t_1} [F_x(t, x, x')h + F_{x'}(t, x, x')h'] \, dt = 0. \tag{3}$$

Integrate the second term by parts, with $F_{x'} = u$ and $h' \, dt = dv$:

$$\int_{t_0}^{t_1} F_{x'} h' \, dt = F_{x'} h \,|_{t_0}^{t_1} - \int_{t_0}^{t_1} (h \, dF_{x'}/dt) \, dt$$

$$= (F_{x'} h) \,|_{t_1} - \int_{t_0}^{t_1} (h \, dF_{x'}/dt) \, dt$$

since $h(t_0) = 0$. The notation $(F_{x'}, h) \,|_{t_1}$ means that the expression in parentheses preceding the vertical line is to be evaluated at $t = t_1$. Substitution into (3) gives

$$\int_{t_0}^{t_1} h(F_x - dF_{x'}/dt) \, dt + (F_{x'} h) \,|_{t_1} = 0 \tag{4}$$

with F_x and $F_{x'}$ in the integrand evaluated along the optimal path $(t, x(t), x'(t))$.

Since (4) must be zero for *all* admissible comparison functions, it must hold, in particular, for functions terminating at the same point as the candidate function x. Therefore, (4) must be zero for all differentiable functions h satisfying $h(t_1) = 0$. This implies that the optimal function x must obey the Euler equation

$$F_x(t, x, x') = dF_{x'}(t, x, x')/dt, \qquad t_0 \le t \le t_1. \tag{5}$$

Since x satisfies (5), the condition that (4) be zero for all admissible functions h imposes the requirement that

$$F_{x'}(t_1, x(t_1), x'(t_1)) h(t_1) = 0$$

for any admissible $h(t_1)$. Since $h(t_1)$ is unrestricted and need not be zero, this in turn implies that

$$F_{x'} = 0 \quad \text{at} \quad t_1 \quad \text{if} \quad x_1 \text{ is free}. \tag{6}$$

Roughly, this means that a slight change in course at the last moment cannot improve the objective value. The boundary requirement (6) arising from optimality considerations is called a *transversality condition*. It is employed along with the given initial condition $x(t_0) = x_0$ to find the particular values of the two arbitrary constants of integration in the solution of the second order differential equation (5). To sum up, *the necessary conditions for a function x to solve problem (1) are that it satisfy the Euler equation (5), the initial condition $x(t_0) = x_0$, and the transversality condition (6).* The Legendre condition is also necessary.

Example 1. Find the shortest distance between the point $x(a) = A$ and the line $t = b$.

SOLUTION. From the formulation in Example 1.4, the problem is

$$\min \quad \int_a^b \left\{1 + \left[x'(t)\right]^2\right\}^{1/2} dt$$

$$\text{subject to} \qquad x(a) = A \quad (a, A, b \text{ fixed}).$$

Since the integrand F depends only on x', the solution to the Euler equation (following Case 5.3) has the form $x(t) = c_1 t + c_2$. Since the transversality condition (6) is $F_{x'} = x'/[1 + x'^2]^{1/2} = 0$, $x'(b) = 0$. Thus the constants c_1, c_2 must obey

$$x(a) = A = c_1 a + c_2, \qquad x'(b) = 0 = c_1,$$

so the extremal sought is

$$x(t) = A, \qquad a \leq t \leq b,$$

the horizontal line from (a, A) to (b, A). Note that the Legendre condition is satisfied since $F_{x'x'} > 0$. Indeed the integrand is convex and therefore this solution is minimizing, as is apparent.

Example 2. Optimum Checking Schedules and Random Failures.

A sophisticated application of our tools is determining a schedule for checking the status of a system that can fail at any time t. The cost per inspection of the status (failed or okay) is c_0. The longer the period T between the failure and its detection, the greater the loss $L(T)$ from failure. The more frequently a system is checked, the lower will be the loss $L(T)$ of undetected failure and the higher will be the total cost of checking. An optimal checking schedule minimizes the expected sum of these costs.

Suppose that inspections are so frequent that they can be described by a smooth density function $n(t)$, giving the number of checks per unit time. Therefore, $1/n(t)$ is the time interval between checks and $1/2n(t)$ is the expected time interval between a failure and the check that detects it. Suppose that the first failure occurs at t. The approximate (exact, if L is linear) expected loss will be $L(1/2n(t))$ and the cost of checking will be $c_0 \int_0^t n(s)\, ds$ (cost per check times number of checks up to t). Let $F(t)$ be the (known) probability of failure between time 0 and time t, so that $F'(t)$ is the probability density of failure. F is a nondecreasing function $F(0) = 0$ and $F(t_1) = 1$; the system will surely fail by t_1. The expected cost incurred over the period from time 0 until the time of detection of the first failure is

$$\int_0^{t_1} \left[c_0 \int_0^t n(s)\, ds + L(1/2n(t)) \right] F'(t)\, dt. \tag{7}$$

This is the cost if failure occurs at t multiplied by the probability density of failure at t, and integrated over all possible failure times. We seek a checking function $n(t)$ to minimize the expected cost (7).

Expression (7) will appear more familiar if we define $x(t) = \int_0^t n(s)\, ds$ so that $x'(t) = n(t)$. With these substitutions, (7) becomes

$$\begin{aligned} &\min \int_0^t [c_0 x(t) + L(1/2x')] F'(t)\, dt \\ &\text{subject to} \quad x(0) = 0, \quad x(t_1) \text{ free.} \end{aligned} \tag{8}$$

Routine computation produces the Euler equation

$$c_0 F'(t) = -d[L'(1/2x')F'(t)/2x'^2]/dt \tag{9}$$

Transversality condition (6) becomes

$$-L'F'/2x'^2 = 0 \quad \text{at} \quad t_1. \tag{10}$$

Separate variables in (9) and integrate:

$$c_0 F(t) = -L'F'/2x'^2 + k$$

where k is the constant of integration. Write $a = k/c_0$, recall that $x' = n$, and rearrange to

$$L'(1/2n)/n^2 = 2c_0[a - F]/F'. \tag{11}$$

In view of (11), $-L'F'/2n^2 = c_0[a - F]$, so that (10) is equivalent to

$$c_0[a - F(t_1)] = 0. \tag{12}$$

Since $F(t_1) = 1$ by assumption, (12) implies that

$$a = 1. \tag{13}$$

Therefore, putting (13) into (11), the rule for determining the optimum checking schedule $n(t)$ is given implicitly by

$$L'(1/2n(t))/n^2(t) = 2c_0[1 - F(t)]/F'(t). \tag{14}$$

In the special case where the loss due to a failed system is just proportional to the duration of the undetected failure, $L(T) = c_1 T$, Equation (14) gives an explicit solution

$$n(t) = \{c_1 F'(t)/2c_0[1 - F(t)]\}^{1/2}. \tag{15}$$

The larger the conditional probability density of failure $F'/(1 - F)$ at t, given survival until t, the smaller the cost c_0 of checking, and the larger the loss c_1 from undetected failure, the more frequently one checks.

EXERCISES

1. Find necessary conditions for the function $x = x^*(t)$, $t_0 \le t \le t_1$, to maximize

$$\int_{t_0}^{t_1} F(t, x, x')\, dt$$

   with only t_0 and t_1 given. Note that both $x(t_0)$ and $x(t_1)$ can be chosen optimally. (Answer: the Euler equation and the conditions $F_{x'}(t_i, x(t_i), x'(t_i)) = 0$, $i = 0, 1$.)

2. What necessary conditions are obeyed by a solution to (1) if "maximize" were replaced by "minimize"?

3. Find extremals for

$$\int_0^1 \{(1/2)[x'(t)]^2 + x(t)x'(t) + x(t)\}\, dt$$

   when $x(0)$ and $x(1)$ may be chosen freely.

4. In the checking model of Example 2, assume that the loss rate is constant ($L(T) = c_1 T$) and the optimal checking schedule (15) is applied.
   a. Show that the expected cost is

$$(2c_0 c_1)^{1/2} \int_0^{t_1} [F'(t)(1 - F(t))]^{1/2}\, dt.$$

   b. In the worst possible case, nature would chose F to maximize the expected cost in *a*. Find the worst possible function $F(t)$, assuming that failure is certain by a specified time t_1 ($F(t_1) = 1$). Then find the associated checking schedule. (Answer: $1 - F(t) = (1 - t/t_1)^{1/2}$, $0 \le t \le t_1$.)

FURTHER READING

Example 2 and Exercise 4 are based on the work of J. B. Keller (1974b).

Section 9

Free Horizon—Transversality Conditions

Consider a problem

$$\max \int_{t_0}^{t_1} F(t, x(t), x'(t))\, dt \qquad \text{subject to} \quad x(t_0) = x_0 \tag{1}$$

in which the initial point (t_0, x_0) is given but neither of the terminal coordinates are necessarily fixed in advance. The function F is twice continuously differentiable, as before.

Let t_1 and $x^*(t)$, $t_0 \le t \le t_1$, be optimal and consider a comparison function $x(t)$, $t_0 \le t \le t_1 + \delta t_1$. The domains of the two functions may differ slightly, with the number δt_1 small in absolute value but of any sign. Both x^* and x are continuously differentiable functions and satisfy the initial condition. Since their domains may not be identical, either x^* (if $\delta t_1 > 0$) or x (if $\delta t_1 < 0$) is extended on the interval $[t_1, t_1 + \delta t_1]$, so that the functions (as extended) do have a common domain. For instance, if $\delta t_1 > 0$, one may let x^* continue along a tangent drawn to x^* at t_1 so

$$x^*(t) = x^*(t_1) + x^{*\prime}(t_1)(t - t_1), \qquad t_1 \le t \le t_1 + \delta t_1.$$

If, on the other hand, $\delta t_1 < 0$, then x can be extended from $t_1 + \delta t_1$ to t_1 by a similar linear extrapolation. The extension is not specified in the sequel, but a workable procedure has been indicated.

Define the function $h(t)$ as the difference between the extended functions at each t on their common domain, so

$$x(t) = x^*(t) + h(t), \qquad t_0 \le t \le \max(t_1, t_1 + \delta t_1). \tag{2}$$

Since $x^*(t_0) = x(t_0) = x_0$ by hypothesis, we have $h(t_0) = 0$. Only comparison functions that are "close" to the candidate function are considered, where

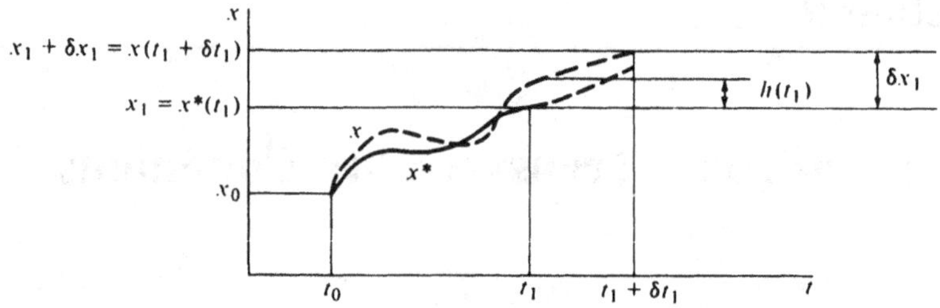

Figure 9.1

the distance between the functions x and x^* is defined by

$$\|x - x^*\| = \max_t |h(t)| + \max_t |h'(t)| + |\delta t_1| + |x(t_1 + \delta t_1) - x^*(t_1)|, \tag{3}$$

with the maximum absolute value taken over the domain of h. The last two terms reflect the difference in terminal coordinates of the two functions. Thus two continuously differentiable functions are close if, at each point of the extended domain, their values are close and their slopes are similar, and if, further, their termination points are close.

Figure 9.1 illustrates some of the notation to be used. In the figure, the distance between paths is exaggerated for clarity. Let $x^*(t)$, $t_0 \le t \le t_1$, optimize (1), let $h(t)$ be an admissible, arbitrary, and fixed function, and let δt_1 be a small fixed number. Define

$$g(a) = \int_{t_0}^{t_1 + a\,\delta t_1} F(t, x^*(t) + ah(t), x^{*\prime}(t) + ah'(t))\, dt. \tag{4}$$

The function g assumes its optimum at $a = 0$, so that $g'(0) = 0$ is required. Application of Leibnitz's rule (see (A1.10)) gives

$$g'(0) = F(t_1, x^*(t_1), x^{*\prime}(t_1))\, \delta t_1 + \int_{t_0}^{t_1} (F_x h + F_{x'} h')\, dt = 0. \tag{5}$$

Integrate the second term in the integral by parts, recalling that $h(t_0) = 0$, to get

$$g'(0) = F(t_1, x^*(t_1), x^{*\prime}(t_1))\, \delta t_1 + F_{x'}(t_1, x^*(t_1), x^{*\prime}(t_1))\, h(t_1) + \int_{t_0}^{t_1} (F_x - dF_{x'}/dt)\, h\, dt, \tag{6}$$

where the integrand is evaluated along the optimal path $(t, x^*(t), x^{*\prime}(t))$.

Call the difference in the value of the functions at their respective terminal points δx_1:

$$\delta x_1 \equiv x(t_1 + \delta t_1) - x^*(t_1). \tag{7}$$

Now approximate $x(t_1 + \delta t_1)$ by a line with intercept $x(t_1)$ and slope $x^{*\prime}(t_1)$ (see Figure 9.1). Then, we have approximately

$$\begin{aligned}\delta x_1 &\approx x(t_1) - x^*(t_1) + x^{*\prime}(t_1)\,\delta t_1 \\ &= h(t_1) + x^{*\prime}(t_1)\,\delta t_1. \end{aligned} \tag{8}$$

Thus, the difference in value of the functions at their respective terminal points is approximated by their difference in value at t_1 plus the change in value over the interval between t_1 and $t_1 + \delta t_1$. Rearrange (8) to

$$h(t_1) \approx \delta x_1 - x^{*\prime}(t_1)\,\delta t_1. \tag{9}$$

Substitute (9) into (6) and collect terms.

$$\int_{t_0}^{t_1} (F_x - dF_{x'}/dt)h\,dt + F_{x'}|_{t_1}\,\delta x_1 + (F - x'F_{x'})|_{t_1}\,\delta t_1 = 0. \tag{10}$$

This is the expression sought. It is the first variation for (1).

Since the comparison curve x *could* terminate at exactly the same point that x^* does, with $\delta t_1 = 0$ and $\delta x_1 = 0$, it follows that

$$\int_{t_0}^{t_1} (F_x - dF_{x'}/dt)h\,dt = 0 \tag{11}$$

must hold for all admissible functions h satisfying $h(t_0) = h(t_1) = 0$. Hence, it is necessary that the Euler equation

$$F_x - dF_{x'}/dt = 0$$

be satisfied. But then (10) reduces to

$$F_{x'}|_{t_1}\,\delta x_1 + (F - x'F_{x'})|_{t_1}\,\delta t_1 = 0. \tag{12}$$

Expression (12) is basic to finding the first order necessary conditions for optimality corresponding to any specification of terminal conditions. These conditions are used, together with the initial condition $x(t_0) = x_0$, to find the constants of integration in the solution of the Euler equation.

Before obtaining new results from (12), we check that old results follow. Of course, if both t_1 and $x(t_1) = x_1$ are given, (12) reduces to the identity $0 = 0$. In Section 8, we considered the case of t_1 fixed and $x(t_1)$ free. In the present context, this means $\delta t_1 = 0$ and δx_1 unrestricted so (12) implies

$$F_{x'}(t_1, x(t_1), x'(t_1))\,\delta x_1 = 0$$

for any δx_1. Therefore,

$$F_{x'}(t_1, x(t_1), x'(t_1)) = 0 \tag{13}$$

if x_1 is free is required, in agreement with Section 8.

Now suppose that t_1 is free but $x(t_1) = x_1$ is fixed. Then $\delta x_1 = 0$ and (12) becomes in this case

$$(F - x'F_{x'})|_{t_1} \delta t_1 = 0$$

for all δt_1. Therefore, we require

$$F - x'F_{x'} = 0 \tag{14}$$

at t_1 if t_1 is free.

Finally, if $x(t_0) = x_0$ and both $x(t_1)$ and t_1 are free, then (13) and (14) must hold. The initial condition and these two transversality conditions determine the two constants of integration in the Euler equation and the terminal time t_1. If x_0 and x_1 are given but t_1 is free, then the two given boundary conditions and (14) determine the two constants of integration and t_1.

We summarize the necessary conditions obtained for

$$\text{max or min} \quad \int_{t_0}^{t_1} F(t, x(t), x'(t))\, dt$$

$$\text{subject to} \quad x(t_0) = x_0.$$

Necessary Conditions

a. Euler equation: $F_x = dF_{x'}/dt$, $t_0 \le t \le t_1$.

b. Legendre condition:

$$\text{(max)} \quad F_{x'x'} \le 0, \quad t_0 \le t \le t_1$$

$$\text{(min)} \quad F_{x'x'} \ge 0, \quad t_0 \le t \le t_1.$$

c. Boundary conditions:
   (i) $x(t_0) = x_0$.
   (ii) If $x(t_1)$ is fixed, then $x(t_1) = x_1$ is known.
   (iii) If t_1 is fixed, then t_1 is known.

d. Transversality conditions:
   (i) If $x(t_1)$ is free, then $F_{x'} = 0$ at t_1.
   (ii) If t_1 is free, then $F - x'F_{x'} = 0$ at t_1.
   If both $x(t_1)$ and t_1 are free, conditions (d) can be written equivalently.

d′. Transversality conditions, if both $x(t_1)$ and t_1 are free:
   (i′) $F_{x'} = 0$ at t_1.
   (ii′) $F = 0$ at t_1.

Example 1. Find extremals for

$$\int_0^T \{c_1[x'(t)]^2 + c_2 x(t)\}\, dt \tag{15}$$

$$\text{subject to} \quad x(0) = 0, \quad x(T) = B, \tag{16}$$

where B is a given constant, but T is free.

From Section 4, the solution to the Euler equation is

$$x(t) = c_2 t^2/4c_1 + K_1 t + K_2, \qquad 0 \le t \le T. \tag{17}$$

Since T is free, the transversality condition is (14); in particular

$$c_1[x'(T)]^2 = c_2 x(T). \tag{18}$$

The three unknown constants K_1, K_2, and T are to be determined from (17) with the aid of the two given boundary conditions (16) and the transversality condition (18).

The three conditions are solved for the three unknowns. Since $x(0) = 0$, $K_2 = 0$. Then (18) yields

$$c_1(c_2 T/2c_1 + K_1)^2 = c_2(c_2 T^2/4c_1 + K_1 T).$$

Expanding and collecting terms gives $K_1 = 0$. Thus

$$x(t) = c_2 t^2/4c_1, \qquad 0 \le t \le T. \tag{19}$$

Also $x(T) = c_2 T^2/4c_1 = B$, so

$$T = 2(Bc_1/c_2)^{1/2}. \tag{20}$$

The extremal is given by (19) and (20).

Example 2. Example 4.7, in which we sought an optimal consumption plan over a definite time horizon, will now be generalized. We assume that the individual's lifetime is unknown so an optimal contingency plan is sought. Let $F(t)$ be the probability of dying by time t, $F'(t)$ the associated probability density function, and T an upper bound on possible lifetime (say 140 years), so that $F(T) = 1$. Then $1 - F(t) = \int_t^T F'(s)\,ds$ is the probability of living at least until t.

The individual derives satisfaction not only from consumption $U(C)$, but also from leaving an estate to beneficiaries. The latter is expressed through a utility-of-bequest function $W(K)$ that is continuously differentiable, nonnegative, and increasing. The function $W(K)$ differs from the utility-of-consumption function $U(C)$; it depends on a stock rather than a flow and reflects consumption permitted to beneficiaries. Let $a(t)$ be a discount term associated with estate utility. The behavior of this function is not specified. It may increase up to some t and decline thereafter, reflecting the individual's assessment of the relative importance of leaving a large estate in the middle years of life when the children are not fully grown, compared to early years before children arrive or later years when the children are on their own.

If the individual dies at time t, the total lifetime utility will consist of the discounted (at rate r) stream of utility from the consumption path up to t plus the discounted (factor $a(t)$) utility from the bequest at death. Hence the individual's problem is to

$$\max \int_0^T F'(t)\left[\int_0^t e^{-rs}U(C(s))\,ds + a(t)W(K(t))\right] dt \tag{21}$$

subject to the budget constraint (4.9),

$$C(t) = iK(t) + w(t) - K'(t), \tag{22}$$

and boundary condition

$$K(0) = K_0. \tag{23}$$

To put the problem in a readily manageable form, integrate by parts the portion of the objective involving the double integral (letting the inner integral be u and its coefficient $F'\,dt$ be dv). Then (21) is equivalent to

$$\int_0^T \{e^{-rt}U(C(t))[1 - F(t)] + a(t)W(K(t))F'(t)\}\,dt. \tag{21'}$$

This alternative form (21′) can be interpreted as follows. If the individual lives at least until t (probability $1 - F(t)$), then utility from consumption $U(C(t))$ is collected. If the individual dies at t (probability $F'(t)$), then utility from the bequest is also received.

Denote the integrand of (21′) by G, and use (22). Compute

$$G_K = e^{-rt}U'(C)i(1 - F) + aW'(K)F',$$
$$G_{K'} = -e^{-rt}U'(C)(1 - F).$$

After simplification, the Euler equation can be written

$$C'(t) = -(i - r)U'(C)/U''(C) + m(U'(C) - e^{rt}aW'(K))/U''(C) \tag{24}$$

where

$$m(t) \equiv F'(t)/[1 - F(t)] \tag{25}$$

is the conditional probability density of dying at t given survival until then. Comparing (24) with (4.13), one sees that the difference in the optimal rate of change of consumption under certainty and uncertainty is reflected in the second term of (24). In particular, if the bequest is not valued, that is, $a(t) = 0$, (24) becomes

$$C'(t) = -(i - r - m)U'(C)/U''(C),$$

from which it can be seen that the effect of uncertainty about one's lifetime is the same as an increase in the discount rate r; that is, as an increase in the rate of impatience. Specifically, uncertainty about one's lifetime raises the "effective'' discount rate at each t from r to $r + F'(t)/[1 - F(t)] = r + m$. Thus one discounts for both impatience and risk of dying. All this holds as well if $a(t) > 0$.

Since T is fixed and $K(T)$ is free, the relevant transversality condition is

$$F_{K'}|_T = -e^{-rT}U'(C(T))[1 - F(T)] = 0. \tag{26}$$

But since $1 - F(T) = 0$ by hypothesis, (26) provides no new information.

EXERCISES

1. Let B be the total quantity of some exhaustible resource, for example, the amount of mineral in a mine, controlled by a monopolist who discounts continuously at rate r and wishes to maximize the present value of profits from the mine. Let $y(t)$ be the cumulative amount sold by time t and $y'(t)$ be the current rate of sales. The net price (gross price less cost of mining) $p(y')$ is assumed to be a decreasing, continuously differentiable function of the current rate of sales:

$$p'(y') < 0.$$

a. Let T denote the time at which the resource will be depleted. Then choose $y(t)$ and T to maximize

$$\int_0^T e^{-rt}p(y'(t))y'(t)\,dt \qquad \text{subject to} \quad y(0) = 0, \quad y(T) = B.$$

Employ the Euler equation, transversality condition, and Legendre condition to show that the optimal plan involves sales decreasing over time, with $y'(T) = 0$.

b. Show that at T, the average profit per unit of resource extraction just equals the marginal profit.

c. Find the solution if

$$p(y') = (1 - e^{-ky'})/y',$$

where $k > 0$ is a given constant. (Partial answer: $T = (2kB/r)^{1/2}$.)

Note that increasing the initial stock by a fraction lengthens the extraction period only by a smaller fraction; that is, $T^*(B)$ is an increasing concave function.

d. Now suppose that the net price depends on the cumulative amount extracted and sold, as well as on the current rate of production—sales. (Mining may become more expensive as the digging goes deeper. Alternatively, the resource may be durable—e.g., aluminum—and the amount previously sold may be available for a second-hand market, affecting the demand for newly mined resource.) In particular, suppose

$$p(y', y) = a - by - cy',$$

where a, b, and c are given positive constants. Find the sales plan to maximize the present value of the profit stream in this case. The constants of integration and T may be stated implicitly as the solution of a simultaneous system of equations.

2. Find necessary conditions for optimizing (1) if either $x(t_0)$ or t_0, or both, may also be chosen freely.
3. Use your answer to Exercise 2 together with transversality conditions d and d' to show that

$$\int_{t_0}^{t_1} F_x(t, x^*(t), x^{*\prime}(t))\, dt = 0$$

if both t_0 and t_1 can be chosen freely (i.e., that F_x must equal zero on average if t_0 and t_1 can be chosen freely). Hint: Recall the form (3.11) of the Euler equation.

FURTHER READING

Example 2 is due to Yaari. Exercise 1 is due to Hotelling (1931).

Section 10

Equality Constrained Endpoint

Thus far terminal coordinates have been either given or else freely chosen. There are intermediate possibilities, with some degree of constrained choice regarding terminal coordinates.

Consider

$$\text{max or min} \quad \int_{t_0}^{t_1} F(t, x(t), x'(t))\, dt \tag{1}$$

$$\text{subject to} \quad x(t_0) = x_0, \tag{2}$$

$$R(t_1) = x_1, \tag{3}$$

where R is some differentiable function. Terminal time is neither fixed nor completely free; any modification in terminal time t_1 must be accompanied by a compensating change in the terminal function value x_1 to end on the terminal curve (3) (see Figure 10.1).

In particular, a slight change in δt_1 in terminal time occasions a change $R'(t_1)\,\delta t_1$ in terminal value:

$$R'(t_1) = \delta x_1 / \delta t_1. \tag{4}$$

The first variation for (1)–(3) is (9.10) with (4). As before, the Euler equation must hold for all $t_0 \le t \le t_1$. Then substituting from (4) into (9.12) gives

$$F + (R' - x')F_{x'} = 0 \quad \text{at} \quad t_1 \tag{5}$$

if $R(t_1) = x_1$ is required. Note that x' is the rate of change along the optimal path while R' is the rate of change along the terminal curve (3). With a fixed initial point and a terminal point to satisfy (3), one must find two constants of integration plus t_1 with the aid of the solution of the Euler equation, the initial condition, the terminal condition (3), and the transversality condition (5).

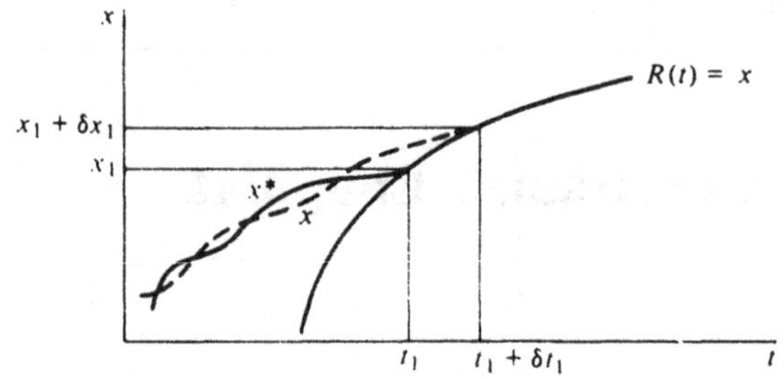

Figure 10.1

The summary at the end of Section 9 may be extended by

d. (iii) If t_1, x_1 must satisfy $R(t_1) = x_1$, then in addition

$$F + (R' - x')F_{x'} = 0 \qquad \text{at} \quad t_1.$$

Example. For functionals of the form

$$\int_{t_0}^{t_1} f(t, x)(1 + x'^2)^{1/2}\, dt$$

with $x(t_0) = x_0$, $R(t_1) = x_1$, the transversality condition (5) takes the form

$$f(1 + x'^2)^{-1/2}(1 + R'x') = 0,$$

so unless $f = 0$, the transversality requirement is

$$x'(t_1) = -1/R'(t_1).$$

The optimal path and terminal curve are orthogonal.

If the terminal curve were written in the implicit form

$$Q(t_1, x_1) = 0 \tag{6}$$

the transversality condition would look different. Since changes $\delta t_1, \delta x_1$ in terminal position must obey, to a linear approximation,

$$Q_t\, \delta t_1 + Q_x\, \delta x_1 = 0$$

we have

$$\delta x_1 / \delta t_1 = -Q_t / Q_x,$$

so the condition appears as

$$F - F_{x'}(x' + Q_t/Q_x) = 0 \qquad \text{at} \quad t_1 \tag{7}$$

if (6) must hold. Note that (5) and (7) are the same requirement; they differ in appearance alone. These equations indicate that

$$F - F_{x'}(x' - \delta x_1/\delta t_1) = 0 \quad \text{at} \quad t_1$$

where x' is the slope of the optimal path and $\delta x_1/\delta t_1$ denotes the slope of the terminal curve, both at the terminal point.

Example. Two groups of people are located within a circular city of radius r (and hence area πr^2). The smaller group occupies an area of size A and the larger group occupies the rest. What spatial configuration of the minority neighborhood provides a border with the majority population of minimal length?

Loury shows that the optimal spatial configuration involves locating the smaller group at the city's fringe, in a lens-shaped area determined by the intersection of two circles, one of which is the city's boundary.

We formulate the problem as follows: The city is of given radius r and may be taken to have center $(r, 0)$. The interior boundary of the minority neighborhood, denoted by $x(t)$, may be taken to begin at the origin $(0, 0)$ and end on the city fringe, so (t_1, x_1) satisfies

$$(t_1 - r)^2 + x_1^2 = r^2. \tag{8}$$

Its length is

$$\int_0^{t_1} \{1 + [x'(t)]^2\}^{1/2} \, dt. \tag{9}$$

Since the height of the circle (8) above t is $[r^2 - (t - r)^2]^{1/2}$, the total area between the circle (8) and the curve $x(t)$ is

$$\int_0^{t_1} \{[r^2 - (t - r)^2]^{1/2} - x(t)\} \, dt = A. \tag{10}$$

The curve starts at the origin.

$$x(0) = 0, \tag{11}$$

and ends on (8):

$$x(t_1) = x_1. \tag{12}$$

In sum, we seek a curve $x(t)$ (Figure 10.2) that minimizes (9) subject to (8), (10)–(12). This is an isoperimetric problem with a restricted endpoint. Appending constraint (10) to the objective (9) with multiplier λ yields the augmented integrand, denoted F, of

$$F = [1 + (x')^2]^{1/2} + \lambda\{[r^2 - (t - r)^2]^{1/2} - x\}.$$

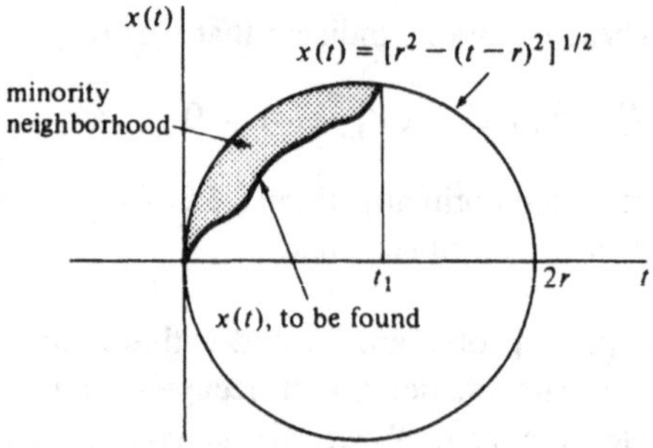

Figure 10.2

The Euler equation

$$-\lambda = d\left(x'/\left[1 + (x')^2\right]^{1/2}\right)/dt$$

may be integrated to

$$k - \lambda t = x'/\left[1 + (x')^2\right]^{1/2},$$

where k is a constant of integration. Square each side and solve algebraically for x'. Separation of variables gives

$$dx = (k - \lambda t)\,dt/\left[1 - (k - \lambda t)^2\right]^{1/2}.$$

Integration yields

$$x - c = \left[1 - (k - \lambda t)^2\right]^{1/2}/\lambda,$$

where c is a constant of integration. Squaring and rearranging terms produces

$$(x - c)^2 + (t - m)^2 = 1/\lambda^2 \tag{13}$$

where $m = k/\lambda$. The optimal path (13) is an arc of a circle. Denoting

$$Q(t_1, x_1) = x_1^2 + (t_1 - r)^2 - r^2$$

the transversality condition (7) becomes

$$\left[(1 + (x'_1)^2\right]^{1/2} - \left[x'_1 + (t_1 - r)/x_1\right]$$

$$x'_1/\left[1 + (x'_1)^2\right]^{1/2} = 0.$$

Simplifying,

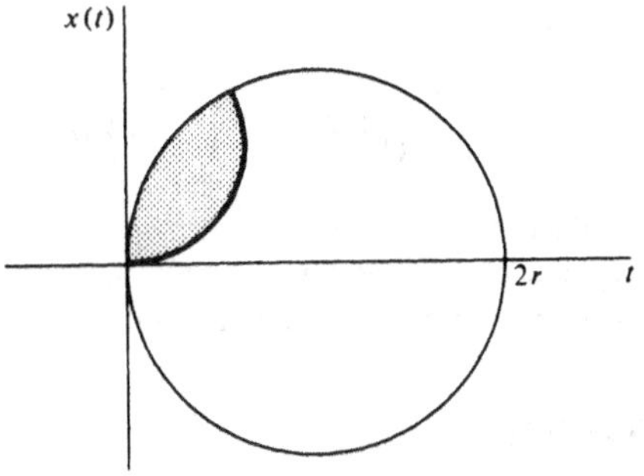

Figure 10.3

$$x'(t_1) = x_1/(t_1 - r).$$

This requires path (13) to be orthogonal to the city boundary at the endpoint. Computing x' from (13), we write this transversality condition as

$$(m - t_1)/(x_1 - c) = x_1/(t_1 - r). \tag{14}$$

Therefore, the solution to the problem posed in (8)–(12) is an arc of a circle that is orthogonal to the city boundary at the endpoints (Figure 10.3). The five constants m, c, λ, t_1, and x_1 that precisely determine this arc are found from the endpoint and transversality conditions summarized in (8), (10)–(12) and (14).

EXERCISES

1. Show that the shortest path from a given point (t_0, x_0) to a curve $R(t) = x$ is a straight line from (t_0, x_0) to $(t_1, R(t_1))$ perpendicular to the tangent to $R(t) = x$ at $(t_1, R(t_1))$, for some t_1.

2. Find necessary conditions for solution to (1), (2) and
$$P(x_1) = t_1. \tag{3'}$$

3. Develop the basic equation for finding transversality conditions in case one or both *initial* coordinates may be chosen freely or may be chosen to lie on a specified curve. Then find the relevant transversality conditions in each case. [Hint:
$$g(a) = \int_{t_0+a\delta t_0}^{t_1+a\delta t_1} F(t, x + ah, x' + ah')\,dt,$$
$$g'(0) = (F - x'F_{x'})|_{t_1}\delta t_1 + F_{x'}|_{t_1}\delta x_1$$
$$-(F - x'F_{x'})|_{t_0}\delta t_0 - F_{x'}|_{t_0}\delta x_0$$
$$+\int_{t_0}^{t_1}(F_x - dF_{x'}/dt)h\,dt.]$$

4. Find the curves for which

$$\int_0^{t_1} \left\{ \left[1 + (x')^2\right]^{1/2} / x \right\} dt$$

can have extrema subject to $x(0) = 0$, if
a. the point (t_1, x_1) must be on the line $x = t - 5$, or
b. the point (t_1, x_1) must be on the circle $(t - 9)^2 + x^2 = 9$.

FURTHER READING

The example is discussed by Loury.

Section 11

Salvage Value

The value of the objective may depend on the terminal position as well as on the path. For example, the reward for completing a task may depend on the speed with which it was completed. There may be a "salvage value" associated with the assets of the firm at terminal time.

Consider choosing t_1 and $x(t)$, $t_0 \le t \le t_1$ to optimize

$$\int_{t_0}^{t_1} F(t, x, x')\, dt + G(t_1, x_1) \qquad \text{subject to} \quad x(t_0) = x_0, \tag{1}$$

where $x_1 = x(t_1)$ is the terminal value of the function. Let $x^*(t)$, $t_0 \le t \le t_1$, be optimal and let $x(t)$, $t_0 \le t_1 \le t_1 + \delta t_1$, be a nearby admissible comparison function. Extend either x^* or x so they share a common domain and define

$$h(t) = x(t) - x^*(t), \qquad t_0 \le t \le \max(t_1, t_1 + \delta t_1). \tag{2}$$

Evaluating (1) along the function $x^* + ah$ on the interval $(t_0, t_1 + a\,\delta t_1)$ gives

$$g(a) = \int_{t_0}^{t_1 + a\,\delta t_1} F(t, x^* + ah, x^{*\prime} + ah')\, dt$$
$$+ G(t_1 + a\,\delta t_1, x_1 + a\,\delta x_1). \tag{3}$$

Since x^* is optimal, we have by Leibnitz's rule

$$g'(0) = \int_{t_0}^{t_1} (F_x h + F_{x'} h')\, dt + (F\,\delta t_1 + G_t\,\delta t_1 + G_x\,\delta x_1)\,|_{t_1} = 0. \tag{4}$$

Integrating by parts and recalling that $h(t_0) = 0$ gives

$$\int_{t_0}^{t_1} (F_x - dF_{x'}/dt) h\,dt + (F_{x'}h + F\,\delta t_1 + G_t\,\delta t_1 + G_x\,\delta x_1)|_{t_1} = 0. \quad (5)$$

Since a comparison curve terminating at (t_1, x_1) is admissible, (5) must hold in case $\delta t_1 = \delta x_1 = h(t_1) = 0$. This implies

$$F_x - dF_{x'}/dt = 0 \quad (6)$$

must hold along $x^*(t)$, $t_0 \le t \le t_1$. Furthermore, we have approximately, according to (9.9)

$$h(t_1) \approx \delta x_1 - x^{*\prime}(t_1)\,\delta t_1. \quad (7)$$

Substituting (6) and (7) into (5) and collecting terms yields

$$(F - x^{*\prime}F_{x'} + G_t)\,\delta t_1 + (F_{x'} + G_x)\,\delta x_1 = 0 \quad (8)$$

at t_1. This is the fundamental result for finding transversality conditions for problem (1).

If terminal time t_1 is free, then δt_1 may have any sign and therefore its coefficient must be zero:

$$F - x^{*\prime}F_{x'} + G_t = 0 \quad \text{at} \quad t_1 \quad \text{if} \quad t_1 \text{ is free}. \quad (9)$$

If the terminal position x_1 is chosen freely, then δx_1 may be of arbitrary sign and its coefficient must be zero.

$$F_{x'} + G_x = 0 \quad \text{at} \quad t_1 \quad \text{if} \quad x_1 \text{ is free}. \quad (10)$$

If the terminal time and position must satisfy a differentiable relation,

$$R(t_1) = x_1. \quad (11)$$

then, since $R'\,\delta t_1 = \delta x_1$, if (11) is required we obtain

$$F + F_{x'}(R' - x') + G_x R' + G_t = 0 \quad \text{at} \quad t_1. \quad (12)$$

The summary of necessary conditions in Sections 9 and 10 may now be replaced by the following summary for

$$\text{max or min} \quad \int_{t_0}^{t_1} F(t, x, x')\,dt + G(t_1, x_1)$$

$$\text{subject to} \quad x(t_0) = x_0.$$

Necessary Conditions

a. Euler equation: $F_x = dF_{x'}/dt$, $t_0 \le t \le t_1$.
b. Legendre condition:

$$\text{(max)} \quad F_{x'x'} \le 0, \quad t_0 \le t \le t_1;$$

$$\text{(min)} \quad F_{x'x'} \ge 0, \quad t_0 \le t \le t_1.$$

c. Boundary conditions:
   (i) $x(t_0) = x_0$.
   (ii) If $x(t_1)$ is fixed, then $x(t_1) = x_1$ is known.
   (iii) If t_1 is fixed, then t_1 is known.
   (iv) If t_1, x_1 must satisfy $R(t_1) = x_1$, then this equation provides a condition.

d. Transversality conditions:
   (i) If $x(t_1)$ is free, then $F_{x'} + G_x = 0$ at t_1.
   (ii) If t_1 is free, then $F - x'F_{x'} + G_t = 0$ at t_1.
   (iii) If t_1, x_1 must satisfy $R(t_1) = x_1$, then
$$F + F_{x'}(R' - x') + G_x R' + G_t = 0 \quad \text{at} \quad t_1.$$

Example. Imagine a research and development project in which there are decreasing returns to spending money faster. The more rapidly a given sum is spent, the less it contributes to total effective effort. (More rapid spending may be used for overtime payments, for less productive factors, for greater use of parallel rather than sequential efforts, etc.)

Let $x(t)$ be the rate of dollar spending at time t and let $z(t)$ be the cumulative effort devoted to the project by time t. The rate of spending $x(t)$ and the growth of cumulative effort are related by

$$z'(t) = x^{1/2}(t). \tag{13}$$

The total effective effort required to complete the project is A:

$$z(0) = 0, \quad z(T) = A, \tag{14}$$

where T denotes the time of completion (to be determined).

A reward of R can be collected when the project is completed. (R might be the value of a patent on the invention or the value of the stream of profits generated by the invention, discounted to the time the project is completed and the profit stream begins.) If the discount rate is r, then the value of the project at time 0 (now) is the profit less the development costs:

$$e^{-rT}R - \int_0^T e^{-rt}x(t)\,dt. \tag{15}$$

We maximize (15), subject to (13) and (14).

To express the problem wholly in terms of z and z', use (13) to eliminate x; then (15) may be written

$$\max e^{-rT}R - \int_0^T e^{-rt}[z'(t)]^2\,dt \tag{16}$$

Since z does not appear in (16), the Euler equation is

$$z'(t) = ce^{rt},$$

to be solved together with the boundary conditions (14) and transversality condition (9), since T may be chosen freely.

Integrating and using $z(0) = 0$ gives

$$z(t) = ce^{rt}/r - c/r. \tag{17}$$

In this problem (9) (after simplifying) requires that

$$z'(T) = (rR)^{1/2}. \tag{18}$$

Using (17) in (18) gives

$$ce^{rT} = (rR)^{1/2}. \tag{19}$$

Finally, setting $z(T) = A$ in (17) gives

$$(e^{rT} - 1)c/r = A. \tag{20}$$

Solve (19) and (20) for the two unknowns c and T:

$$c = (rR)^{1/2} - rA \tag{21}$$

and

$$T = -r^{-1}\ln(1 - A(r/R)^{1/2}). \tag{22}$$

Since the ln function is defined for positive numbers only, (22) is sensible only if

$$rA^2 < R. \tag{23}$$

Substituting from (21) into (17) gives

$$z(t) = \left[(R/r)^{1/2} - A\right](e^{rt} - 1), \qquad 0 \le t \le T, \tag{24}$$

as the optimal path of accumulation of effective R&D effort provided that (23) is satisfied. This proviso can be roughly interpreted as requiring that the effective effort A be sufficiently small in relation to the reward R. If (23) is not satisfied, then the project should not be undertaken. From (22), the optimal duration of development T varies directly with the required effort A and inversely with reward R; easier development or a bigger reward accelerate development. Combining (24) and (13) gives the optimal spending path: If (23) holds, then

$$x(t) = \left[(rR)^{1/2}\text{-}rA\right]^2 e^{2rt}, \qquad 0 \le t \le T,$$

where T is given in (22). Otherwise,

$$x(t) = 0, \qquad 0 \le t.$$

EXERCISES

1. Provide transversality conditions for (1) in the case where
   a. t_1, x_1 must satisfy $Q(t_1, x_1) = 0$,
   b. t_1, x_1 must satisfy $P(x_1) = t_1$.

2. Find transversality conditions for

$$\max \int_{t_0}^{t_1} F(t, x(t), x'(t))\, dt + G(t_0, x_0, t_1, x_1)$$

in the case where t_0 is fixed, x_0 may be freely chosen and the terminal point (t_1, x_1) must be on the curve $R(t_1) = x_1$.

3. Many firms may be doing R&D on similar projects. Assume that the first firm to complete its R&D claims the reward of R, and that no other firm gets a reward. (For instance, the first firm gets an exclusive patent that prevents others from selling similar inventions.) Suppose ABC Co. believes that the probability of rival invention by time t is

$$F(t) = 1 - e^{-ht}, \tag{25}$$

where $h > 0$ is constant. Assume ABC's R&D technology is given by (13) and (14). Find ABC's optimal spending plan and time of completion. How does the possibility of rival preemption affect the optimal development plan?

(Partial Solution: Since ABC spends money on R&D at t only if no rival has claimed the reward by t, and since it will collect the reward at T only if no rival appears by T, (15) is replaced by

$$e^{-rT}R(1 - F(T)) - \int_0^T e^{-rt}x(t)[1 - F(t)]\, dt,$$

where $F(t)$ is specified in (25).

4. Suppose the situation described in Exercise 3 is changed so that the first firm to complete its invention claims the reward of R, but a lesser reward of R_2, where $A^2/r < R_2 < R$, is available to a firm completing its R&D after the first prize of R has been claimed. Suppose that ABC has accumulated effective effort of z_0 by the time t_0, a rival, claims the reward R. Find ABC's optimal development program from the moment t_0 forward. Also find the value of that optimal program.

(Partial Solution: Since the problem is autonomous (defined in Section 15), we may count time from the moment a rival claims the reward R; therefore the optimal value from the moment t_0 forward is

$$r\left[(R_2/r)^{1/2} - A + z_0\right]^2 = \max e^{-rT_2}R_2 - \int_0^{T_2} e^{-rt}(z')^2\, dt$$

$$\text{subject to} \quad z(0) = z_0, \quad z(T_2) = A.$$

5. Suppose the R&D rewards are the same as in Exercise 4 and that ABC's R&D technology is (13) and (14). The probability of rival invention by t is (25) as expressed in Exercise 3. Find ABC's optimal development plan.
[Hints: If no one has completed invention by t, ABC spends at rate $(z')^2$ (to be found). ABC collects R at T (to be found) if no rival has claimed that reward by then. If a rival completes invention and claims the major reward R at t (probability density $F'(t) = he^{-ht}$) when ABC's accumulated effective effort is z, then

ABC can modify its development program. The value from t forward of that optimal program is that found in Exercise 4.

$$\max \quad Re^{-(r+h)T} + \int_0^T e^{-(r+h)t}\left\{hr\left[(R_2/r)^{1/2} - A + z(t)\right]^2 - \left[z'(t)\right]^2\right\} dt$$

subject to $\quad z(0) = 0, \quad z(T) = A.$

FURTHER READING

The R&D problem of this section and Exercises 3–5 have been explored in a series of papers by Kamien and Schwartz. See especially Kamien and Schwartz (1982). A differential game approach to the R&D problem is presented in Example 2 of Section II.23.

Section 12

Inequality Constraint Endpoints and Sensitivity Analysis

Optimization may be subject to endpoint inequality constraints. For example, the terminal time t_1 may be freely chosen so long as it does not exceed a given upper bound T:

$$t_1 \leq T. \tag{1}$$

Or the terminal value x_1 may be selected freely provided that

$$x_1 \geq a. \tag{2}$$

Frequently the lower bound a is zero. To cope with such possibilities, we make two changes in our procedures. First, we deal exclusively with maximization problems. (Analogous results for minimization problems are to be supplied in the exercises.) Second, we look more closely at how changes in the solution function x affect the value of the solution since feasible changes may be restricted.

Suppose the initial coordinates are fixed and requirements on the terminal coordinates are to be specified later. Thus,

$$\max \quad \int_{t_0}^{t_1} F(t, x(t), x'(t))\, dt, \tag{3}$$

$$\text{subject to} \qquad x(t_0) = x_0. \tag{4}$$

Let $x^*(t)$, $t_0 \leq t \leq t_1$, be the optimal function and let F^* denote the value of F along this function; that is, $F^*(t) = F(t, x^*(t), x^{*\prime}(t))$. Let J^* be the maximum value achieved in (3).

Let $x(t)$, $t_0 \le t \le t_1 + \delta t_1$ be a feasible comparison function that is close to the optimal function $x^*(t)$. Extend x or x^* so that they have a common domain. Let J denote the value of the integral in (3) for this comparison function. Then

$$\begin{aligned} J - J^* &= \int_{t_0}^{t_1+\delta t_1} F(t, x, x')\,dt - \int_{t_0}^{t_1} F(t, x^*, x^{*\prime})\,dt \\ &= \int_{t_1}^{t_1+\delta t_1} F(t, x, x')\,dt + \int_{t_0}^{t_1} [F(t, x, x') - F(t, x^*, x^{*\prime})]\,dt. \end{aligned} \tag{5}$$

Since J^* is the maximum, (5) must be nonpositive. We seek a linear approximation to (5). Since δt_1 is small and x is close to x^*, the first integral in the second line of (5) is approximately equal to

$$\int_{t_1}^{t_1+\delta t_1} F(t_1, x^*(t_1), x^{*\prime}(t_1))\,dt = F^*(t_1)\,\delta t_1.$$

Further, the second integrand in the second line of (5) can be expanded by Taylor's theorem (A2.9) around $(t, x^*, x^{*\prime})$. Hence

$$J - J^* = F^*(t_1)\,\delta t_1 + \int_{t_0}^{t_1} ((x - x^*)F_x^* + (x' - x^{*\prime})\,F_{x'}^*)\,dt + h.o.t., \tag{6}$$

where $F_x^*(t) = F_x(t, x^*(t), x^{*\prime}(t))$, $F_{x'}^*$ is defined analogously, and $h.o.t.$ denotes higher-order terms. Writing

$$h(t) = x(t) - x^*(t) \tag{7}$$

gives

$$J - J^* = F^*(t_1)\,\delta t_1 + \int_{t_0}^{t_1} (F_x^* h + F_{x'}^* h')\,dt + h.o.t. \tag{8}$$

The linear terms of (8) comprise the *first variation* of J, written

$$\delta J = F^*(t_1)\,\delta t_1 + \int_{t_0}^{t_1} (F_x^* h + F_{x'}^* h')\,dt. \tag{9}$$

Expression (9) is familiar from earlier calculations; compare it with (9.5). The recalculation highlights the appropriate interpretation of (9): the first variation (9) is the rate of change in the optimal value J^* due to a slight shift in path h, or shift in terminal coordinates. Its importance will be seen after transforming (9) in the now-familiar way.

Integrate the last term in (9) by parts with $h(t_0) = 0$ (since x_0 is fixed):

$$\delta J = F^*(t_1)\,\delta t_1 + F^*_{x'}|_{t_1} h(t_1) + \int_{t_0}^{t_1} (F^*_x - dF^*_{x'}/dt)\,h\,dt.$$

Dropping the asterisks for convenience and using (9.9) for $h(t_1)$ gives the form sought for the first variation of J, in agreement with (9.10)

$$\delta J = (F - x'F_{x'})|_{t_1}\,\delta t_1 + F_{x'}|_{t_1}\delta x_1 + \int_{t_0}^{t_1} (F_x - dF_{x'}/dt)\,h\,dt. \quad (10)$$

Since x^* provides the maximum, any feasible modificatons must lead to a solution with value no greater than J^*, that is, it must produce $\delta J \le 0$.

By a familiar argument that the comparison path was arbitrary and could surely terminate at the same point as the optimal path ($\delta t_1 = \delta x_1 = 0$), the integral in (10) must be nonpositive for all differentiable functions h with $h(t_0) = h(t_1) = 0$. This implies that the Euler equation

$$F_x - dF_{x'}/dt = 0 \quad (11)$$

must hold along the optimal path x^*.

Since the comparison path could have $x(t_1) = x_1$, we take $\delta x_1 = 0$ for this and the next paragraph. Then (10) reduces to

$$\delta J = (F - x'F_{x'})|_{t_1}\delta t_1 \le 0. \quad (12)$$

If t_1 is fixed in the problem specification, then $\delta t_1 = 0$ is required and (12) gives no further restriction. If t_1 is free, then δt_1 may have any sign and the coefficient of δt_1 must be zero. This gives the same transversality condition (9.14) as before.

Now imagine that t_1 is neither fixed nor completely free; rather, (1) is imposed. Then either the optimal $t_1 < T$ or else $t_1 = T$. In the first case, the comparison path may end before or after t_1, and therefore modifications δt_1 may be either positive or negative; hence

$$F - x'F_{x'} = 0 \quad \text{at} \quad t_1 \quad \text{if} \quad t_1 = T \text{ when (1) is imposed.} \quad (13)$$

The upper bound T is not an active constraint and the result is the same as if t_1 were unconstrained. In the second case, $t_1 = T$ and feasible modifications involve either the same or an earlier termination time, thus $\delta t_1 \le 0$ is required. But then, if (12) is to hold for all $\delta t_1 \le 0$, we must have

$$F - x'F_{x'} \ge 0 \quad \text{at} \quad t_1 \quad \text{if} \quad t_1 = T \text{ when (1) is imposed.} \quad (14)$$

This is the new result. Conditions (13) and (14) may be combined into

$$T \ge t_1, \qquad F - x'F_{x'} \ge 0, \qquad (T - t_1)(F - x'F_{x'}) = 0 \quad \text{at} \quad t_1. \quad (15)$$

Next, the appropriate transversality conditions are sought in case x is constrained. We require

$$F_{x'}|_{t_1}\,\delta x_1 \le 0 \qquad \text{for feasible} \quad \delta x_1. \tag{16}$$

If x_1 is fixed, then feasible $\delta x_1 = 0$ and (16) yields no information. If x_1 may be chosen freely, then feasible δx_1 may be positive or negative and (16) is assured only if $F_{x'}(t_1) = 0$. These are familiar results. If (2) is imposed, either $x_1 > a$ or else $x_1 = a$. If $x_1 > a$, then feasible modifications δx_1 may be positive or negative and therefore

$$F_{x'}(t_1) = 0 \qquad \text{if} \quad x_1 > a \text{ when (2) is imposed.} \tag{17}$$

On the other hand, if $x_1 = a$ in an optimal solution, then any comparison path must have the same or larger terminal value; that is, feasible modifications involve $\delta x_1 \ge 0$. In this case, (16) holds for all feasible modifications only if

$$F_{x'}(t_1) \le 0 \qquad \text{in case} \quad x_1 = a \text{ when (2) is imposed.} \tag{18}$$

Combining (17) and (18) gives the equivalent requirement that

$$x_1 \ge a, \qquad F_{x'}(t_1) \le 0, \qquad (x_1 - a)F_{x'}(t_1) = 0 \tag{19}$$

if (2) is required.

The first variation of (3), in case neither the initial nor terminal coordinates are necessarily fixed, can be developed in similar fashion. It is

$$\delta J = (F - x'F_{x'})|_{t_1}\,\delta t_1 + F_{x'}|_{t_1}\,\delta x_1 - (F - x'F_{x'})|_{t_0}\,\delta t_0 - F_{x'}|_{t_0}\,\delta x_0$$
$$+ \int_{t_0}^{t_1} (F_x - dF_{x'}/dt)\,h\,dt. \tag{20}$$

This is the fundamental expression for finding all the first order necessary conditions for (3) whichever boundary conditions are imposed.

The first variation, (20), makes it possible to conduct sensitivity analysis. That is, it can be asked how the optimized value of the integral (3) changes if its upper limit of integration t_1 changes or the terminal value x_1 changes. The changes in t_1 and x_1 are assumed to be independent. We begin by observing that when t_0 and $x(t_0)$ are fixed, the value of (3) along an extremal may be regarded as a function of t_1 and x_1. That is, let

$$V(t_1, x_1) = \int_{t_0}^{t_1} F(t, x^*, x^{*\prime})\,dt \tag{21}$$

$$\text{s.t.} \qquad x(t_0) = x_0, \quad x_1 \text{ free}.$$

Similarly, let

$$V(t_2, x_2) = \int_{t_0}^{t_2} F(t, \hat{x}, \hat{x}')\,dt \tag{22}$$

$$\text{s.t.} \qquad x(t_0) = x_0, \quad x_2 \text{ free},$$

where $t_1 \neq t_2$, $x_1 \neq x_2$ and $\hat{x}(t)$ refers to the extremal for the integral (22). Thus,

$$V(t_2, x_2) - V(t_1, x_1) = \int_{t_0}^{t_2} F(t, \hat{x}, \hat{x}')\, dt - \int_{t_0}^{t_1} F(t, x^*, x^{*\prime})\, dt, \quad (23)$$

which is just $J - J^* = \delta J$ in (5) if we let $t_2 = t_1 + \delta t_1$ and regard $\hat{x}(t) = x(t)$ as the comparison path to $x^*(t)$. But then, from (20) we get

$$V(t_2, x_2) - V(t_1, x_1) = \delta J = (F - x'F_{x'})|_{t_1} \delta t_1 + F_{x'}|_{t_1} \delta x_{1'} \quad (24)$$

where we have taken into account that the Euler equation must be satisfied. For t_1, t_2, x_1, x_2 sufficiently close, by the definition of a total differential of a function of two variables

$$V(t_2, x_2) - V(t_1, x_1) = V_t\, \delta t + V_x\, \delta x, \quad (25)$$

where V_t and V_x refer to the partial derivatives of V with respect to t and x, respectively. It then follows from (24) and (25) that

$$V_t = F - x'F_{x'} = -H, \quad (26a)$$

$$V_x = F_{x'} = p \quad (26b)$$

where H refers to the Hamiltonian and p the generalized momenta defined in (3.13). V_t and V_x represent the changes in the value of the integral (3) with respect to small changes in t_1 and x_1, respectively, when the path $x^*(t)$ is adjusted optimally.

Now if there are constraints $t_1 \leq T$, $x_1 \geq a$, and they are tight, i.e., $t_1 = T$, $x_1 = a$, for the extremal of (3), then V_t and V_x indicate by how much the optimal value of (3) would change with respect to a relaxation of each constraint separately. This means that the Hamiltonian H and the generalized momenta p may be regarded as the respective shadow prices associated with the constraints on the upper limit of integration t_1 and the terminal value x_1. That is, the Hamiltonian H and the generalized momenta p indicate the most one would be willing to pay to have the respective constraints on t_1 and x_1 relaxed. Obviously, if a constraint is not binding for an extremal of (3)—for example, if the optimal $t_1 < T$—then the most one would be willing to pay to relax this constraint is zero. The interpretation of H and p as shadow prices provide intuitive meanings to the transversality conditions (15) and (19). According to (15),

$$(T - t_1)(F - x'F_{x'}) = -(T - t_1)H = 0. \quad (27)$$

What this means is that if $t_1 < T$, then the shadow price of T, the Hamiltonian $H = 0$. While if the shadow price $-H > 0$, then $T = t_1$, the constraint must be binding. Similarly, from (19) we have

$$(x_1 - a)F_{x'} = (x_1 - a)p = 0 \quad \text{at} \quad t_1. \quad (28)$$

This means that either the constraint is not binding, $x_1 > a$, and its shadow price $p = 0$, or its shadow price $p < 0$ and the constraint is binding. (Note that the shadow prices H and p are both nonpositive here, a counter intuitive way of thinking of them. Thus, in optimal control theory the Hamiltonion is defined so that both shadow prices are nonnegative.)

Expressions (27) and (28) are referred to as complementary slackness conditions. Combining (26a, b) gives

$$V_t = F - V_x x', \tag{29}$$

a partial differential equation that must be satisfied along $x^*(t)$. It is known as the *Hamilton-Jacobi* equation (see II.21.7).

Example. Land Use in a Long, Narrow City.

A city is confined to a rectangular strip of given width W and maximum length L. A fixed amount of area A is to be a business district, with the remainder for roadway. Each square yard of business area generates g tons of traffic per unit time, with destinations uniformly distributed over the remaining business area. W is assumed sufficiently small for transport cost in the breadthwise direction to be neglected. The cost per ton-yard of moving traffic lengthwise a small distance at point t is an increasing function $f(v/x)$,

$$f(v/x) = b(v/x)^k, \qquad k \geq 1, \quad b > 0, \tag{30}$$

of the traffic density v/x at that point, with $v(t)$ the total volume of traffic passing lengthwise coordinate t and $x(t)$ the road width at t. Hence the total transport cost in the city per unit time is

$$\int_0^L v(t) f(v(t)/x(t))\, dt. \tag{31}$$

The problem is to choose the width of the road $x(t)$ and the length $L \leq \bar{L}$ of the city to minimize (31). (Note that t is distance, not time, in this problem).

We introduce a new variable y in terms of which both v and x will be given. Let $y'(t)$ denote the width of the business district at t (see Figure 12.1).

$$y'(t) = W - x(t) \tag{32}$$

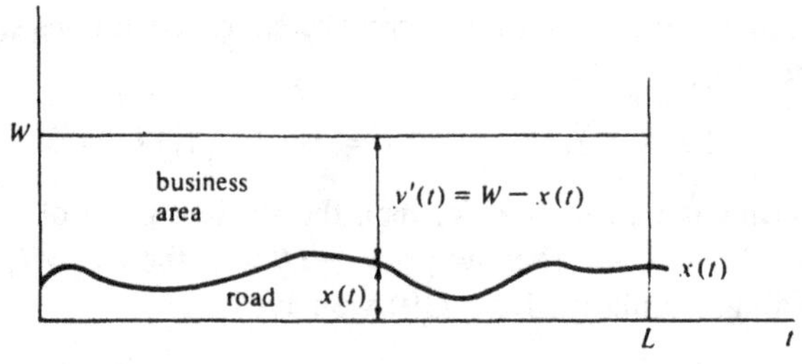

Figure 12.1

and $y(t)$ be the total business area to the left of t

$$y(t) = \int_0^t [W - x(s)]\, ds, \qquad y(0) = 0, \quad y(L) = A. \tag{33}$$

To calculate $v(t)$, first note that the volume of traffic originating to the left of t is $gy(t)$, of which a fraction $[A - y(t)]/A$ has destination to the right of t. Thus $gy(t)[A - y(t)]/A$ passes t from left to right. Second, a similar calculation shows that $g[A - y(t)]y(t)/A$ passes t from right to left. Hence the total volume $v(t)$ passing t is

$$v(t) = 2g[A - y(t)]y(t)/A. \tag{34}$$

Substituting from (30), (32), and (34) into (31) gives

$$\begin{aligned} &\min \quad (2g/A)^{k+1} b \int_0^L y^{k+1}(A - y)^{k+1}(W - y')^{-k}\, dt \\ &\text{subject to} \quad L \le \bar{L}, \quad y(0) = 0, \quad y(L) = A. \end{aligned} \tag{35}$$

The additional requirement $0 \le y'(t) \le W$ turns out to be satisfied automatically. Once the function $y(t)$ that minimizes (35) is known, the optimal x can be found through (32).

The positive constant in front of the integral in (35) may be ignored. The integrand $F(y, y')$ does not contain t, and therefore it is of the form discussed earlier in Case 2 in Section 5. The Euler equation is $F - y'F_{y'} = C$. Thus,

$$[y(A - y)/(W - y')]^{k+1}[W - (k + 1)y'] = C \tag{36}$$

for some constant C.

In the optimal solution, either $L < \bar{L}$ or else $L = \bar{L}$. We consider each case in turn. As is to be shown in Exercise 3 below, if the constraint on the length of the city is not tight, then the transversality condition $F - y'F_{y'} = 0$ applies; therefore $C = 0$. Since neither $y = 0$ nor $y = A$ for all t is feasible, $C = 0$ implies that $W - (k + 1)y' = 0$, or that

$$y' = W/(k + 1). \tag{37}$$

Putting (37) into (32) gives the road width at t:

$$x(t) = W - y'(t) = kW/(k + 1). \tag{38}$$

Optimal land use involves a road of constant width that is a fraction $k/(k + 1)$ of the city's width if $L < \bar{L}$.

To find out when $L < \bar{L}$, note that (37) implies that $y(t) = Wt/(k + 1)$, so that $y(L) = WL/(k + 1)$. Using the boundary conditions of (35), $WL/(k + 1) = A$ and

$$L = A(k + 1)/W \le \bar{L}. \tag{39}$$

Thus (39) is necessary to have road width given by (38).

If $A > \bar{L}W/(k+1)$, (39) is not satisfied. $L = \bar{L}$ and, as is to be shown in Exercise 3, $F - y'F_{y'} \le 0$ at $\bar{L}$, so that, in view of (36), $C \le 0$. In this case, a closed-form solution to the differential equation (36) is not available. From the symmetry of the problem, y will be symmetric about the midpoint $L/2$. This can be verified formally by showing that if $y(t)$ satisfies (36), then so does $z(t) = A - y(L - t)$.

To be meaningful, the first square bracketed expression in (36) must be positive so that the second term has the sign of C, which is negative:

$$W - (k+1)y' < 0. \tag{40}$$

Then (40) and (32) imply that

$$x(t) < kW/(k+1). \tag{41}$$

From (41), when the city is constricted, the road will be narrower at every point than would be optimal if the city length were unconstrained. (Recall that the business area is fixed while the total area is "too small.") Thus not only is the total road area smaller than when L is unrestricted, but also the road width is narrower at each t in the present case. To determine the relative shape or width of the road through the city, we seek y''. Take the logarithm of (36),

$$\begin{aligned}(k+1)\ln y + (k+1)\ln(A-y) - (k+1)\ln(W-y')\\ +\ln((k+1)y' - W) = \ln(-C),\end{aligned}$$

differentiate totally

$$\begin{aligned}(k+1)y'/y - (k+1)y'/(A-y) + (k+1)y''/(W-y')\\ +(k+1)y''/[(k+1)y' - W] = 0,\end{aligned}$$

and solve for y'' algebraically:

$$y'' = (2y - A)(W - y')[y'(k+1) - W]/ky(A-y). \tag{42}$$

From (40) and (42), y'' has the sign of $y - A/2$. By symmetry, $y'' < 0$ when $t < L/2$, $y''(L/2) = 0$, and $y'' > 0$ for $t > L/2$. Since $y'' = -x'$, the road

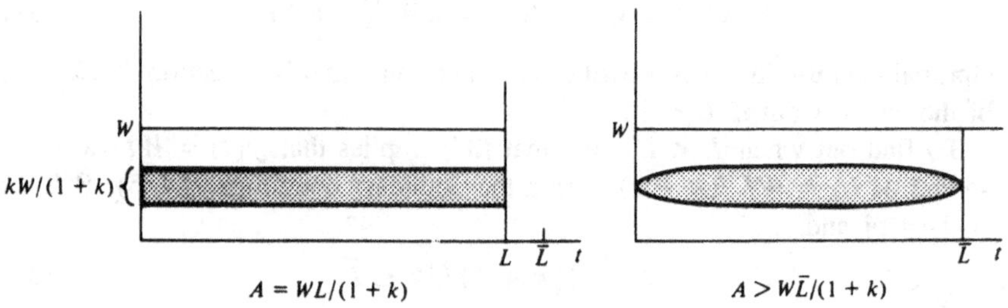

Figure 12.2

widens steadily from $t = 0$ to $t = L/2$ and then narrows symmetrically for $t > L/2$. Since $y'(0) = W$, $x(0) = W - y'(0) = 0$ and, by symmetry, $x(L) = 0$ (see Figure 12.2).

EXERCISES

1. Find necessary conditions for solution of (3) and (4) if at the same time t_1 is fixed and $a \le x(t_1) \le b$ is required.

2. Find necessary conditions for solution of

$$\min \int_{t_0}^{t_1} F(t, x, x')\, dt$$

$$\text{subject to} \quad x(t_0) = x_0, \quad a \le x(t_1) \le b, \quad t_0 \text{ fixed}, \quad t_1 \text{ free}.$$

3. Find necessary conditions for solution of

$$\min \int_{t_0}^{t_1} F(t, x, x')\, dt$$

$$\text{subject to} \quad x(t_0) = x_0, \quad x(t_1) = x_1, \quad t_1 \le T.$$

4. Show that (20) is the first variation of

$$\int_{t_0}^{t_1} F(t, x, x')\, dt,$$

where neither the initial nor terminal coordinates are necessarily fixed.

FURTHER READING

The example of land use in a long, narrow city was taken from Solow and Vickery.

Section 13

Corners

In the problems discussed so far, the admissible functions had to be continuously differentiable. It may be desirable to admit functions that are continuous throughout and are continuously differentiable except possibly at one or several points. Such functions are said to be *piecewise smooth*. A point at which the derivative is discontinuous is called a *corner*.

Example 1. The value of

$$\min \int_0^3 (x-2)^2(x'-1)^2\, dt$$

$$\text{subject to} \qquad x(0)=0, \quad x(3)=2,$$

is bounded below by zero. This lower bound is attained if either $x = 2$ or $x' = 1$ at each t, $0 \le t \le 3$. Thus

$$x(t) = \begin{cases} t, & 0 \le t \le 2, \\ 2, & 2 \le t \le 3, \end{cases}$$

must be minimizing. This function (Figure 13.1) is piecewise smooth, with a corner at $t = 2$.

We seek necessary conditions for a function x to optimize

$$\int_{t_0}^{t_1} F(t, x, x')\, dt \qquad \text{subject to} \quad x(t_0) = x_0, \quad x(t_1) = x_1, \tag{1}$$

where x must be piecewise smooth. Therefore, it must be continuous, with a continuous derivative everywhere on $[t_0, t_1]$ except possibly at a few points.

In particular, suppose the optimizing function $x^*(t)$ is continuously differ-

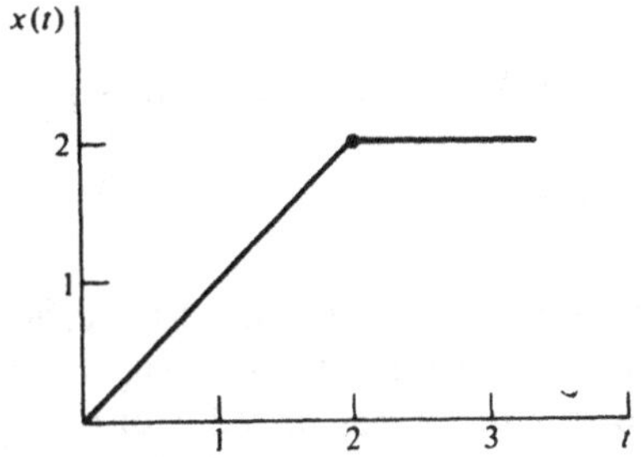

Figure 13.1

entiable on $[t_0, t_1]$ except at some point t_2, $t_0 < t_2 < t_1$. Write

$$\int_{t_0}^{t_1} F(t, x, x')\,dt = \int_{t_0}^{t_2} F(t, x, x')\,dt + \int_{t_2}^{t_1} F(t, x, x')\,dt. \quad (2)$$

On each interval, $[t_0, t_2]$, and $[t_2, t_1]$, $x^*(t)$ satisfies an Euler equation. To prove this, imagine that $x(t_2) = x_2$. Then $x^*(t)$, $t_0 \le t \le t_2$, must optimize

$$\int_{t_0}^{t_2} F(t, x, x')\,dt \qquad \text{subject to} \quad x(t_0) = x_0, \quad x(t_2) = x_2, \quad (3)$$

while $x^*(t)$, $t_2 \le t \le t_1$, optimizes

$$\int_{t_2}^{t_1} F(t, x, x')\,dt \qquad \text{subject to} \quad x(t_2) = x_2, \quad x(t_1) = x_1. \quad (4)$$

If the value of either of these subproblems could be improved, then the sum could be enhanced by replacing the relevant segment of x^* by the improved solution of the subproblem. Thus, if x^* solves (1), it also solves (3) and (4). Although x^* satisfies the same differential equation $F_x = dF_{x'}/dt$ on each interval, the constants of integration may differ with the boundary conditions appropriate to each interval employed. The conditions that hold at (t_2, x_2) are called the *Weierstrass-Erdmann Corner Conditions*. These are developed as follows.

Both t_2 and x_2 must be selected optimally, so no modification can improve the value in (2). To consider its implication, select arbitrary modifications $\delta t_2, \delta x_2$ in the optimal coordinates t_2, x_2. Changing the endpoint of (3) also changes the initial point of (4) since the solution is continuous on $t_0 \le t \le t_1$. To find the impact on (2) of slightly changing the switching point, we compute the changes in the values of (3) and (4).

By (12.20), the variation δJ of

$$J = \int_a^b F(t, x, x')\,dt \quad (5)$$

is

$$\delta J = (F - x'F_{x'})|_b \, \delta b + F_{x'}|_b \delta x_0 - (F - x'F_{x'})|_a \, \delta_a - F_{x'}|_a \delta x_a + \int_a^b (F_x - dF_{x'}/dt) h \, dt. \tag{6}$$

Evaluating (6) along x^* gives the linear part of the difference $J - J^*$, where J^* corresponds to x^* and J corresponds to a comparison curve x. We use (6) with $a = t_0$, $b = t_2$ to find the change in value of (3) due to the move of the terminal coordinates t_2, x_2 by $\delta t_2, \delta x_2$. Since the initial coordinates are unchanged, $\delta a = \delta t_0 = 0$ and $\delta x_a = \delta x_0 = 0$. The Euler equation will also be obeyed; therefore the integral is zero. Thus (6) applied to (3) reduces to

$$(F - x'F_{x'})|_{t_2^-} \, \delta t_2 + F_{x'}|_{t_2^-} \, \delta x_2 \tag{7}$$

as the change in value of (3) resulting from the move of the terminal coordinates. The superscript minus on t_2 indicates that the values are to be taken as t_2 is approached from below.

Moving the point t_2, x_2 not only shifts the terminal point of (3) but also shifts the initial conditions for problem (4). To find the resulting changes in value of (4), again we use (6), this time identifying $a = t_2$, $x_a = x_2$, $b = t_1$, $x_b = x_1$. The Euler equation is obeyed by x^*, so the integral in (6) equals zero. Since the comparison curve also ends at t_1, x_1, the terms δb and δx_b are zero. The change in (4), using (6) is thus

$$-(F - x'F_{x'})|_{t_2^+} \, \delta t_2 - F_{x'}|_{t_2^+} \, \delta x_2. \tag{8}$$

The plus superscripts indicate evaluation as t_2 is approached from above.

The net change in (2) from moving t_2, x_2 is the sum of changes in (3) and (4), namely, (7) plus (8):

$$\left[(F - x'F_{x'})|_{t_2^-} - (F - x'F_{x'})|_{t_2^+}\right] \delta t_2 + \left(F_{x'}|_{t_2^-} - F_{x'}|_{t_2^+}\right) \delta x_2. \tag{9}$$

If t_2, x_2 is optimal, no change in these coordinates can improve the value in (2). Since feasible changes in δt_2 and δx_2 are independent and may have any sign, their coefficients in (9) must each equal zero. Therefore,

$$F_{x'}|_{t_2^-} = F_{x'}|_{t_2^+} \tag{10}$$

and

$$(F - x'F_{x'})|_{t_2^-} = (F - x'F_{x'})|_{t_2^+}. \tag{11}$$

Conditions (10) and (11) which must hold at any discontinuity of $x^{*\prime}$ are the *Weierstrass-Erdmann Corner Conditions* and state that along an optimal path x^* the functions $F_{x'}$ and $F - x'F_{x'}$ are continuous. These functions must be continuous even where $x^{*\prime}$ is not continuous.

The two (identical) Euler equations are solved with the aid of the three boundary conditions $x(t_i) = x_i$, $i = 0, 1, 2$, and the two corner conditions (10) and (11) to determine the four constants of integration and t_2.

The corner conditions for Example 1 are that

$$F_{x'} = 2(x-2)^2(x'-1)$$

and

$$F - x'F_{x'} = -(x-2)^2(x'-1)(x'+1)$$

be continuous throughout. The proposed solution renders both expressions equal to zero on $[0, 3]$. Recall that if t does not appear in the integrand, then the Euler equation takes the form

$$F - x'F_{x'} = \text{const}.$$

We have just shown this to be satisfied (with constant = 0).

Example 2. Find the extremals with corners, if any, for

$$\int_0^T \left(c_1 x'^2 + c_2 x\right) dt \qquad \text{subject to} \quad x(0) = 0, \quad x(T) = B.$$

The expressions

$$F_{x'} = 2c_1 x'$$

and

$$F - x'F_{x'} = c_2 x - c_1 x'^2$$

must be continuous throughout. But continuity of $F_{x'}$ in this case directly implies continuity of x', so there can be no corners (no points of discontinuity of x').

EXERCISE

Find extremals of

$$\min \int_0^4 (x'-1)^2(x'+1)^4\, dt$$

$$\text{subject to} \qquad x(0) = 0, \quad x(4) = 2$$

that have just one corner.

FURTHER READING

Corners are often more readily handled by optimal control; see Section II12.

Section 14

Inequality Constraints in (t, x)

Consider optimizing

$$\int_{t_0}^{t_1} F(t, x, x')\, dt \tag{1}$$

$$\text{subject to} \quad R(t) \geq x(t), \quad t_0 \leq t \leq t_1,$$
$$x(t_0) = x_0, \quad x(t_1) = x_1, \tag{2}$$

where $R(t)$ is a given continuous function. The solution of the Euler equation for (1), with the given boundary points, may satisfy (2); the problem is then solved. Alternatively, there is some interval in which the solution to the Euler equation is not feasible and then constraint (2) is followed with $x(t) = R(t)$. One might think of solving (1) and (2) by finding the extremal that satisfies the endpoint conditions and replacing infeasible portions of the extremal by segments of the boundary $x(t) = R(t)$. This is generally *not* optimal, as will be illustrated later.

Suppose that the solution of the Euler equation satisfies (2) over $[t_0, t_2]$ and that (2) is tight over $[t_2, t_3]$, with $t_0 < t_2 < t_3 < t_1$ (see Figure 14.1). Then boundary conditions determining the constants of integration for the Euler equation $F_x = dF_{x'}/dt$, $t_0 \leq t \leq t_2$ are $x(t_0) = x_0$ and $x(t_2) = R(t_2)$. The optimal path continues, following $x(t) = R(t)$, $t_2 \leq t \leq t_3$ and a similar Euler equation is obeyed over the final interval $[t_3, t_1]$, with constants of integration determined in part by $x(t_3) = R(t_3)$ and $x(t_1) = x_1$. We must still determine t_2 and t_3. Since these are found in a similar manner, the selection of t_2 is emphasized.

The integral (1) can be expressed as the sum

$$\int_{t_0}^{t_2} F(t, x, x')\, dt + \int_{t_2}^{t_1} F(t, x, x')\, dt. \tag{3}$$

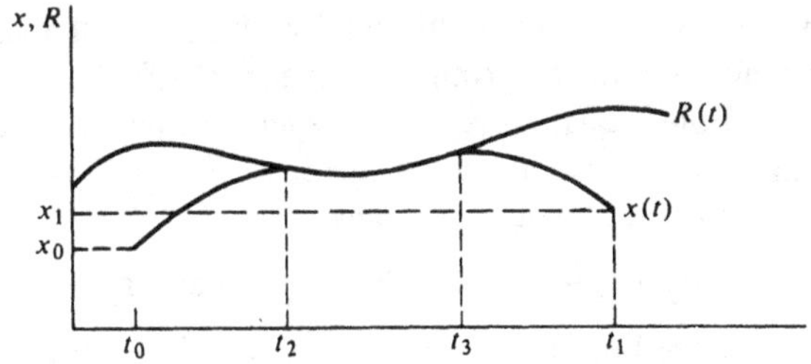

Figure 14.1

If $x^*(t)$, $t_0 \le t \le t_1$ is optimal for (1), then for given t_2, x_2 (with $R(t_2) = x_2$), the function $x^*(t)$, $t_0 \le t \le t_2$, must be optimal for

$$\int_{t_0}^{t_2} F(t, x, x')\, dt \qquad (4)$$

$$\text{subject to} \qquad x(t_0) = x_0, \quad x(t_2) = x_2, \quad R(t) \ge x(t),$$

while $x^*(t)$, $t_2 \le t \le t_1$, is optimal for

$$\int_{t_2}^{t_1} F(t, x, x')\, dt \qquad (5)$$

$$\text{subject to} \qquad x(t_2) = x_2, \quad x(t_1) = x_1, \quad R(t) \ge x(t).$$

This is obvious, since if the value of either of these subproblems could be improved, the value of the sum could clearly be enhanced by replacing the relevant segment of x^* with the optimal solution to the subproblem.

Suppose the optimal path x^* reaches the curve $R(t)$ at t_2. If t_2 were modified to $t_2 + \delta t_2$, the following changes would ensue. First, the Euler equation would be followed to point C instead of point B in Figure 14.2. Thus, the first portion of the solution, following the extremal to the curve $R(t)$ changes the endpoint slightly. Second, the path follows the constraint from time $t_2 + \delta t_2$ (point C) to t_3, rather than from t_2 to t_3.

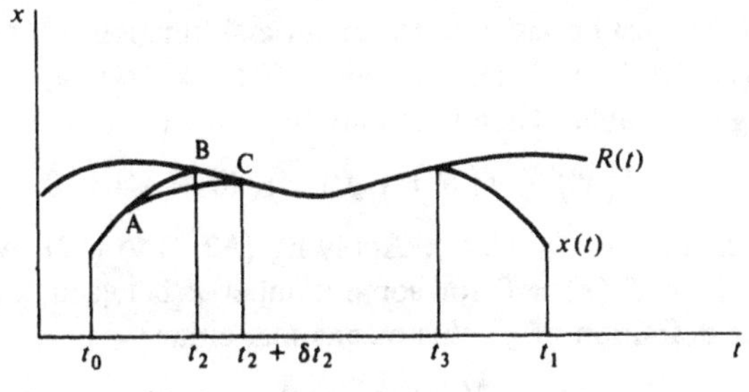

Figure 14.2

The change in the value of (1) is the sum of changes in the value of (4) and (5) due to the altering of t_2. Apply Equation (13.6) to (4) with $a = t_0$, $b = t_2$. Since the Euler equation is to be obeyed, the integral term in (13.6) is zero. The initial coordinates are unchanged, so $\delta t_0 = 0$ and $\delta x_0 = 0$. The change in (4) due to adjustment δt_2 is thus

$$F_{x'} \delta x_2 + (F - x' F_{x'})\, \delta t_2 \qquad \text{at} \quad t_2. \tag{6}$$

But since B and C are both on the constraint, δt_2, δx_2 must obey

$$R'(t_2)\, \delta t_2 = \delta x_2. \tag{7}$$

Combining (6) and (7) yields

$$\left[F + F_{x'}(R' - x')\right] \delta t_2. \tag{8}$$

Thus (8) is (the linear part of) the change in the value of (4) corresponding to δt_2. In Figure 14.2, it is the value of modifying the path to the constraining curve from AB to AC.

The value of the integral (5), the value of continuing from C rather than from B, is also affected. Since the path from B to C follows the curve $R(t)$, the change in (5) is

$$\int_{t_2}^{t_2 + \delta t_2} F(t, R(t), R'(t))\, dt \approx +F(t_2, R(t_2), R'(t_2))\, \delta t_2 \tag{9}$$

i.e., the value of F at t_2 times the length of the interval. This is the value foregone by omitting the segment BC from (5).

Subtracting (9) from (8) gives the net change in (1):

$$\left[F(t, x, x') + F_{x'}(t, x, x')(R' - x') - F(t, R, R')\right]\Big|_{t_2} \delta t_2. \tag{10}$$

Since feasible δt_2 may have any sign, its coefficient in (10) must be zero if no improvement is to be possible:

$$F(t, x_2, x') - F(t, x_2, R') + F_{x'}(t, x_2, x')(R' - x') = 0 \qquad \text{at} \quad t_2. \tag{11}$$

In (11), $x'(t_2)$ is the left-hand derivative, the rate of change of the extremal as it approaches the constraint; $R'(t_2)$ is the slope of the constraint at t_2; and $R(t_2) = x_2$.

Equation (11) must be satisfied by an optimal function $x^*(t)$ and number t_2. With $t_2, x^*(t_2)$ held fixed, one can view $F(t_2, x^*(t_2), y) = f(y)$ as a function of a single variable. Then (11) may be written

$$f(x') - f(R') + f'(x')(R' - x') = 0 \tag{12}$$

where $x' = x'(t_2)$, $R' = R'(t_2)$. Applying (A2.3) to (12), we can show that either $x' = R'$ or $f''(r) = 0$ for some number r between $x'(t_2)$ and $R'(t_2)$. Recalling our definition of f, this means that either

$$F_{x'x'}(t_2, x^*(t_2), r) = 0 \qquad \text{for some} \quad r$$

or

$$R'(t_2) = x^{*\prime}(t_2) \qquad \text{where } t_2 \text{ and } x^*(t) \text{ are optimal.} \tag{13}$$

A similar argument shows that (13) holds with t_2 replaced by t_3. Since R' is the slope of the constraint and x' is the slope of the extremal (solution of the Euler equation), the extremal is tangent to the constraining curve at juncture points if $F_{x'x'} \neq 0$ everywhere.

To sum up, an optimal solution to (1) is characterized by the following conditions:

a. On every interval such that $R(t) > x(t)$, the Euler equation $F_x = dF_{x'}/dt$ must be obeyed.
b. On every other interval, the function x is determined by $x(t) = R(t)$.
c. Let t^0 denote a time of switching from an interval of type a to an interval of type b or vice versa. Boundary conditions for determining the constants of integration in the solution to the Euler equation include $R(t^0) = x(t^0)$ and $R'(t^0) = x'(t^0)$ (provided $F_{x'x'} \neq 0$).

Example. A firm must meet an exogenous cyclical shipments schedule $S'(t)$, $0 \le t \le T$. Production cost is proportional to the square of the production rate and inventory cost is linear. Inventory must be nonnegative. Find the production plan to make the required shipments at minimum cost.

Let $x(t)$ be cumulative output by time t, so that $x'(t)$ is the output rate at t. Inventory is the cumulative output $x(t)$ less the cumulative shipments to date $S(t)$. Thus

$$\min \int_0^T \{c_1[x'(t)]^2 + c_2[x(t) - S(t)]\}\, dt \tag{14}$$

$$\text{subject to} \qquad x(t) \ge S(t), \quad 0 \le t \le T, \quad x(0) = x_0. \tag{15}$$

The solution involves time intervals on which $x(t) = S(t)$ and time intervals for which x obeys the Euler equation. While the constraint is tight, current production x' just matches current shipments S'. Extremals are of the form

$$x(t) = c_2 t^2/4c_1 + k_1 t + k_2. \tag{16}$$

Suppose an extremal is followed for $t_1 \le t \le t_2$. The four numbers t_1, t_2, k_1, and k_2 are determined by

$$\begin{aligned} x(t_1) &= S(t_1), & x(t_2) &= S(t_2), \\ x'(t_1) &= S'(t_1), & x'(t_2) &= S'(t_2), \end{aligned} \tag{17}$$

since the optimal path x must be continuous, the constraint $x \ge S$ is tight just before t_1 and after t_2, and $F_{x'x'} = 2c_1 \neq 0$. Therefore, by condition (c), $x' = S'$ at the moment between intervals of the two types. Since S is a known function, t_1, t_2, k_1, k_2 can be found for each interval on which the constraint is not tight.

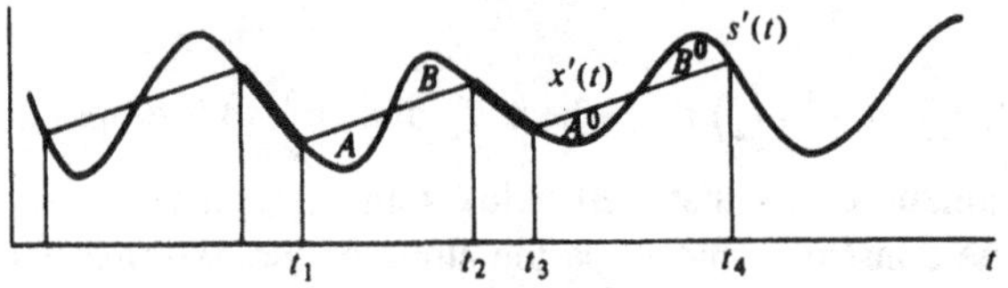

Figure 14.3

Note in Figure 14.3 that the slope of x', that is, the rate of increase in production, is $c_2/2c_1$ on each free interval, on which the constraint is not tight. In the early part of the free interval, such as (t_1, t_2) or (t_3, t_4), the production rate exceeds the shipment rate and inventory accumulates. The peak inventory on each free interval occurs when $x' = S'$. For the remainder of the free interval, shipments exceed production with the difference supplied from inventory. Area A = area B and area A^0 = area B^0. Inventory reaches zero at the end of the free interval, t_2 or t_4. For $t_2 \le t \le t_3$, production satisfies shipping requirements exactly.

We remark that if $\max_t S''(t) < c_2/2c_1$, then it would never be optimal to produce for inventory. Output equals sales throughout.

It is apparent from Figure 14.3 that the optimal solution does *not* consist of feasible portions of a single extremal joined by boundary segments, for such a solution would involve portions of a *single* straight line, $x' = c_2 t/2c_1 + k_1$, whereas Figure 14.3 involves portions of several straight lines (same slopes, different intercepts). One does *not* delete infeasible portions of the unconstrained solution to get the constrained optimal solution.

EXERCISE

$$\min \int_0^5 \{[x'(t)]^2 + 4x\}\,dt$$

$$\text{subject to} \qquad x(0) = 10, \quad x(5) = 0, \quad x(t) \ge 6 - 2t.$$

Sketch the optimal path and compare it with the path obtained by ignoring the inequality constraint.

FURTHER READING

The example is adopted from Anderson.

Section 15

Infinite Horizon Autonomous Problems

A problem of the form

$$\max \int_0^\infty e^{-rt} F(x(t), x'(t))\, dt \qquad \text{subject to} \quad x(0) = x_0 \tag{1}$$

has an infinite planning time or horizon. A problem is said to be *autonomous* if it does not depend on time explicitly; t is not an argument. But economists also say that (1) is *autonomous* since its dependence on time is merely through the discount term, which is constant. (The Euler equation for (1) will be autonomous; t is not an argument. See (2) following.) These two definitions differ but the context should make clear which is being applied.

In infinite horizon problems, there may be no necessary transversality condition. Even if a transversality condition holds, it may be of little help in determining constants of integration. The needed condition is often obtained from the observation that if such problems depend on time only through the discount term, it may be reasonable to expect the solution $x(t)$ to tend toward a stationary level x_s in the long run. A *steady state* or *stationary state* is one in which $x' = x'' = 0$. The steady state x_s for (1) is found by putting $x' = x'' = 0$ in the Euler equation

$$F_x = -rF_{x'} + F_{x'x}x' + F_{x'x'}x''. \tag{2}$$

The number x_s is therefore implicitly specified by

$$F_x(x_s, 0) + rF_{x'}(x_s, 0) = 0. \tag{3}$$

The second boundary condition for problem (1) is thus frequently taken to be

$$\lim_{t \to \infty} x(t) = x_s. \tag{4}$$

For example

$$\min \int_0^\infty e^{-rt}(x^2 + ax + bx' + cx'^2)\,dt$$

$$\text{subject to} \qquad x(0) = x_0.$$

where $c > 0$ and $r > 0$ gives rise to the Euler equation

$$x'' - rx' - x/c = (a + rb)/2c,$$

that has the particular solution

$$x_s = -(a + rb)/2.$$

The general solution to the Euler equation is

$$x(t) = Ae^{w_1 t} + Be^{w_2 t} + x_s$$

where $w_1, w_2 = r/2 \pm [(r/2)^2 + 1/c)]^{1/2}$ are the roots of the associated characteristic equation. These roots are real and opposite in sign, so $w_1 > 0 > w_2$. Selecting the constants of integration A and B so that the initial condition and the terminal condition (4) are satisfied yields

$$x(t) = (x_0 - x_s)e^{w_2 t} + x_s.$$

The steady state will not be attained in finite time but will be approached as t grows (recall that $w_2 < 0$). Differentiating gives

$$x'(t) = w_2[x(t) - x_s],$$

which has the form of a common "adjustment" equation. The value of $x(t)$ moves toward x_s at a rate proportional to the gap between the current value $x(t)$ and the "desired" or steady state value x_s. The rate of adjustment w_2 depends on the discount rate r and the parameter c.

Section 16

Most Rapid Approach Paths

Consider the infinite horizon autonomous problem that is linear in x':

$$\int_0^\infty e^{-rt}[M(x) + N(x)x']\,dt \tag{1}$$

$$\text{subject to} \qquad x(0) = x_0, \quad A(x) \le x' \le B(x). \tag{2}$$

The rate of change of x may be bounded, as noted. The Euler equation,

$$M'(x) + rN(x) = 0, \tag{3}$$

in an ordinary equation in a single variable x, not a differential equation (recall Case 5 of Section 5). Suppose (3) has a *unique* solution x_s; then x_s is the stationary solution. But this will not be feasible for all t unless $x_0 = x_s$. The optimal solution to (1) to (2) is to move from x_0 to x_s as quickly as possible and then remain at x_s. This is called a *most rapid approach path* (MRAP).

The claim can be made plausible as follows. (A more rigorous proof is in the appendix to this section.) Define $S(x) = \int_{x_0}^{x} N(y)\,dy$. Then $S'(x) = N(x)$, so (1) is equivalent to

$$\int_0^\infty e^{-rt}[M(x) + S'(x)x']\,dt. \tag{4}$$

Integration by parts (with $u = e^{-rt}$ and $dv = S'(x)x'\,dt$) gives

$$\int_0^\infty e^{-rt}[M(x) + rS(x)]\,dt, \tag{5}$$

which depends on x but not x'. (We assume $\lim_{t\to\infty} e^{-rt}S(x(t)) = 0$.) The problem has been written in such a way that the reward or payoff depends only on the state variable; one wants to get to a desirable state as soon as possible

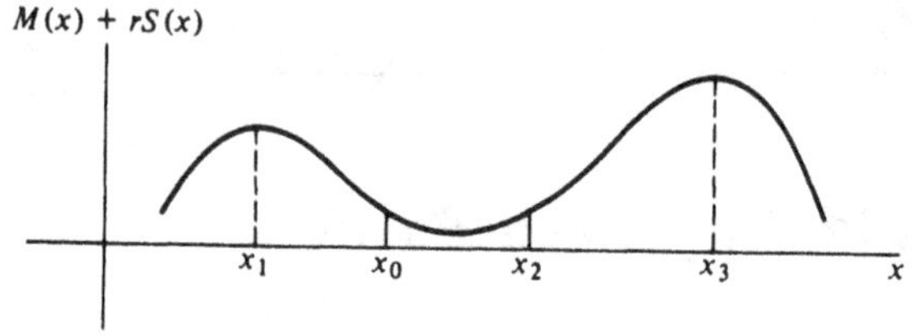

Figure 16.1

and stay there. There is the unique value of x that maximizes $M(x) + rS(x)$ and therefore satisfies $M'(x) + rS'(x) = M'(x) + rN(x) = 0$. This is x_s; see (3).

Recall that x_s satisfies (3) and suppose

$$M'(x) + rN(x) > 0 \quad \text{for} \quad x < x_s; \tag{6}$$
$$M'(x) + rN(x) < 0 \quad \text{for} \quad x > x_s.$$

Then the result can be stated as follows. If $x_0 < x_s$, then the optimal path involves the fastest possible rate of increase, $x' = B(x)$, until x_s is reached. Similarly, if $x_0 > x_s$, then the optimal path involves the fastest possible rate of decrease with $x' = A(x)$, until x_s is reached. Then x_s is maintained thereafter. If the inequalities of (6) are reversed, the MRAP is minimizing.

If (3) has more than one solution, then the optimal solution will be a MRAP to some local maximum of $M(x) + rS(x)$. To determine which local maximum is to be approached, it is necessary to evaluate (1) along the MRAP from x_0 to each local maximum x_s. Consider Figure 16.1. Starting at x_0, it clearly pays to move away from x_0, though it is not readily apparent which direction is best. There are two local maxima, x_1 and x_3. Certainly movement to the left as far as x_1 would be superior to staying at x_0 since it involves an improved path with $M(x) + rS(x) > M(x_0) + rS(x_0)$ at all times. Thus we might move from x_0 to x_1 as quickly as possible and then remain there.

Alternatively, it might be preferable to move to x_3 as quickly as possible and then remain there. The desirability of this alternative has to be closely evaluated, since to attain the higher peak at x_3, it is necessary to take a short-term decrease in payoff while passing from x_0 to x_2 before enjoying values of $M + rS$ in excess of that available at x_0. If there is more than one stationary state, as in this example, the one to approach may depend on the initial state.

Example 1. Let $R(x)$ be the maximum revenue a firm can earn with "goodwill" of x. Assume $R'(0) > 0$ and $R'' < 0$. Goodwill increases with advertising I and decays at constant proportional rate b: $x'(t) = I(t) - bx(t)$. The firm is to choose advertising spending $I(t)$ to maximize the present value of the stream of profits, $R(x) - I$:

$$\int_0^\infty e^{-rt}[R(x) - x' - bx]\,dt \tag{7}$$

$$\text{subject to} \qquad x(0) = x_0 > 0, \tag{8}$$

$$-bx \le x' \le \bar{I} - bx. \tag{9}$$

The lower bound on x' is achieved when spending on advertising is zero. The upper bound reflects a maximum (possibly infinite) permissible spending rate $\bar{I}$.

Equation (3) defining x_s becomes

$$R'(x_s) = r + b. \tag{10}$$

Since $R'' < 0$, (10) has at most one solution (we assume it has one) and (6) is satisfied. Therefore, the solution to (7)–(9) is a MRAP to x_s. Specifically, if $x_0 < x_s$, then $x' + bx = \bar{I}$ (so $x(t) = \bar{I}/b + (x_0 - \bar{I}/b)e^{-bt}$) until $x(t) = x_s$. Once x_s is achieved, it is maintained by keeping $I(t) = bx_s$.

Example 2. Modify the investment problem in Section 5 by making the terminal time infinite and the terminal value free:

$$\max \quad \int_0^\infty e^{-rt}[pf(K) - c(K' + bK)]\,dt$$

$$\text{subject to} \qquad K(0) = K_0,$$

where f is output from capital stock K, $f'(0) > 0$, $f'' < 0$; p is the constant price at which output can be sold; c is the constant cost per unit investment; and b is the rate of capital depreciation. The solution is the MRAP to the unique capital stock defined by

$$pf'(K_s) = (r + b)c.$$

If K' is unconstrained and $K_s \ge K_0$, K_s can be achieved immediately by a jump in K. If the growth in the firm must be self-financed so that $K' \le pf(K)/c - bK$, then the MRAP involves setting K' as large as possible until K_s is attained and choosing K' to maintain K_s thereafter. On the other hand, if $K_s < K_0$, the solution is to make no investment, $K' + bK = 0$, until K falls to K_s. Then K_s is maintained by investing at the constant rate bK_s to offset depreciation.

APPENDIX TO SECTION 16

The optimality of a MRAP can be proven by writing (1) as a line integral:

$$\int_0^\infty e^{-rt}[M(x) + N(x)x']\,dt = \int_C e^{-rt}M\,dt + e^{-rt}N\,dx,$$

where C is the curve $x = x(t)$, $0 \le t$ (see Section A7).

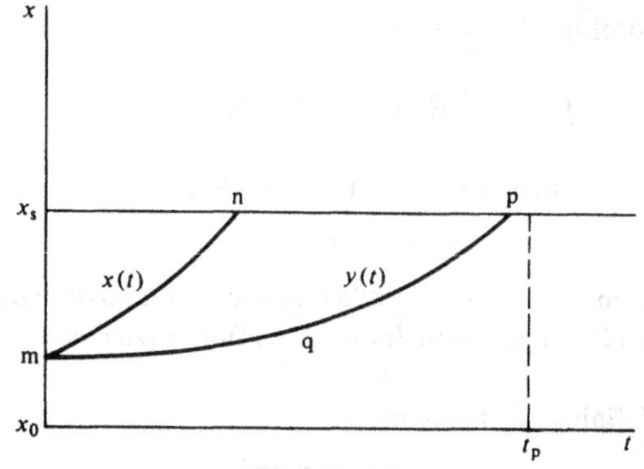

Figure 16.2

Let $x(t)$ be the path that reaches x_s as rapidly as possible and let $y(t)$ be some other feasible path (it reaches x_s more slowly). Let t_p be the time y gets to x_s. Then x and y coincide for $t > t_p$. We must show that x gives a higher value to the integral from 0 to t_p than y does (see Figure 16.2):

$$\int_0^{t_p} e^{-rt}[M(x) + N(x)x']\,dt - \int_0^{t_p} e^{-rt}[M(y) + N(y)y']\,dt > 0. \quad (11)$$

Each of these integrals can be written as a corresponding line integral, so the left side of (11) equals

$$\int_{mnp} e^{-rt}M\,dt + e^{-rt}N\,dx - \int_{mqp} e^{-rt}M\,dt + e^{-rt}N\,dx,$$

where the line integrals are evaluated along the optimal path of x and the comparison path mqp of y. Reversing the orientation of the optimal path, we write equivalently

$$-\int_{pnm} e^{-rt}M\,dt + e^{-rt}N\,dx - \int_{mqp} e^{-rt}M\,dt + e^{-rt}N\,dx$$

$$= -\int_{pnmqp} e^{-rt}M\,dt + e^{-rt}N\,dx. \quad (12)$$

This is a line integral around a single closed curve, oriented so the enclosed region is on the left as one proceeds along the curve. Apply Green's theorem to write (12) as the double integral (see Section A7)

$$\iint e^{-rt}[M'(x) + rN(x)]\,dx\,dt, \quad (13)$$

where the double integral is to be evaluated over the region bounded by the curve $mqpnm$. Throughout the region $x < x_s$ the integrand in (13) by hypothesis (6) is positive throughout the region and hence the integral (13) is positive. Consequently (12), and thereby the left side of (11), is positive, completing the demonstration.

If the inequalities in (6) were reversed, then the sign of (13) would be negative and the *MRAP* $x(t)$ would be minimizing.

EXERCISE

The population of a fish species in a lake at time t is $x(t)$. The natural growth rate of the population is $g(x)$, where $g(x)$ is a concave function with $g(0) = g(x_m) = 0$ and $g(x) > 0$ for $0 < x < x_m$. The fish are caught for sale at rate $h(t)$. Hence, the rate of change of the fish population is

$$x'(t) = g(x(t)) - h(t).$$

Let p be the price at which fish are sold and $c(x)$ be the cost of catching a fish when the population is x; $c(x)$ is a nonincreasing function. Why?

a. Show that the present value of profits derived from fishing can be written

$$\int_0^\infty e^{-rt}[p - c(x(t))][g(x(t)) - x'(t)]\, dt.$$

b. Find and characterize the fish population that maximizes the value of profits from fishing, found in *a*. (What are the bounds on $x'(t)$?)

FURTHER READING

See Spence and Starrett for a discussion of *MRAP* and Colin Clark (1976) and Sethi (1977b) for complementary discussions in terms of Green's theorem. Example 1 is due to Nerlove and Arrow. The exercise is due to Colin Clark, who has extensively studied the dynamics of fishing and fish populations. Also see V. L. Smith (1977). A differential game approach to the fish harvesting problem is presented in Example 3 of Section II.23.

Section 17

Diagrammatic Analysis

Consider the problem

$$\min \int_0^T e^{-rt}[f(x'(t)) + g(x(t))]\, dt \tag{1}$$

$$\text{subject to} \qquad x(0) = x_0, \quad x(T) = x_T, \tag{2}$$

where $f'' > 0$ and $g'' > 0$. The functions f and g are twice continuously differentiable and strictly convex, but not further specified. The Euler equation implies

$$x'' = [rf'(x') + g'(x)]/f''(x'). \tag{3}$$

The Legendre condition is always satisfied. Indeed the integrand of (1) is convex in x, x', so a solution to (2) and (3) will be minimizing.

Equation (3) cannot be solved without further specification of f and g. Since the signs of f' and g' have not been restricted, the sign of x'' is not obvious. Yet the solution can be characterized qualitatively by diagrammatic analysis. This is done by deducing how a path $x(t)$ that satisfies (3) can proceed through time. And this, in turn, is done by constructing a phase diagram in the x-x' plane.

Note that t is not an argument of (3); it is an autonomous differential equation. To ascertain directions of movement in the x-x' plane consistent with a solution to (3), note that whenever $x' > 0$, x increases through time. Therefore, from any point x, x' above the $x' = 0$ axis, the x coordinate of a solution to (3) must be increasing. Similarly, from any point below the $x' = 0$ line, the x coordinate of a solution to (3) must decrease. Thus, the $x' = 0$ line divides the plane into two regions, one in which x is increasing and the other in which x is decreasing (see Figure 17.1).

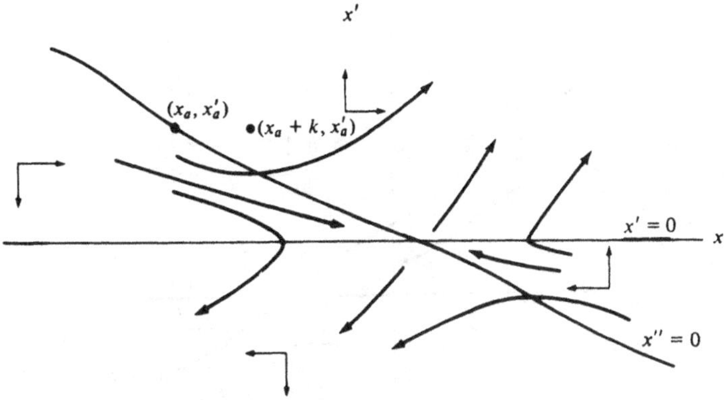

Figure 17.1

Next, we consider the locus of points x, x' such that $x'' = 0$. From (3), such points satisfy

$$rf'(x') + g'(x) = 0 \tag{4}$$

The slope of the curve implicitly described by (4) is

$$dx'/dx = -g''(x)/rf''(x') < 0 \quad \text{along} \quad x'' = 0. \tag{5}$$

Along the $x'' = 0$ locus, a small increase in the x coordinate must be accompanied by a small decrease in the x' coordinate in order to stay on the locus.

The $x'' = 0$ locus, (4), divides the x-x' plane into two regions. In one, the right side of (3) is positive so $x'' > 0$, and therefore x' is increasing. In the other, the right side of (3) is negative, so $x'' < 0$ and x' is decreasing. To determine the direction of movement to the right of the $x'' = 0$ locus, let (x_a, x_a') satisfy (4) (i.e., be on the locus), so $(x_a + k, x_a')$ is to the right of it for $k > 0$. Since $g'' > 0$, g' is an increasing function and therefore $rf'(x_a') + g'(x_a + k) > rf'(x_a') + g'(x_a) = 0$.

From (3), this means that $x'' > 0$ and thus x' is increasing at $(x_a + k, x_a')$. Similarly, from any point to the left of the $x'' = 0$ locus, we have $x'' < 0$ and the x' coordinate decreasing. The $x' = 0$ and $x'' = 0$ loci each divide the x-x' plane into two regions, giving four regions altogether. The directions of movement through time in the plane consistent with the analysis thus far are indicated in Figure 17.1. Both x and x' depend on t, but the argument has been suppressed. Typical paths consistent with the directional arrows are also illustrated. Thus, a point on a path indicates values of x, x' that might be realized at a moment t. The values x, x' advance along the path in the direction indicated as time elapses. Since the solution to the differential equation with given boundary conditions is unique, each point in the plane lies on exactly one path. Note that each path is horizontal (x' stationary) as it

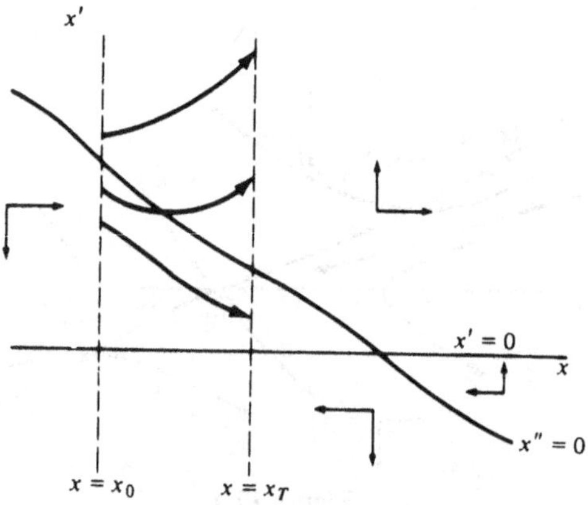

Figure 17.2

crosses the $x'' = 0$ locus and is vertical (x stationary) as it crosses the $x' = 0$ locus.

We have found diagrammatically the class of solutions to the Euler equations (3). It is clear from Figure 17.1 that the solution must be either monotonic, single-peaked, or single-troughed in $x(t)$: $x'(t)$ changes sign at most once.

From all the solutions to (3), we select the one that satisfies boundary conditions (2); this will be the path that begins on the vertical line $x = x_0$ and terminates on the vertical line $x = x_T$ with an elapsed time of T to make the trip. Each of the three paths illustrated in Figure 17.2 is the solution for some T.

If x_T in (2) is not specified, so that $x(T)$ may be selected freely, then the transversality condition is

$$f'(x'(T)) = 0. \tag{6}$$

Since f' is strictly monotone, there is at most one value of x' that satisfies (6). The Euler equation (3) must be satisfied, as must the initial condition $x(0) = x_0$. In this case, we select from among all the paths satisfying (3) the path that begins on the vertical line $x = x_0$ and goes to the horizontal line implicitly specified by (6) in a total elapsed time of T (see Figure 17.3).

Finally, modify problem (1) and (2) by assuming that $T \to \infty$ and $\lim_{t\to\infty} x(t)$ is free. In this case, the transversality condition is typically replaced by the requirement that x approach a steady state or stationary solution. A stationary solution x_s for which $x' = 0$ and $x'' = 0$, lies at the intersection of the two curves that divide the plane. It is given implicitly, from (3) by

$$rf'(0) + g'(x_s) = 0. \tag{7}$$

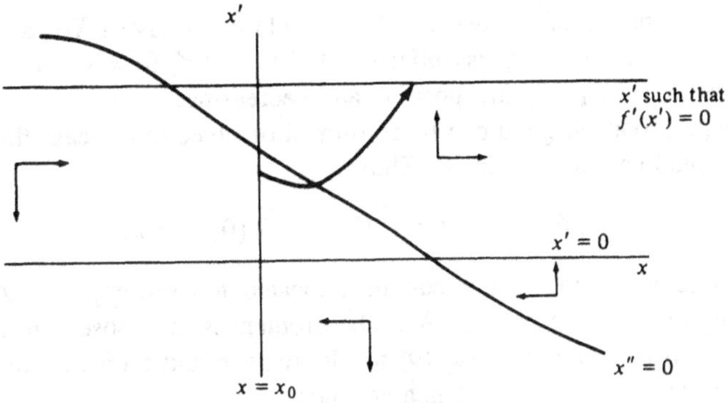

Figure 17.3

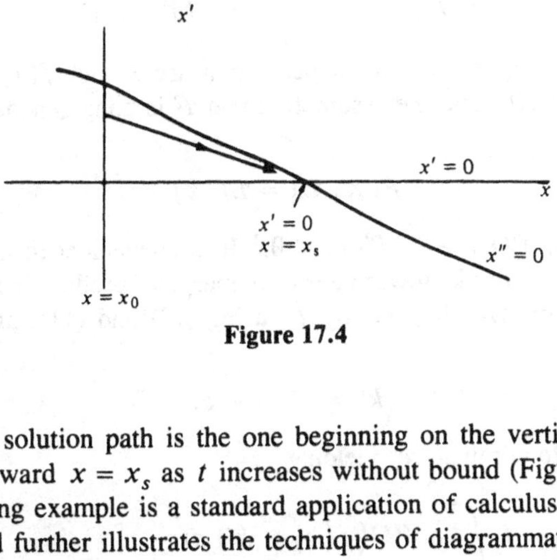

Figure 17.4

Therefore the solution path is the one beginning on the vertical line $x = x_0$ and tending toward $x = x_s$ as t increases without bound (Figure 17.4).

The following example is a standard application of calculus of variations to economics and further illustrates the techniques of diagrammatic analysis.

Example. The Neoclassical Growth Model.

In a constant population of L identical workers, per capita consumption is $c(t)$. Each worker derives utility $U(c)$ from consumption, where $U' > 0$ and $U'' < 0$. We suppose that marginal utility increases without bound as consumption shrinks to zero: $\lim_{c\to 0} U'(c) = \infty$. (For example, $U(c) = c^{1/2}$ has this property.) A central planner is to select a consumption plan to maximize aggregate discounted utility

$$\max \int_0^\infty e^{-rt} L U(c(t))\, dt \tag{8}$$

taking into account the production possibilities of the economy. A single output is produced using capital K and labor L in a production function $F(K, L)$

that is homogeneous of degree 1. (See A1.11) – (A1.14).) We assume that both productive factors are essential ($F(0, L) = F(K, 0) = 0$) and that their marginal products F_K, F_L are positive and decreasing.

The single produced good can be consumed or saved to increase the stock of capital. Capital can be consumed. Thus

$$F(K, L) = Lc + K', \qquad K(0) = K_0. \tag{9}$$

According to (9), total output can be allocated to consumption, Lc, or to augmenting the stock of capital, K'. The problem is to choose c to maximize (8), subject to (9). One can use (9) to eliminate c and then find the optimal function $K(t)$, recalling that L is a constant.

A change of variable is especially useful when L is growing exponentially. We make it now, in anticipation of later use. Define

$$k \equiv K/L, \qquad f(k) \equiv F(K/L, 1). \tag{10}$$

Thus capital per capita and output per capita are k and $f(k)$, respectively. Using definitions (10) and the assumption that F is homogeneous of degree 1, we have

$$F(K, L) = Lf(k) \tag{11}$$

where $f(0) = 0$, $f'(k) > 0$, $f''(k) < 0$.[1] It is convenient to assume that, as the use of a factor shrinks toward zero, its marginal product increases without bound: $f'(0) = \infty$. Dividing (9) by L, using (10) and (11), and rearranging gives

$$k' = f(k) - c. \tag{12}$$

Use (12) in (8) to eliminate c, yielding

$$\begin{aligned} &\max_k \quad L\int_0^\infty e^{-rt}U(f(k) - k')\,dt \\ &\text{subject to} \qquad k(0) = k_0, \quad k \geq 0, \quad f(k) - k' \geq 0. \end{aligned} \tag{13}$$

The Euler equation for (13) is

$$[dU'(c)/dt]/U'(c) = U''(c)c'/U' = -[f'(k) - r]. \tag{14}$$

The assumptions on f' and U' assure that the nonnegativity restrictions will be satisfied.

---

[1] Differentiate (11) using definitions (10) and the chain rule:

$$0 < F_K = Lf'(k)\,\partial k/\partial K = Lf'(k)/L = f'(k),$$
$$0 > F_{KK} = f''(k)\,\partial k/\partial K = f''(k)/L.$$

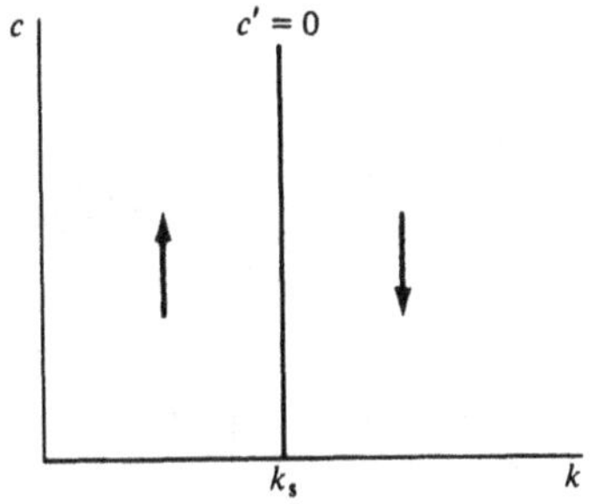

Figure 17.5

Rule (14) should be compared with (4.13). The proportionate rate of change of marginal utility equals the difference between the earnings rate on invested capital and the discount rate. In the present case, the earnings rate on capital is endogenous; that is, the marginal productivity of capital is not a given parameter but rather is determined within the model.

The optimal solution satisfies both the basic balance equation (12) and the Euler equation (14). To develop qualitative properties of the optimal solution, we construct a phase diagram in the nonnegative c-k plane. From (14) $c' = 0$ if

$$f'(k) = r. \tag{15}$$

Thus, (15) defines the locus of points in the c-k plane for which $c' = 0$. Since the left side is monotone decreasing, (15) has a unique solution. Call it k_s. If $k > k_s$, then

$$f'(k) - r < f'(k_s) - r = 0$$

since $f'' < 0$ by assumption. Hence, from (14) (recall $-U''/U' > 0$ always), it follows that $c' < 0$ when $k > k_s$. Similarly, $c' > 0$ when $k < k_s$. The directional arrows in Figure 17.5 reflect these conclusions. From a point in k-c space to the right (left) of k_s, the c coordinate must fall (rise).

Next, from (12), $k' = 0$ whenever

$$c = f(k). \tag{16}$$

Thus, (16) defines the locus of points for which $k' = 0$. The curve described by (16) passes through the origin and has derivatives

$$dc/dk = f'(k), \qquad d^2c/dk^2 = f''(k).$$

In view of our assumptions about f, the curve described by (16) is increasing and concave (see Figure 17.6). It divides the k-c plane into two sections. Let (k_a, c_a) satisfy (16) and consider the point $(k_a, c_a + m)$ where $m > 0$. Thus $(k_a, c_a + m)$ lies above (k_a, c_a). Then, from (12),

$$k' = f(k_a) - (c_a + m) = -m < 0,$$

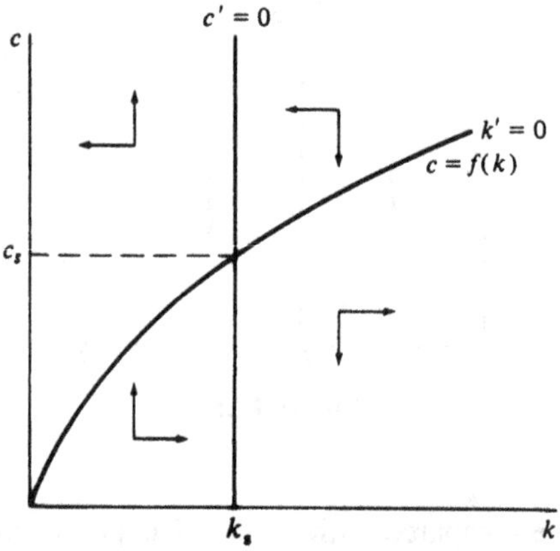

Figure 17.6

and so k declines at $(k_a, c_a + m)$. Therefore k is falling at every point above the $k' = 0$ locus. Similarly, from any point below the $k' = 0$ locus, $k' > 0$ and so k is rising. The directional arrows in Figure 17.6 reflect these conclusions. The diagram now indicates the general direction of movement that (k, c) would take from any location. Note in each case that c is momentarily stationary along a path as the $c' = 0$ locus is crossed and that k is stationary as the $k' = 0$ locus is crossed.

Is there a steady-state level of per capita consumption and capital that is sustainable forever? If so, is there a path tending toward this point? The answer to both questions is "yes." A steady-state is one in which $c' = 0$ and also $k' = 0$, namely, (k_s, c_s) where $c_s = f(k_s)$. Furthermore, from the theorem regarding the existence of a solution to a differential equation, there is at most one path from the line $k = k_0$ to the point (k_s, c_s).

Suppose that $k_0 < k_s$. Then the approach to (k_s, c_s) must be from below, with c and k both rising monotonically from their initial values to their stationary values along the unique path 1 in Figure 17.7. If initial consumption were chosen too large (such as in path 2), then capital will be accumulated for a while and later diminish, eventually becoming depleted and crossing into $k < 0$, an infeasible solution. On the other hand, if initial consumption were chosen too small, then consumption will rise for a while, peak, and then fall, as shown in path 3. Such a path cannot be optimal since, after a time, one could increase consumption, jumping vertically to path 4 leading toward k_s, c_s from above. Path 1 is the only optimal plan if $k_0 < k_s$. If $k_0 > k_s$, then the optimal path to the steady state has both k and c decreasing monotonically along path 4. Other paths are consistent with the directional arrows and hence the two differential equations, but if the horizon is infinite, they can be ruled out as either infeasible or inferior, by arguments similar to those just used. The paths 1 and 4 tending to k_s, c_s are the only optimal ones. If $k_0 < k_s$, path 1 is

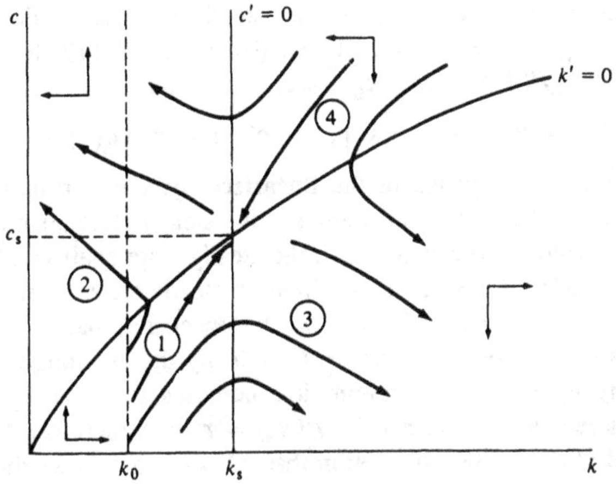

Figure 17.7

followed with both c and k increasing monotonically with advancing time. If $k_0 > k_s$, path 4 is followed with k and c each decreasing over time.

The behavior of the system (12) and (14) can be approximated in the neighborhood of the steady state (k_s, c_s) by a related system of linear differential equations. Expand the right sides of (12) and (14) in a Taylor series around (k_s, c_s), retaining only the linear terms. For example,

$$f(k) - c = [f(k_s) - c_s] + f'(k_s)(k - k_s) - (c - c_s) + h.o.t.$$

In view of (15), the square-bracketed term on the right is zero. Thus, approximately,

$$k' = f'(k_s)(k - k_s) - (c - c_s) \tag{17}$$

and, similarly,

$$c' = -f''(k_s)(k - k_s)U'(c_s)/U''(c_s). \tag{18}$$

The homogeneous system corresponding to (17) and (18) has the characteristic equation (See Section B5.)

$$m^2 - mf'(k_s) - f''(k_s)U'(c_s)/U''(c_s) = 0 \tag{19}$$

with roots

$$m_1, m_2 = \left\{f'(k_s) \pm \left[f'(k_s)^2 + 4f''(k_s)U'(c_s)/U''(c_s)\right]^{1/2}\right\}/2. \tag{20}$$

These roots are real and of opposite signs. Let $m_1 > 0 > m_2$. The general form of the solution to (17) and (18) is

$$k(t) = A_1e^{m_1t} + B_1e^{m_2t} + k_s, \qquad c(t) = A_2e^{m_1t} + B_2e^{m_2t} + c_s. \tag{21}$$

Since the solution is to converge to the steady state (i.e., $\lim_{t\to\infty} k_t = k_s$, $\lim_{t\to\infty} c(t) = c_s$), $A_1 = A_2 = 0$. Use $k(0) = k_0$ and (17) to find $B_1 = k_0 - k_s$ and $B_2 = [f'(k_s) - m_2](k_0 - k_s)$. Note that

$$k'(t) = m_2[k(t) - k_s], \qquad c'(t) = m_2[c(t) - c_s].$$

Thus the rate of convergence of the linearized system is proportional to the deviation between the current position and the steady state position.

A question asked in connection with the steady-state analysis of the neoclassical growth model is: What is the highest steady-state level of per capita consumption, c_s. The answer is provided by observing that in the steady-state $c_s = f(k_s)$, and therefore $dc_s/dk_s = f' > 0$, by the assumption that $f' > 0$. That is, steady-state per capita assumption increases as the steady-state optimal stock, k_s, increases. But from (15) $f'(k_s) = r$, and therefore from the total differential of (15) and the assumption that $f'' < 0$, it follows that $dk_s/dr = 1/f''(k_s) < 0$. Thus, the steady-state capital stock and therefore the steady-state level of consumption , c_s, increase as the discount rate r declines. The highest level of steady-state per capita consumption occurs if $r = 0$. The path of capital accumulation that leads to this highest steady-state level of per capita consumption is known as the *golden rule* of capital accumulation. The golden rule, "Do unto others as you would have them do unto you," in the context of optimal growth means that the present generation should not discount the utility of consumption of future generations (you would not like them to do that if you were the future generation). If the present generation discounts future generations utility, then it consumes more today and leaves a lower capital stock for the future that ultimately leads to lower steady-state per capita consumption. However, if the discount rate $r = 0$, the integral in (8) does not converge, and the subsequent analysis does not apply. This difficulty can be avoided by replacing the integrand in (8) by $u(c(t)) - u(\bar{c})$, where $\bar{c}$ is the level of per capita consumption in which the individual is completely satisfied, i.e., above this level the marginal utility of consumption declines and minimizing the integral. $\bar{c}$ is sometimes referred to as the bliss point. Ramsey used this formulation in his original analysis of the optimal growth problem.

If the problem had a finite terminal time T, then the transversality condition would be

$$-e^{-rT}U'(c(T)) = 0 \qquad \text{if} \quad K(T) \text{ is free},$$

which implies $c(T) = \infty$. This is clearly infeasible. The nonnegativity constraint on $K(T)$ has been ignored. The best that can be done at T is to consume all the capital that remains. Hence, the boundary condition must be

$$K(T) = 0.$$

The phase diagram already drawn is appropriate. The optimal path will be the one beginning on the line $k = k_0$ and terminating on the line $k = 0$ at T.

Analysis of the relationship between the optimal finite horizon and infinite horizon capital accumulation paths are the subject matter of turnpike theorems.

EXERCISES

1. **a.** Verify that the solution to (1) in the case where

$$f(x') = bx' + cx'^2, \qquad g(x) = x^2 + ax$$

is consistent with the general behavior determined by diagrammatic analysis.

b. Repeat part a, in the case where $x(T)$ may be chosen freely.

c. Repeat part a, in the case where $T \to \infty$ and $\lim_{t\to\infty} x(t)$ is not constrained but assumed to tend to a steady state. What is that steady state ($x = x_s$)?

2. Approximate Equation (3) in the neighborhood of the stationary state by a linear differential equation. Describe its behavior in the neighborhood of the stationary state. (Partial solution: $x'' = rx' + (x - x_s)g''(x_s)/f''(0)$ where x_s satisfies $rf'(0) + g'(x_s) = 0$.)

3. Sketch the optimal solution to the optimal growth problem above under the alternative supposition that the population (or labor force) is growing at a constant proportionate rate n; that is $L'(t)/L(t) = n$, so that $L(t) = L(0)e^{nt}$. Assume also that the capital stock depreciates at a constant proportionate rate b. [Hint: In this case

$$K' = F(K, L) - Lc - bK,$$

and therefore

$$k' = d(K/L)/dt = K'/L - KL'/L^2 = f(k) - c - bk - kn.]$$

FURTHER READING

There is very extensive literature on the neoclassical growth model, with a myriad of generalizations and extensions. See, for example, Arrow and Kurz (1970), Shell, Intriligator, and Takayama. Phelps (1966) is the source for the golden rule literature and Samuelson (1965) and Takayama for turnpike theorems.

See Appendix B, Section 5 regarding pairs of differential equations and linear approximations to them in the neighborhood of a point.

Section 18

Several Functions and Double Integrals

The procedures developed thus far can be extended to problems involving determination of more than one function. Suppose that in the class of differentiable functions $x^*(t)$ and $y^*(t)$ optimize

$$\int_{t_0}^{t_1} F(t, x(t), y(t), x'(t), y'(t))\, dt$$
$$\text{subject to} \qquad x(t_0) = x_0, \quad y(t_0) = y_0 \tag{1}$$

with the terminal time and terminal values of the functions not necessarily fixed. Let $x(t)$, $y(t)$, $t_0 \le t \le t_1 + \delta t_1$, be admissible comparison functions close to x^*, y^*, respectively, in the sense of (9.3). Extend the functions so the candidate and comparison functions all have the same domain. Define

$$h(t) = x(t) - x^*(t), \qquad k(t) = y(t) - y^*(t),$$
$$t_0 \le t \le \max(t_1, t_1 + \delta t_1). \tag{2}$$

Since x_0 and y_0 are fixed, $h(t_0) = k(t_0) = 0$. With x^*, y^*, h, k all held fixed, the value of the objective (1) when evaluated at $x^* + ah$, $y^* + ak$ depends solely on the parameter a. Write

$$g(a) = \int_{t_0}^{t_1 + a\delta t_1} F(t, x^* + ah, y^* + ak, x^{*\prime} + ah', y^{*\prime} + ak')\, dt. \tag{3}$$

Since $g(a)$ assumes its extremum at $a = 0$,

$$g'(0) = \int_{t_0}^{t_1} (F_x h + F_{x'} h' + F_y k + F_{y'} k')\, dt + F|_{t_1} \delta t_1 = 0 \tag{4}$$

with the integral evaluated along $(t, x^*(t), x^{*\prime}(t), y^*(t), y^{*\prime}(t))$. Integrating the terms in h', k' by parts and recalling that $h(t_0) = k(t_0) = 0$, one can rewrite (4)

$$\int_{t_0}^{t_1} [(F_x - dF_{x'}/dt)h + (F_y - dF_{y'}/dt)k]\, dt$$
$$+ (F_{x'}h)|_{t_1} + (F_{y'}k)|_{t_1} + F|_{t_1}\,\delta t_1 = 0. \tag{5}$$

We have approximately, according to (9.9),

$$h(t_1) \approx \delta x_1 - x^{*\prime}(t_1)\delta t_1$$
$$k(t_1) \approx \delta y_1 - y^{*\prime}(t_1)\,\delta t_1, \tag{6}$$

where

$$\delta x_1 \equiv x(t_1 + \delta t_1) - x^*(t_1),$$
$$\delta y_1 \equiv y(t_1 + \delta t_1) - y^*(t_1).$$

Substitute into (5) to get the expression sought,

$$\int_{t_0}^{t_1} [(F_x - dF_{x'}/dt)h + (F_y - dF_{y'}/dt)k]\, dt$$
$$+ (F - x^{*\prime}F_{x'} - y^{*\prime}F_{y'})|_{t_1}\,\delta t_1 + F_{x'}|_{t_1}\,\delta x_1 + F_{y'}|_{t_1}\,\delta y_1 = 0. \tag{7}$$

Now (7) holds for any admissible choice of functions h and k. In particular, it holds for h arbitrary but with $h(t_0) = h(t_1) = \delta x_1 = \delta t_1 = 0$ and with $k \equiv 0$ on $t_0 \le t \le t_1$. It then follows from (7) that x^*, y^* satisfy

$$\int_{t_0}^{t_1} (F_x - dF_{x'}/dt)h\, dt = 0,$$

and therefore

$$F_x - dF_{x'}/dt = 0. \tag{8}$$

Similarly, x^* and y^* must satisfy

$$F_y - dF_{y'}/dt = 0. \tag{9}$$

This system of two interdependent Euler equations (8) and (9) must be satisfied simultaneously by x^* and y^*. The solution of this pair of second order differential equations contains four arbitrary constants of integration to be determined from the boundary and transversality conditions.

If terminal time is freely chosen, then it follows from (7)–(9) that

$$(F - x'F_{x'} - y'F_{y'})|_{t_1} = 0 \quad \text{if} \quad t_1 \text{ is free}. \tag{10}$$

If the terminal values of x and y are free, then we must have

$$F_{x'}|_{t_1} = 0 \qquad \text{if } x(t_1) \text{ is free}$$
$$F_{y'}|_{t_1} = 0 \qquad \text{if } y(t_1) \text{ is free} \tag{11}$$

Suppose, instead that $x(t_1)$, $y(t_1)$ must lie on a curve

$$Q(x, y) = 0. \tag{12}$$

Then, since

$$Q_x \delta x_1 + Q_y \delta y_1 = 0$$

for admissible changes in terminal values, we must have

$$F_{x'}/F_{y'} = Q_x/Q_y \qquad \text{at} \quad t_1 \qquad \text{if} \quad (12) \text{ is required.} \tag{13}$$

Finally, the terminal position may be required to satisfy a relation of the form

$$Q(t, x, y) = 0. \tag{14}$$

Then modification of the endpoint must obey, to a linear approximation,

$$Q_t \, \delta t_1 + Q_x \, \delta x_1 + Q_y \, \delta y_1 = 0. \tag{15}$$

Suppose $\delta y_1 = 0$; then δt_1, δx_1 must obey

$$Q_t \, \delta t_1 + Q_x \, \delta x_1 = 0,$$

which, combined with (7), implies

$$F - F_{x'}(x' + Q_t/Q_x) - F_{y'} y' = 0 \qquad \text{at} \quad t_1. \tag{16}$$

Similarly, setting $\delta x_1 = 0$ and allowing δy_1, δt_1 to vary according to (15) leads to

$$F - F_{x'} x' - F_{y'}(y' + Q_t/Q_y) = 0 \qquad \text{at} \quad t_1. \tag{17}$$

Therefore, if the terminal point must obey (14), the transversality conditions are (16) and (17).

Second order conditions for problems involving n unknown functions are analogous to those in Section 6 with one unknown function. In particular, a second order *necessary* condition for functions x^*, y^* to maximize (1) is the Legendre condition that F be concave in x', y' along x^*, y^*. If $F(t, x, y, x', y')$ is jointly concave in its last four arguments, then functions x^*, y^* satisfying the Euler equations and relevant boundary conditions will maximize (1). Straightforward extensions obtain for minimization and for the case that n unknown functions are to be found.

Isoperimetric constraints may be appended, each with its own constant multiplier, as in Section 7. A finite constraint, such as

$$g(t, x(t), y(t)) = 0, \tag{18}$$

is appended to the integrand with a multiplier function $\lambda(t)$. If $x^*(t)$, $y^*(t)$ optimize (1) subject to (18), and if g_x and g_y are not both zero anywhere, then there is a function $\lambda(t)$ such that $x^*(t)$, $y^*(t)$ satisfy the Euler equations for

$$\int_{t_0}^{t_1} (F + \lambda g)\, dt. \tag{19}$$

Similarly, a differential constraint

$$g(t, x(t), y(t), x'(t) y'(t)) = 0 \tag{20}$$

may be appended to the integrand with a multiplier function; if x^*, y^* optimize (1) subject to (20) and if $g_{x'}$ and $g_{y'}$ are not both zero anywhere, then there is a function $\lambda(t)$ such that $x^*(t)$, $y^*(t)$ satisfy the Euler equations for (19), where g is as defined in (20).

Example.
Find the extremal for

$$\int_0^{\pi/2} (x'^2 + y'^2 + 2xy)\, dt \qquad \text{subject to} \quad x(0) = 0, \quad x(\pi/2) = 1,$$

$$y(0) = 0, \quad y(\pi/2) = -1.$$

SOLUTION. Since $F_x = 2y$, $F_{x'} = 2x'$, $F_y = 2x$, and $F_{y'} = 2y'$, the system of Euler equations is

$$y = x'', \qquad x = y''.$$

Differentiating the first equation twice and combining it with the second yields

$$x^{(4)}(t) = x(t),$$

which has characteristic equation $r^4 = 1$ with roots $r = \pm 1, \pm i$. Therefore,

$$x(t) = c_1 e^t + c_2 e^{-t} + c_3 \cos t + c_4 \sin t,$$

$$y(t) = x''(t) = c_1 e^t + c_2 e^{-t} - c_3 \cos t - c_4 \sin t.$$

The boundary conditions give

$$c_1 = c_2 = c_3 = 0, \qquad c_4 = 1,$$

and so the extremal sought is

$$x(t) = \sin t, \qquad y(t) = -\sin t.$$

An Euler equation can be derived for a double integral,

$$\int_A \int F(t, s, x(t, s), x_t(t, s), x_s(t, s))\, dt\, ds, \tag{21}$$

where t and s are the independent variables of which x is a continuous function with continuous partial derivatives x_t, x_s with respect to t and s, respectively. A represents the area of the fixed region in the t-s plane, bounded by a simply connected closed curve C (i.e., one which does not cross itself and has no holes in its interior) over which the integration is performed. Corresponding to the boundary conditions $x(t_0) = x_0$, $x(t_1) = x_1$, in the case of a single integral is the condition that $x(t, s)$ taken on specified values for all t, s along the curve C.

Suppose that $x^*(t, s)$ optimizes (21) and let

$$x(t, s) = x^*(t, s) + ah(t, s) \tag{22}$$

be a comparison path that satisfies the same boundary conditions as x^*, so that $h(t, s) = 0$ for all t, s along C, and is continuous as are its partial derivatives. As before, let

$$g(a) = \int_A\int F(t, s, x^* + ah, x_t^* + ah_t, x_s^* + ah_s)\, dt\, ds. \tag{23}$$

Thus,

$$g'(0) = \int_A\int (F_x h + F_{x_t} h_t + F_{x_s} h_s)\, dt\, ds = 0. \tag{24}$$

At this stage of the derivation of the Euler equation in the case of a single integral the term involving $h'(t)$ would be integrated by parts. The counterpart trick here is to appeal to Green's theorem (see Section A7).

$$\int_C P(t, s)\, dt + Q(t, s)\, ds = \int_A\int (Q_t - P_s)\, dt\, ds \tag{25}$$

by letting $P = -h(t, s)F_{x_s}$, $Q = h(t, s)F_{x_t}$, where Q_t and P_s refer to partial derivatives with respect to t and s of P and Q, respectively. So, then

$$\int_C h(t, s)[-F_{x_s}\, dt + F_{x_t}\, ds]$$

$$= \int_A\int (\partial(hF_{x_t})/\partial t + \partial(hF_{x_s})/\partial s]\, dt\, ds \tag{26}$$

$$= \int_A\int (h_t F_{x_t} + h\,\partial F_{x_t}/\partial t + h_s F_{x_s} + h\,\partial F_{x_s}/\partial s]\, dt\, ds = 0. \tag{27}$$

The reason (27) equals zero is that $h(t, s)$, which appears in the integral on the left side of (26) must be zero for all t, s along the curve c. Thus, from (27) it follows that

$$\iint_A (h_t F_{x_t} + h_s F_{x_s})\, dt\, ds = -\iint_A (h\,\partial F_{x_t}/\partial t + h\,\partial F_{x_s}/ds)\, dt\, ds. \tag{28}$$

Now, substituting from (28) into (24) yields

$$\iint_A (F_x - \partial F_{x_t}/\partial t - \partial F_{x_s}/\partial s)\, h\, dt\, ds = 0, \tag{29}$$

from which it follows that the Euler equation is

$$F_x - \partial F_{x_t}/\partial t - \partial F_{x_s}/\partial s = 0. \tag{30}$$

This derivation of the Euler equation can be extended to multiple integrals.

Example. Find the Euler equation for

$$\min \iint_A (x_t^2 + x_s^2)\, dt\, ds,$$

$F_x = 0$, $F_{x_t} = 2x_t$, $F_{x_s} = 2x_s$. Thus, the Euler equation is

$$x_{tt} + x_{ss} = 0.$$

This partial differential equation is referred to as Laplace's equation and arises in many physics problems.

EXERCISES

1. Show that the necessary conditions for continuously differentiable functions $x_1(t), \ldots, x_n(t)$ to maximize

$$\int_{t_0}^{t_1} F(t, x_1(t), \ldots, x_n(t), x_1'(t), \ldots, x_n'(t))\, dt$$

$$\text{subject to} \quad x_i(t_0) = a_i, \quad x_i(t_1) = b_i, \quad i = 1, \ldots n,$$

are that these n functions satisfy the endpoint conditions and the simultaneous system of Euler equations

$$F_{x_i} = dF_{x_i'}/dt, \qquad i = 1, \ldots, n.$$

2. Show that if $F(t, x, y, x' y')$ is a concave function of (x, y, x', y') and if the functions $x^*(t)$, $y^*(t)$ satisfy the boundary conditions in (1) and the Euler equations (8) and (9), then the functions x^*, y^* maximize (1).

3. State the Legendre necessary condition for the problem posed in Exercise 1.

4. Find extremals for the fixed endpoint problems

   a. $\int_{t_0}^{t_1} [2xy - 2x^2 - (x')^2 + (y')^2]\, dt,$

   b. $\int_{t_0}^{t_1} [2xy - 2x^2 + (x')^2 - (y')^2]\, dt,$

   c. $\int_{t_0}^{t_1} (x^2 + x'y' + y^2)\, dt.$

5. Show that the Euler equation obeyed by an optimal solution to

$$\max \int_{t_0}^{t_1} F(t, x(t), x'(t), x''(t))\, dt$$

$$\text{subject to} \quad x(t_0) = x_0, \quad x(t_1) = x_1, \quad t_1 \text{ fixed}$$

is

$$F_x - dF_{x'}/dt + d^2F_{x''}/dt^2 = 0.$$

[Hint: Find

$$g'(0) = \int_{t_0}^{t_1} (F_x h + F_{x'} h' + F_{x''} h'')\, dt = 0$$

for any admissible function h. Integrate the middle term by parts as usual. Integrate the last term by parts twice.]

FURTHER READING

Section B4 regards solution of nth order linear differential equations.

If the objective function is quadratic, the differential equations arising from optimization have a known solution. See Connors and Teichroew for applications to management science. Other functional forms often lead to systems of differential equations that are difficult to solve. If the functional forms are not fully specified, as frequently happens in economics, then the analysis uses phase diagrams (recall Section 17). But for a problem with two functions to be analyzed, a four-dimensional diagram would be required, an impractical situation. Thus analysts frequently content themselves with a formal statement and interpretation of the necessary conditions. See Treadway for an exceptional example of exploiting the information in the necessary conditions. Finally, see Example II6.2; while it is more readily solved in the optimal control format, the calculus of variations is also applicable. H. T. Davis is a good source for further study of double integral problems.

PART II

OPTIMAL CONTROL

Section 1

Introduction

The classical calculus of variations has been generalized. The *maximum principle* for optimal control, developed in the late 1950s by L. S. Pontryagin and his co-workers, applies to all calculus of variations problems. In such problems, optimal control gives equivalent results, as one would expect. The two approaches differ, however, and the optimal control approach sometimes affords insights into a problem that might be less readily apparent through the calculus of variations.

Optimal control also applies to problems for which the calculus of variations is not convenient, such as those involving constraints on the derivatives of functions sought. For instance, one can solve problems in which net investment or production rates are required to be nonnegative. While proof of the maximum principle under full generality is beyond our scope, the now-familiar methods are used to generate some of the results of interest and to lend plausibility to others.

In optimal control problems, variables are divided into two classes, *state* variables and *control* variables. The movement of state variables is governed by first order differential equations. The simplest control problem is one of selecting a piecewise continuous control function $u(t)$, $t_0 \leq t \leq t_1$, to

$$\max \int_{t_0}^{t_1} f(t, x(t), u(t))\, dt \tag{1}$$

$$\text{subject to} \quad x'(t) = g(t, x(t), u(t)), \tag{2}$$

$$t_0, t_1, \quad x(t_0) = x_0 \text{ fixed}; \quad x(t_1) \text{ free}. \tag{3}$$

Here f and g are assumed to be continuously differentiable functions of three independent arguments, none of which is a derivative. The *control* variable $u(t)$ must be a piecewise continuous function of time. The *state* variable $x(t)$

changes over time according to the differential equation (2) governing its movement. The control u influences the objective (1), both directly (through its own value) and indirectly through its impact on the evolution of the state variable x (that enters the objective (1)). The highest derivative appearing in the problem formulation is the first derivative, and it appears *only* as the left side of the *state equation* (2). (Equation (2) is sometimes also called the *transition* equation.) A problem involving higher derivatives can be transformed into one in which the highest derivative is the first, as will be shown later.

The prototypical calculus of variations problem of choosing a continuously differentiable function $x(t)$, $t_0 \le t \le t_1$, to

$$\begin{aligned} &\max \quad \int_{t_0}^{t_1} f(t, x(t), x'(t))\, dt \\ &\text{subject to} \quad x(t_0) = x_0 \end{aligned} \tag{4}$$

is readily transformed into an equivalent problem in optimal control. Let $u(t) = x'(t)$. Then the equivalent optimal control problem is

$$\begin{aligned} &\max \quad \int_{t_0}^{t_1} f(t, x(t), u(t))\, dt \\ &\text{subject to} \quad x'(t) = u(t), \quad x(t_0) = x_0. \end{aligned} \tag{5}$$

The state variable is x, while u is the control. For instance, our production planning Example I1.1 appears as

$$\begin{aligned} &\min \quad \int_0^T \left[c_1 u^2(t) + c_2 x(t)\right] dt \\ &\text{subject to} \quad x'(t) = u(t), \quad x(0) = 0, \quad x(T) = B, \quad u(t) \ge 0, \end{aligned} \tag{6}$$

where the production rate $u(t)$ is the control and the current inventory on hand $x(t)$ is the state variable. In this case, the objective is minimization rather than maximization, and the terminal point is fixed rather than free. These are typical variants from the initial format of (1) or (5).

While a calculus of variations problem (4) can always be put into an optimal control format (5), it is not always the most natural or useful form. For instance, Example I1.2 can be readily expressed as an optimal control problem:

$$\begin{aligned} &\max \quad \int_0^T e^{-rt} U(C(t))\, dt \\ &\text{subject to} \quad K' = F(K(t)) - C(t) - bK(t), \\ &\qquad\qquad K(0) = K_0, \quad K(T) \ge 0, \quad C(t) \ge 0. \end{aligned} \tag{7}$$

Here $K(t)$ is the sole state variable; its rate of change is given in the differential equation. There is one control variable, the rate of consumption

$C(t)$. Choice of $C(t)$ determines the rate of capital accumulation and also the value of the objective function.

Likewise, Example I1.3 is readily expressed as a problem of optimal control:

$$\begin{aligned} \max \quad & \int_0^T e^{-rt}[P(K(t)) - C(I(t))]\, dt \\ \text{subject to} \quad & K'(t) = I(t) - bK(t), \\ & K(0) = K_0, \quad K(T) \geq 0, \quad I(t) \geq 0. \end{aligned} \tag{8}$$

The objective is maximization of the discounted stream of profits, namely, revenues attainable with capital stock K less the cost of capital investment. Capital is augmented by gross investment but decays at exponential rate b. The state variable is the capital stock K; the control variable is net investment I.

An optimal control problem may have several state variables and several control variables. Each state variable evolves according to a differential equation. The number of control variables may be greater or smaller than the number of state variables.

The optimal control results are developed in the next sections for problems already solved by calculus of variations. This will develop familiarity with the new notations and tools. New problem will then be solved and their use illustrated.

FURTHER READING

References on the techniques of optimal control theory include Pontryagin et al. (1962), Berkovitz (1974, 1976), Bryson and Ho, Fleming and Rishel, Hestenes, and Lee and Markus. In addition, there are a number of books that provide an introduction to the theory as well as discussion of applications in economics and management science; these include books by Hadley and Kemp, Intriligator (1971); Takayama, Sethi and Thompson; Feichtinger and Hartl; and Seierstad and Sydsaeter. For further surveys of applications in management science, see Bensoussan, Hurst, and Naslund; Bensoussan, Kleindorfer and Tapiero; and Sethi (1978). The references at the back of this book provide an overview of other applications to economics and management science; the list is merely suggestive of the range of work that has appeared.

Section 2

Simplest Problem—Necessary Conditions

The simplest problem in calculus of variations had the values of the state variable at both endpoints fixed. But the simplest problem in optimal control involves a free value of the state variable at the terminal point. To find necessary conditions that a maximizing solution $u^*(t)$, $x^*(t)$, $t_0 \le t \le t_1$, to problem (1.1)–(1.3) must obey, we follow a procedure reminiscent of solving a nonlinear programming problem with Lagrange multipliers (see Section A5). Since the constraining relation (1.2) must hold at each t over the entire interval $t_0 \le t \le t_1$, we have a multiplier function $\lambda(t)$, rather than a single Lagrange multiplier value as would be associated with a single constraint. For now, let $\lambda(t)$ be any continuously differentiable function t on $t_0 \le t \le t_1$; shortly, a convenient specification for its behavior will be made.

For any functions x, u satisfying (1.2) and (1.3) and any continuously differentiable function λ, all defined on $t_0 \le t \le t_1$, we have

$$\int_{t_0}^{t_1} f(t, x(t), u(t))\, dt = \int_{t_0}^{t_1} [f(t, x(t), u(t)) + \lambda(t) g(t, x(t), u(t)) - \lambda(t) x'(t)]\, dt \quad (1)$$

since the coefficients of $\lambda(t)$ must sum to zero if (1.2) is satisfied, as we assume. Integrate the last term on the right of (1) by parts

$$-\int_{t_0}^{t_1} \lambda(t) x'(t)\, dt = -\lambda(t_1) x(t_1) + \lambda(t_0) x(t_0) + \int_{t_0}^{t_1} x(t) \lambda'(t)\, dt. \quad (2)$$

Substituting from (2) into (1) gives

$$\int_{t_0}^{t_1} f(t, x(t) u(t))\, dt = \int_{t_0}^{t_1} [f(t, x(t), u(t)) + \lambda(t) g(t, x(t), u(t)) + x(t)\lambda'(t)]\, dt - \lambda(t_1) x(t_1) + \lambda(t_0) x(t_0). \quad (3)$$

A control function $u(t)$, $t_0 \le t \le t_1$, together with the initial condition (1.3) and the differential equation (1.2) determine the path of the corresponding state variable $x^*(t)$, $t_0 \le t \le t_1$. Thus we may speak of finding the control function, since a corresponding state function is implied. Since selection of the control function $u(t)$ determines the state variable $x(t)$, it determines the value of (3) as well.

To develop the necessary conditions for solution of the calculus of variations problem (1.4), we constructed a one-parameter family of comparison curves $x^*(t) + ah(t)$, $x^{*\prime}(t) + ah'(t)$, where $h(t)$ was arbitrary but fixed. In the current notation (1.5), $x' = u$ and a modified control function $u(t) + ah'(t)$ produces, via integration, a modified state function $x(t) + ah(t)$. However, for the implicit state equation (1.2), one cannot give an explicit expression for the modified control. Hence the modified state function will be expressed implicitly. Since the previous h, h' notation is not helpful here, we depart from the previous usage of h and now let $h(t)$ represent a fixed modification in the *control* $u(t)$.

We consider a one-parameter family of comparison controls $u^*(t) + ah(t)$, where $u^*(t)$ is the optimal control, $h(t)$ is some fixed function, and a is a parameter. Let $y(t, a)$, $t_0 \le t \le t_1$, denote the state variable generated by (1.2) and (1.3) with control $u^*(t) + ah(t)$, $t_0 \le t \le t_1$. We assume that $y(t, a)$ is a smooth function of both its arguments. The second argument enters parametrically. Clearly $a = 0$ provides the optimal path x^*. Further, all comparison paths satisfy the initial condition. Hence

$$y(t, 0) = x^*(t), \quad y(t_0, a) = x_0. \tag{4}$$

With the functions u^*, x^* and h all held fixed, the value of (1.1) evaluated along the control function $u^*(t) + ah(t)$ and the corresponding state $y(t, a)$ depends on the single parameter a. Thus we write

$$J(a) = \int_{t_0}^{t_1} f(t, y(t, a), u^*(t) + ah(t))\, dt.$$

Using (3),

$$\begin{aligned} J(a) = \int_{t_0}^{t_1} [&f(t, y(t, a), u^*(t) + ah(t)) \\ &+\lambda(t) g(t, y(t, a), u^*(t) + ah(t)) \\ &+y(t, a)\lambda'(t)]\, dt - \lambda(t_1) y(t_1, a) + \lambda(t_0) y(t_0, a). \end{aligned} \tag{5}$$

Since u^* is a maximizing control, the function $J(a)$ assumes its maximum at $a = 0$. Hence $J'(0) = 0$. Differentiating with respect to a and evaluating at

$a = 0$ gives, on collecting terms,

$$J'(0) = \int_{t_0}^{t_1} \left[(f_x + \lambda g_x + \lambda') y_a + (f_u + \lambda g_u) h \right] dt - \lambda(t_1) y_a(t_1, 0), \tag{6}$$

where f_x, g_x and f_u, g_u denote the partial derivatives of the functions f, g with respect to their second and third arguments, respectively; and y_a is the partial derivative of y with respect to its second argument. Since $a = 0$, the functions are evaluated along $(t, x^*(t), u^*(t))$. The last term of (5) is independent of a—that is, $y_a(t_0, a) = 0$—since $y(t_0, a) = x_0$ for all a.

To this point, the function $\lambda(t)$ was only required to be differentiable. Since the precise impact of modifying the control variable on the course of the state variable (i.e., y_a) is difficult to determine, $\lambda(t)$ is selected to eliminate the need to do so. Let λ obey the linear differential equation.

$$\lambda'(t) = -\left[f_x(t, x^*, u^*) + \lambda(t) g_x(t, x^*, u^*) \right], \quad \text{with } \lambda(t_1) = 0. \tag{7}$$

(Recall that x^* and u^* are fixed functions of t.) With λ given in (7), (6) holds provided that

$$\int_{t_0}^{t_1} \left[f_u(t, x^*, u^*) + \lambda g_u(t, x^*, u^*) \right] h \, dt = 0 \tag{8}$$

for an arbitrary function $h(t)$. In particular, it must hold for $h(t) = f_u(t, x^*, u^*) + \lambda(t) g_u(t, x^*, u^*)$, so that

$$\int_{t_0}^{t_1} \left[f_u(t, x^*(t), u^*(t)) + \lambda(t) g_u(t, x^*(t), u^*(t)) \right]^2 dt = 0. \tag{9}$$

This, in turn, implies the necessary condition that

$$f_u(t, x^*(t), u^*(t)) + \lambda(t) g_u(t, x^*(t), u^*(t)) = 0, \quad t_0 \le t \le t_1 \tag{10}$$

To sum up, we have shown that if the functions $u^*(t)$, $x^*(t)$ maximize (1.1), subject to (1.2) and (1.3), then there is a continuously differentiable function $\lambda(t)$ such that u^*, x^*, λ simultaneously satisfy the *state equation*

$$x'(t) = g(t, x(t), u(t)), \quad x(t_0) = x_0, \tag{11}$$

the *multiplier equation*

$$\lambda'(t) = -\left[f_x(t, x(t), u(t)) + \lambda(t) g_x(t, x(t), u(t)) \right], \quad \lambda(t_1) = 0, \tag{12}$$

and the *optimality condition*

$$f_u(t, x(t), u(t)) + \lambda(t) g_u(t, x(t), u(t)) = 0. \tag{13}$$

for $t_0 \le t \le t_1$. The multiplier equation (12) is also known as the *costate*, *auxiliary*, *adjoint*, or *influence* equation.

The device for remembering, or generating these conditions (similar to solving a nonlinear programming problem by forming the Lagrangian, differentiating, etc.) is the *Hamiltonian*

$$H(t, x(t), u(t), \lambda(t)) \equiv f(t, x, u) + \lambda g(t, x, u). \tag{14}$$

Now

$$\partial H/\partial u = 0 \quad \text{generates (13):} \quad \partial H/\partial u = f_u + \lambda g_u = 0; \tag{13'}$$

$$-\partial H/\partial x = \lambda' \quad \text{generates (12):} \quad \lambda'(t) = -\partial H/\partial x = -(f_x + \lambda g_x); \tag{12'}$$

$$\partial H/\partial \lambda = x' \quad \text{recovers (11):} \quad x' = \partial H/\partial \lambda = g. \tag{11'}$$

In addition, we have $x(t_0) = x_0$ and $\lambda(t_1) = 0$. At each t, u is a stationary point of the Hamiltonian for the given values of x and λ. One can find u as a function of x and λ from (13) and substitute into (12) and (11) to get a system of two differential equations in x and λ. These conditions (11)–(13) are also necessary for a minimization problem. It will be shown in Section 7 that the Hamiltonion as defined in (14) is just the negative of the Hamiltonion of expression (14) in Section I.3.

For a maximization problem, it is also necessary that $u^*(t)$ maximize $H(t, x^*(t), u, \lambda(t))$ with respect to u. Thus, $H_{uu}(t, x^*, u^*, \lambda) \le 0$ is necessary for maximization. In a minimization problem, $u^*(t)$ must minimize $H(t, x^*(t), u, \lambda(t))$ with respect to u, and therefore $H_{uu}(t, x^*, u^*, \lambda) \ge 0$ is necessary. These two results have not yet been proven but they will be discussed later.

Note that the Hamiltonian as defined in (14) is similar to the Hamiltonian defined in (I.3.14). Indeed, if we let $\lambda = -p$, then they differ only by a minus sign. Moreover, the necessary conditions $-\partial H/\partial x = \lambda'$, $\partial H/\partial \lambda = x'$ resemble the canonical form of the Euler equation (I.3.15) and $\partial H/\partial u = 0$ corresponds to the definition of p in (I.3.13). In fact, if $g(t, x, u) = u$ so that $x' = u$, then the necessary conditions (11′) and (12′) are exactly the canonical form of the Euler equation and (13′) is the definition of p.

Example 1. Show that the necessary conditions for optimality in (1.5) are equivalent to the Euler equation

$$f_x = df_{x'}/dt \tag{15}$$

and transversality condition

$$f_{x'} = 0 \quad \text{at} \quad t_1 \tag{16}$$

that must be obeyed for the equivalent calculus of variations problem (1.4). What are the second order necessary conditions?

Form the Hamiltonian, following (14),

$$H = f(t, x, u) + \lambda u.$$

Then,

$$\partial H/\partial u = f_u + \lambda = 0, \tag{17}$$

$$\lambda' = -\partial H/\partial x = -f_x, \qquad \lambda(t_1) = 0. \tag{18}$$

Thus, if x^*, u^* are optimal, there must be a continuously differentiable function λ such that x^*, u^*, λ simultaneously satisfy $x' = u$, $x(t_0) = x_0$ and (17)–(18) over the interval $t_0 \le t \le t_1$.

To show that these conditions are equivalent to (15)–(16), differentiate (17) with respect to time:

$$df_u/dt + \lambda' = 0$$

and use the result to eliminate λ' from (18):

$$f_x = df_u/dt,$$

which is (15). Also, the boundary condition $\lambda(t_1) = 0$ is the same as in (16). Finally, the necessary condition $H_{uu} = f_{uu}(t, x^*, u^*) \le 0$ corresponds to the necessary Legendre condition $f_{x'x'}(t, x^*, x^{*\prime}) \le 0$.

Thus we have no new result. Optimal control yields, as necessary conditions, a system of two first order differential equations instead of the Euler equation a single second order differential equation. The transversality conditions and second order necessary conditions under each formulation are likewise equivalent. In each case, the boundary conditions for solution of the differential equation are split, with one holding at the initial moment and the other holding at the final moment.

To show that (15) and (16) imply (17) and (18), we need only reverse the process. That is, define

$$\lambda(t) = -f_{x'}(t, x(t), x'(t)). \tag{19}$$

Differentiate (19) with respect to t and substitute into (15);

$$-f_x = \lambda'(t). \tag{20}$$

Putting (19) into (16) gives

$$\lambda(t_1) = 0. \tag{21}$$

But (19)–(21) correspond exactly to conditions (17) and (18). Thus the two approaches yield equivalent necessary conditions for optimality, as claimed.

Since optimal control is equivalent to calculus of variations for all problems to which the latter applies, one may wonder why it is useful to learn about optimal control. One response is that it applies to a wider class of problems, to be studied later. Another answer is that optimal control may be more convenient for certain problems and may also suggest economic interpretations that are less readily apparent in solving by the calculus of variations. Each of these points will be illustrated; see Example 2 and 3 below.

Example 2.

$$\max \int_0^1 (x + u)\, dt \tag{22}$$

$$\text{subject to} \quad x' = 1 - u^2, \quad x(0) = 1. \tag{23}$$

Form the Hamiltonian

$$H(t, x, u, \lambda) = x + u + \lambda(1 - u^2).$$

Necessary conditions are (23) and

$$H_u = 1 - 2\lambda u = 0, \qquad H_{uu} = -2\lambda \le 0, \tag{24}$$

$$\lambda' = -H_x = -1, \qquad \lambda(1) = 0. \tag{25}$$

Integrate (25) and use the boundary condition to find

$$\lambda = 1 - t. \tag{26}$$

Then $H_{uu} = -2(1 - t) \le 0$ for $0 \le t \le 1$. Also, from (24),

$$u = 1/2\lambda = 1/2(1 - t). \tag{27}$$

Substituting (27) into (23) gives

$$x' = 1 - 1/4(1 - t)^2, \qquad x(0) = 1.$$

Integrating, using the boundary condition, and drawing the results together yields the solution:

$$x(t) = t - 1/4(1 - t) + 5/4,$$
$$\lambda(t) = 1 - t,$$
$$u(t) = 1/2(1 - t).$$

Example 3. The rate at which a new product can be sold at any time t is $f(p(t))g(Q(t))$ where p is the price and Q is the *cumulative* sales. We assume $f'(p) < 0$; sales vary inversely with price. Also $g'(Q) \gtreqless 0$ for $Q \lesseqgtr Q_1$. For a given price, current sales grow with past sales in the early stages as people learn about the good from past purchasers. But as cumulative sales increase, there is a decline in the number of people who have not yet purchased the good. Eventually the sales rate for any given price falls, as the market becomes saturated. The unit production cost c may be constant or may decline with cumulative sales if the firm learns how to produce less expensively with experience: $c = c(Q)$, $c'(Q) \le 0$. Characterize the price policy

$p(t)$, $0 \le t \le T$, that maximizes profits from this new "fad" over a fixed horizon T.

The problem is

$$\max \int_0^T [p - c(Q)] f(p) g(Q)\, dt \tag{28}$$

$$\text{subject to} \quad Q' = f(p) g(Q), \quad Q(0) = Q_0 > 0. \tag{29}$$

Price p is the control variable and cumulative sales Q is the state variable.

Form the Hamiltonian

$$H = f(p) g(Q)[p - c(Q) + \lambda]. \tag{30}$$

The optimal solution must satisfy (29) and

$$H_p = g(Q)\{f'(p)[p - c(Q) + \lambda] + f(p)\} = 0, \tag{31}$$

$$H_{pp} = g(Q)\{f''(p)[p - c(Q) + \lambda] + 2f'(p)\} \le 0, \tag{32}$$

$$\lambda' = -H_Q = f(p)\{g(Q)c'(Q) - g'(Q)[p - c(Q) + \lambda]\}, \tag{33}$$

$$\lambda(T) = 0. \tag{34}$$

We use these conditions to characterize the solution qualitatively. Since $g > 0$, we know from (31) that

$$\lambda = -f/f' - p + c. \tag{35}$$

Differentiating (35) totally with respect to t gives

$$\lambda' = -p'[2 - ff''/(f')^2] + c'Q'. \tag{36}$$

Substituting (35) into (32) and (33) gives

$$gf'[2 - f''f/(f')^2] \le 0, \tag{32'}$$

$$\lambda' = f[gc' + g'f/f']. \tag{33'}$$

Equate (36) and (33'), using (29):

$$[2 - ff''/(f')^2]p' = -g'f^2/f', \tag{37}$$

from which we conclude that

$$\text{sign } p' = \text{sign } g' \tag{38}$$

since $f' < 0$ and (32') holds. Result (38) tells us that in marketing a good its optimal price rises while the market expands ($Q < Q_1$) and falls as the market matures ($Q > Q_1$).

EXERCISES

1. Use optimal control to find the shortest distance between the point $x(a) = A$ and the line $t = b$.

2. Solve by optimal control

$$\min \int_0^T [x^2(t) + ax(t) + bu(t) + cu^2(t)]\, dt$$

$$\text{subject to} \quad x'(t) = u(t), \quad x(0) = x_0 \text{ fixed}, \quad T \text{ fixed}, \quad x(T) \text{ free}, \quad c > 0.$$

3.

$$\max \int_1^5 (ux - u^2 - x^2)\, dt$$

$$\text{subject to} \quad x' = x + u, \quad x(1) = 2.$$

4. Find necessary conditions for solution of

$$\max \int_{t_0}^{t_1} f(t, x, u)\, dt$$

$$\text{subject to} \quad x' = g(t, x, u), \quad t_0, t_1 \text{ fixed}, \quad x(t_0), x(t_1) \text{ free}.$$

5.

$$\min \int_0^1 u^2(t)\, dt$$

$$\text{subject to} \quad x'(t) = x(t) + u(t), \quad x(0) = 1.$$

6. Show that necessary conditions for the solution of

$$\max \int_{t_0}^{t_1} f(t, x, u)\, dt + \phi(x_1)$$

$$\text{subject to} \quad x'(t) = g(t, x, u), \quad x(t_0) = x_0, \quad t_0, t_1 \text{ fixed}, \quad x(t_1) = x_1 \text{ free}.$$

are (11)–(13) except that $\lambda(t_1) = \phi'(x_1)$. Relate to the corresponding transversality condition in the calculus of variations for the case $g = u$.

7. Using the results of exercise 6,

$$\min \int_0^1 u^2(t)\, dt + x^2(1)$$

$$\text{subject to} \quad x'(t) = x(t) + u(t), \quad x(0) = 1.$$

8. The problem

$$\max \int_{t_0}^{t_1} f(x, u)\, dt$$

$$\text{subject to} \qquad x' = g(x, u), \quad x(t_0) = x_0 \text{ fixed}, \quad x(t_1) = x_1 \text{ free},$$

is *autonomous* since there is no explicit dependence on t. Show that the Hamiltonian is a constant function of time along the optimal path. [Hint: Compute using the chain rule,

$$dH/dt = H_x x' + H_u u' + H_\lambda \lambda'$$

and substitute from the necessary conditions for the partial derivatives of H.] Autonomous problems and their advantages are discussed further in Section 8.

9. The Euler equation for the calculus of variations problem

$$\max \int_{t_0}^{t_1} F(x, x')\, dt$$

$$\text{subject to} \qquad x(t_0) = x_0 \text{ fixed}, \quad x(t_1) = x_1 \text{ free}.$$

can be written

$$F - x'F_{x'} = \text{const}, \qquad t_0 \le t \le t_1.$$

Show that this is equivalent to the condition that the Hamiltonian for the related control problem is constant.

10. Use the results of Exercise 8 to show that for Example 3,

a. if $f(p) = e^{-ap}$, then the optimal sales rate fg is constant.

b. if $f(p) = p^{-a}$, then revenue (pfg) is constant in an optimal program.

11. Discuss how the calculus of variations could be used to analyze Examples 2 and 3.

FURTHER READING

See (B4.7)–(B4.9) on the equivalence between a single second order differential equation and a pair of first order differential equations.

Example 3 was stimulated by Robinson and Lakhani, who also provided numerical results for a special case with discounting.

Compare the results mentioned here with those above, especially of Section I.8 and I.11. Note that while the simplest problem of calculus of variations had a fixed endpoint, the simplest problem of optimal control has a free end value.

Section 3

Sufficiency

When are the necessary conditions for optimality both necessary and sufficient? In nonlinear programming, the Kuhn-Tucker necessary conditions are also sufficient provided that a concave (convex) objective function is to be maximized (minimized) over a closed convex region. In the calculus of variations, the necessary conditions are also sufficient for optimality if the integrand $F(t, x, x')$ is concave (convex) in x, x'. Analogous results obtain for optimal control problems.

Suppose that $f(t, x, u)$ and $g(t, x, u)$ are both differentiable concave functions of x, u in the problem

$$\max \quad \int_{t_0}^{t_1} f(t, x, u)\, dt \tag{1}$$

$$\text{subject to} \qquad x' = g(t, x, u), \quad x(t_0) = x_0. \tag{2}$$

The argument t of $x(t)$ and $u(t)$ will frequently be suppressed. Suppose that the functions x^*, u^*, λ satisfy the necessary conditions

$$f_u(t, x, u) + \lambda g_u(t, x, u) = 0, \tag{3}$$

$$\lambda' = -f_x(t, x, u) - \lambda g_x(t, x, u), \tag{4}$$

$$\lambda(t_1) = 0, \tag{5}$$

and the constraints (2) for all $t_0 \le t \le t_1$. Suppose further that x and λ are continuous with

$$\lambda(t) \ge 0 \tag{6}$$

for all t if $g(t, x, u)$ is nonlinear in x or u, or both. Then the functions x^*, u^* solve the problem given by (1) and (2). Thus if the functions f and g

are both jointly concave x, u (and if the sign restriction in (6) holds), then the necessary conditions (2)–(5) are also sufficient for optimality.

The assertion can be verified as follows. Suppose x^*, u^*, λ satisfy (2)–(6). Let x, u be functions satisfying (2). Let f^*, g^*, and so on denote functions evaluated along (t, x^*, u^*) and let f, g, and so on denote functions evaluated along the feasible path (t, x, u). Then we must show that

$$D \equiv \int_{t_0}^{t_1} (f^* - f)\, dt \geq 0. \tag{7}$$

Since f is a concave function of (x, u), we have

$$f^* - f \geq (x^* - x) f_x^* + (u^* - u) f_u^*, \tag{8}$$

and therefore (reasons to follow)

$$\begin{aligned} D &\geq \int_{t_0}^{t_1} [(x^* - x) f_x^* + (u^* - u) f_u^*]\, dt \\ &= \int_{t_0}^{t_1} [(x^* - x)(-\lambda g_x^* - \lambda') + (u^* - u)(-\lambda g_u^*)]\, dt \\ &= \int_{t_0}^{t_1} \lambda [g^* - g - (x^* - x) g_x^* - (u^* - u) g_u^*]\, dt \\ &\geq 0. \end{aligned} \tag{9}$$

as was to be shown. The second line of (9) was obtained by substituting from (4) for f_x^* and from (3) for f_u^*. The third line of (9) was found by integrating by parts the terms involving λ', recalling (2) and (5). The last line follows from (6) and the assumed concavity of g in x and u.

If the function g is linear in x, u, then λ may assume any sign. The demonstration follows since the last square bracket in (9) will equal zero. Further, if f is concave while g is convex and $\lambda \leq 0$, then the necessary conditions will also be sufficient for optimality. The proof proceeds as shown above, except that in the next to last line λ and its coefficients are each nonpositive, therefore making their product nonnegative.

EXERCISES

1. Show that if f and g are both concave functions of x and u, if (6) holds, and if x^*, u^*, λ satisfy (2)–(5), then $u^*(t)$ does maximize the Hamiltonian $H(t, x^*(t), u, \lambda(t))$ at each t, $t_0 \leq t \leq t_1$, with respect to u.

2. Show that if minimization was required in problem (1) and (2), and if the functions f and g are both jointly convex in x, u, then functions x^*, u^*, λ satisfying (2)–(6) will solve the problem. Also show that $u^*(t)$ will minimize the Hamiltonian $H(t, x^*(t), u, \lambda(t))$ at each t.

3. Investigate whether the solutions obtained in the exercises of Section 2 minimize or maximize.

4. Suppose $\phi(x_1)$ is a concave function and that $f(t, x, u)$ and $g(t, x, u)$ are differentiable concave functions of (x, u). State and prove a sufficiency theorem for

$$\max \quad \int_{t_0}^{t_1} f(t, x, u)\, dt + \phi(x_1)$$

$$\text{subject to} \qquad x' = g(t, x, u), \quad x(t_0) = x_0, \quad t_0, t_1 \text{ fixed}, \quad x(t_1) = x_1 \text{ free}.$$

FURTHER READING

Mangasarian provided the basic sufficiency theorem for optimal control. See also Section 15 and Seierstad and Sydsaeter (1977, 1987) for extensions to more complex control problems.

Compare the present results with those of Section I.6.

Section 4

Interpretations

The multiplier λ in optimal control problems has an interesting and economically meaningful interpretation. In nonlinear programming the Lagrange multiplier is interpreted as a marginal valuation of the associated constraint. (See Section A5.) Here $\lambda(t)$ is the marginal valuation of the associated state variable at t.

Consider

$$\max \int_{t_0}^{t_1} f(t, x, u)\, dt$$
$$\text{subject to} \quad x'(t) = g(t, x, u), \quad x(t_0) = x_0. \tag{1}$$

Let $V(x_0, t_0)$ denote the maximum of (1), for a given initial state x_0 at initial time t_0. Let x^*, u^* be the state and control functions providing this maximum, and let λ be the corresponding multiplier. Suppose u^* is a continuous function of t.

We also consider a modification to problem (1) in which the initial state is $x_0 + a$, where a is a number close to zero. The maximum for the modified problem is $V(x_0 + a, t_0)$. Let x and u denote the state and control functions providing this maximum.

Appending the differential equation in (1) with a continuously differentiable multiplier function $\lambda(t)$ gives

$$\begin{aligned} V(x_0, t_0) &= \int_{t_0}^{t_1} f(t, x^*, u^*)\, dt \\ &= \int_{t_0}^{t_1} \left[f(t, x^*, u^*) + \lambda g(t, x^*, u^*) - \lambda x' \right] dt. \end{aligned} \tag{2}$$

Integrate the last term by parts (recalling 2.2)):

$$V(x_0, t_0) = \int_{t_0}^{t_1} (f^* + \lambda g^* + \lambda' x^*)\, dt - \lambda(t_1)x^*(t_1) + \lambda(t_0)x(t_0), \tag{3}$$

where asterisks label functions evaluated along (t, x^*, u^*). Similarly, one finds that (using the same λ)

$$\begin{aligned} V(x_0 + a, t_0) &= \int_{t_0}^{t_1} f\, dt \\ &= \int_{t_0}^{t_1} (f + \lambda g + \lambda' x)\, dt - \lambda(t_1)x(t_1) + \lambda(t_0)[x(t_0) + a], \end{aligned}$$

where x, u are optimal for this problem. Subtracting,

$$\begin{aligned} V(x_0, + a, t_0) - V(x_0, t_0) &= \int_{t_0}^{t_1} [f(t, x, u) - f(t, x^*, u^*)]\, dt \\ &= \int_{t_0}^{t_1} (f + \lambda g + \lambda' x - f^* - \lambda g^* - \lambda' x^*)\, dt \\ &\quad + \lambda(t_0)a - \lambda(t_1)[x(t_1) - x^*(t_1)]. \end{aligned} \tag{4}$$

Substitute for the integrand its Taylor series around (t, x^*, u^*):

$$\begin{aligned} V(x_0 + a, t_0) - V(x_0, t_0) = \int_{t_0}^{t_1} &[(f_x^* + \lambda g_x^* + \lambda')(x - x^*) \\ &+ (f_u^* + \lambda g_u^*)(u - u^*)]\, dt \\ &+ \lambda(t_0)a - \lambda(t_1)[x(t_1) - x^*(t_1)] \\ &+ h.o.t. \end{aligned} \tag{5}$$

Let λ be the multiplier satisfying the necessary conditions for (1). Since x^*, u^*, λ satisfy the necessary conditions (2.11)–(2.13) for optimality,

$$\lambda' = -(f_x^* + \lambda g_x^*), \qquad f_u^* + \lambda g_u^* = 0, \qquad \lambda(t_1) = 0,$$

(5) reduces to

$$V(x_0 + a, t_0) - V(x_0, t_0) = \lambda(t_0)a + h.o.t. \tag{6}$$

Divide (6) by the parameter a and then let a approach zero:

$$\lim_{a \to 0}[V(x_0 + a, t_0) - V(x_0, t_0)]/a = V_x(x_0, t_0) = \lambda(t_0) \tag{7}$$

provided the limit exists. The first equation of (7) constitutes the definition of derivative of V with respect to x. We assume that this derivative exists. Thus the multiplier $\lambda(t_0)$ is the marginal valuation in the optimal program of the state variable at t_0.

The discussion thus far has only considered the initial time. However, $\lambda(t)$ is the marginal valuation of the associated state variable at time t. If there were an exogenous, tiny increment to the state variable at time t and if the problem were modified optimally thereafter, the increment in the total value of the objective would be at the rate $\lambda(t)$.

To verify this assertion, recall that the objective in (1) is additive. Any portion of an optimal program is itself optimal.

For instance, suppose we follow the solution x^*, u^* to (1) for some period $t_0 \le t \le t^{\#}$ and then stop and reconsider the optimal path from that time forward:

$$\begin{aligned} &\max \quad \int_{t^{\#}}^{t_1} f(t, x, u)\, dt \\ &\text{subject to} \qquad x'(t) = g(t, x, u), \quad x(t^{\#}) = x^*(t^{\#}). \end{aligned} \tag{8}$$

A solution to (8) must be $x^*(t)$, $u^*(t)$, $t^{\#} \le t \le t_1$, namely, the same as the original solution to (1) on the interval from $t^{\#}$ forward. To see this, suppose it is untrue. Then there is solution to (8) providing a larger value than does x^*, u^* on $t^{\#} \le t \le t_1$. The value of (1) could then be improved by following x^*, u^* to $t^{\#}$ and switching to the solution to (8). But this contradicts the assumed optimality of x^*, u^* for (1). Therefore, x^*, u^*, $t^{\#} \le t \le t_1$ must solve (8).

We return to the question of the interpretation of λ. Application of the methods used to reach (7) to problem (8) leads to the result

$$V_x(x(t^{\#}), t^{\#}) = \lambda(t^{\#}), \tag{9}$$

provided that this derivative exists, where λ is the function associated with problem (1) (since the solutions to (1) and (8) coincide on $t^{\#} \le t \le t_1$). Then $\lambda(t^{\#})$ is the marginal valuation of the state variable at $t^{\#}$. But the time $t^{\#}$ was arbitrary, so for any t, $t_0 \le t \le t_1$,

$$V_x(x(t), t) = \lambda(t), \qquad t_0 \le t \le t_1, \tag{10}$$

is the marginal valuation of the state variable at time t, whenever this derivative exists.

It is easy to confirm the interpretation at t_1. If there is no salvage term, the marginal value of the state at terminal time is zero: $\lambda(t_1) = 0$. And if there is a salvage term, the marginal value of the state is the marginal contribution of the state to the salvage term: $\lambda(t_1) = \phi'(x_1)$ (recall Exercise 2.16).

For ease in discussion, let x be the stock of an asset and $f(t, x, u)$ current profit. It is an identity that

$$\lambda(t_1)x(t_1) = \lambda(t_0)x(t_0) + \int_{t_0}^{t_1}(x'\lambda + x\lambda')\,dt$$
$$= \lambda(t_0)x(t_0) + \int_{t_0}^{t_1}[d(x\lambda)/dt]\,dt. \tag{11}$$

Recall that $\lambda(t)$ is the marginal valuation of the state variable at t. Thus the value of the terminal stock of assets is the value of the original stock plus the change in the value of assets over the control period $[t_0, t_1]$. The total rate of change in the value of assets

$$d(x\lambda)/dt = x'\lambda + x\lambda'$$

is composed of the value of additions (or reductions) in the stock of assets (the first term on the right side) plus the change in the value of existing assets (the second term on the right side). That is, both changes in amount of assets held and in the unit value of assets contribute to the change in the value of all assets. From (3), the rate at which the total value accumulates is

$$f + \lambda g + x\lambda' = H + x\lambda' \qquad \text{where} \qquad H = f + \lambda g. \tag{12}$$

The first term is the direct gain at t of $f(t, x, u)$, say the current cash flow. The second term is an indirect gain through the change in the state variable. One can think of $\lambda g = \lambda x'$ as the increase in future profitability attributable to the increase in the stock of assets. The third and remaining term $x\lambda'$ represents the changed valuation in current assets, the capital gains. Thus, (12) represents the contribution rate at t, both direct and indirect, toward the total value.

At each moment, one chooses the control u to maximize the net contribution (12) toward total value. For given state variable $x(t)$ and marginal valuation of the state $\lambda(t)$, this means choosing $u(t)$ to maximize H, and hence to satisfy

$$\partial H/\partial u = f_u + \lambda g_u = 0, \qquad t_0 \le t \le t_1, \tag{13}$$

and also

$$\partial^2 H/\partial u^2 = f_{uu} + \lambda g_{uu} \le 0. \tag{14}$$

Note also that if one were free to choose x to maximize (12), then one would set

$$f_x + \lambda g_x + \lambda' = 0. \tag{15}$$

Of course, the choice of u completely determines x. A sufficient condition for x^*, u^*, λ to be optimal is that they be feasible with $\lambda(t_1) = 0$ and that the problem

$$\max_{x,u}[H(t, x, u, \lambda(t)) + \lambda'(t)x] \tag{16}$$

have $x = x^*(t), u = u^*(t)$ as its solution for all $t_0 \le t \le t_1$.

Example. Let $P(x)$ be the profit rate that can be earned with a stock of productive capital x, where $P'(0) > 0$ and $P'' < 0$. The capital stock decays at a constant proportionate rate $b \geq 0$. Investment cost is an increasing convex function of the gross investment rate u, with $C'(0) = 0$ and $C'' > 0$. We seek the investment rate $u(t)$ that maximizes the present value of the profit stream over the fixed planning period $0 \leq t \leq T$:

$$\max \quad \int_0^T e^{-rt}\big[P(x) - C(u)\big]\, dt \tag{17}$$

$$\text{subject to} \qquad x' = u - bx, \quad x(0) = x_0 > 0, \quad u \geq 0. \tag{18}$$

We assume that the nonnegativity condition on u will be met automatically. Therefore, since the capital stock is initially positive, it cannot become negative; review (18). (This is Problem 1.8).

The Hamiltonian is

$$H = e^{-rt}\big[P(x) - C(u)\big] + \lambda(u - bx).$$

Optimal functions x, u and λ satisfy (18) and

$$H_u = -e^{-rt}C'(u) + \lambda = 0, \tag{19}$$

$$H_{uu} = -e^{-rt}C''(u) < 0, \tag{20}$$

$$\lambda' = -H_x = -e^{-rt}P'(x) + b\lambda, \qquad \lambda(T) = 0. \tag{21}$$

Condition (20) is satisfied by our assumptions on C. Equation (19) states that, at each t, the marginal cost of investment must equal the marginal value λ of a unit of capital. Both terms are discounted back to the initial moment $t = 0$ of planning. Equivalently,

$$C'\big(u(t)\big) = e^{rt}\lambda(t) \tag{22}$$

requires that marginal cost equal marginal benefit contemporaneously at each t along the optimal investment path.

Differential equation (21) can be manipulated to show the composition of the marginal value of a unit of capital. Subtract $b\lambda$ from each side, multiply by the integrating factor e^{-bt}, and integrate using the boundary conditions on (21):

$$e^{-bt}\lambda(t) = \int_t^T e^{-(r+b)s}P'\big(x(s)\big)\, ds.$$

Therefore, the value at time t of a marginal unit of capital is the discounted stream of marginal profits it generates from the present to the end of the planning horizon:

$$e^{rt}\lambda(t) = \int_t^T e^{-(r+b)(s-t)}P'\big(x(s)\big)\, ds. \tag{23}$$

The calculation reflects the fact that capital decays, and therefore at each time $s > t$ a unit of it contributes only a fraction e^{-bs} of what it contributed originally at time t. Combining (22) and (23) yields the marginal cost = marginal benefit condition for optimal investment

$$C'(u(t)) = \int_t^T e^{-(r+b)(s-t)} P'(x(s))\, ds. \tag{24}$$

EXERCISE

Derive condition (24) using the calculus of variations.

FURTHER READING

These interpretations are greatly elaborated by Dorfman. The sufficiency condition related to (16) is due to Seierstad and Sydsaeter (1977); see also Section 15. Benveniste and Scheinkman (1979) give sufficient conditions for the value function to be differentiable.

Section 5

Several Variables

Our procedures are readily extended to problems with several control and state variables. We illustrate with a problem in two states and two controls (but note that the number of states and the number of controls need not be equal in general):

$$\max \quad \int_{t_0}^{t_1} f(t, x_1(t), x_2(t), u_1(t), u_2(t)\, dt \tag{1}$$

$$\text{subject to} \qquad x_i'(t) = g_i(t, x_1(t), x_2(t), u_1(t), u_2(t)), \quad i = 1,2,$$
$$x_1(t_0), x_2(t_0) \text{ fixed}, \quad x_1(t_1), x_2(t_1) \text{ free}, \tag{2}$$

where the functions f, g_1, and g_2 are continuously differentiable.

Suppose the optimal solution to (1) and (2) is $x_1^*, x_2^*, u_1^*, u_2^*$. Let $h_1(t)$ and $h_2(t)$ be arbitrary fixed feasible modifications in the controls and consider the one parameter family of comparison controls $u_i(t) = u_i^*(t) + ah_i(t)$, $i = 1,2$, where a is a scalar. The solution of the differential equations (2) with this specification of u_1, u_2 is denoted by $x_1 = y_1(t, a)$, $x_2 = y_2(t, a)$. We suppose that y_1 and y_2 are smooth functions of t and a. Since $a = 0$ must yield the optimal function x_1^*, x_2^*, we have $y_i(t, 0) = x_i^*(t)$, $i = 1,2$. Also, since the initial conditions of (2) must be obeyed by any feasible path,

$$y_i(t_0, a) = x_i(t_0), \qquad i = 1,2. \tag{3}$$

Append differential equations (2) to the objective (1) with continuously differentiable multiplier functions $\lambda_1(t), \lambda_2(t)$, as in (2.1), to get

$$\int_{t_0}^{t_1} f\, dt = \int_{t_0}^{t_1} (f + \lambda_1 g_1 - \lambda_1 x_1' + \lambda_2 g_2 - \lambda_2 x_2')\, dt, \tag{4}$$

where f, g_1, and g_2 have arguments (t, x_1, x_2, u_1, u_2) and x_i, u_i, λ_i, $i = 1, 2$, are functions of t. Integration by parts gives

$$-\int_{t_0}^{t_1} \lambda_i x_i'\, dt = -\lambda_i(t_1)x_i(t_1) + \lambda_i(t_0)x_i(t_0) + \int_{t_0}^{t_1} x_i \lambda_i'\, dt, \qquad i = 1, 2. \tag{5}$$

Substituting from (5) into (4) gives

$$\int_{t_0}^{t_1} f\, dt = \int_{t_0}^{t_1} (f + \lambda_1 g_1 + \lambda_2 g_2 + x_1 \lambda_1' + x_2 \lambda_2')\, dt$$
$$-(\lambda_1 x_1 + \lambda_2 x_2)\,|_{t_1} + (\lambda_1 x_1 + \lambda_1 x_2)\,|_{t_0}, \tag{6}$$

Equation (6) holds for any feasible functions x_1, x_2, u_1, u_2 and any differentiable functions λ_1, λ_2. Now evaluate (6) using the control variables $u_1 = u_1^* + ah_1$, $u_2 = u_2^* + ah_2$ and the corresponding state variables y_1, y_2 developed above. The value of the integral is a function of the parameter a, since u_1^*, u_2^*, h_1, h_2 are fixed:

$$J(a) = \int_{t_0}^{t_1} f(t, y_1(t, a), y_2(t, a), u_1^* + ah_1, u_2^* + ah_2)\, dt$$
$$= \int_{t_0}^{t_1} (f + \lambda_1 g_1 + \lambda_2 g_2 + y_1 \lambda_1' + y_2 \lambda_2')\, dt$$
$$-(\lambda_1 y_1 + \lambda_2 y_2)\,|_{t_1} + (\lambda_1 y_1 + \lambda_2 y_2)\,|_{t_0}, \tag{7}$$

where f, g_1, and g_2 have arguments $(t, y_1(t, a), y_2(t, a), u_1^*(t) + ah_1(t), u_2^*(t) + ah_2(t))$ and λ_i, λ_i', $i = 1, 2$ have argument t. Since u_1^*, u_2^* maximize (1) by hypothesis, (7) must achieve its maximum at $a = 0$. This means $J'(0) = 0$. Computing from (7), recalling (3), gives

$$J'(0) = \int_{t_0}^{t_1} [(\partial f/\partial x_1 + \lambda_1 \partial g_1/\partial x_1 + \lambda_2 \partial g_2/\partial x_1 + \lambda_1')\, \partial y_1/\partial a$$
$$+(\partial f/\partial x_2 + \lambda_1 \partial g_1/\partial x_2 + \lambda_2 \partial g_2/\partial x_2 + \lambda_2')\, \partial y_2/\partial a$$
$$+(\partial f/\partial u_1 + \lambda_1 \partial g_1/\partial u_1 + \lambda_2 \partial g_2/\partial u_1) h_1$$
$$+(\partial f/\partial u_2 + \lambda_1 \partial g_1/\partial u_2 + \lambda_2 \partial g_2/\partial u_2) h_2]\, dt$$
$$-\lambda_1(t_1)\, \partial y_1(t_1, a)/\partial a - \lambda_2(t_1)\, \partial y_2(t_1, a)/\partial a = 0. \tag{8}$$

Since (8) is evaluated at $a = 0$, the partial derivatives of f, g_1 and g_2 in (8) are evaluated along $(t, x_1^*(t), x_2^*(t), u_1^*(t), u_2^*(t))$.

Equation (8) holds for any continuously differentiable multiplier functions $\lambda_1(t), \lambda_2(t)$. We specify these multipliers for the coefficients of $\partial y_1/\partial a$ and

$\partial y_2/\partial a$ to be zero:

$$\lambda_1' = -(\partial f/\partial x_1 + \lambda_1 \partial g_1/\partial x_1 + \lambda_2 \partial g_2/\partial x_1),$$
$$\lambda_2' = -(\partial f/\partial x_2 + \lambda_1 \partial g_1/\partial x_2 + \lambda_2 \partial g_2/\partial x_2), \tag{9}$$

with boundary conditions

$$\lambda_1(t_1) = 0, \quad \lambda_2(t_1) = 0, \tag{10}$$

where the partial derivatives in (9) are evaluated along $(t, x_1^*(t), x_2^*(t), u_1^*(t), u_2^*(t))$. Specifying (9) and (10), (8) reduces to

$$\int_{t_0}^{t_1} [(\partial f/\partial u_1 + \lambda_1 \partial g_1/\partial u_1 + \lambda_2 \partial g_2/\partial u_1) h_1$$
$$+ (\partial f/\partial u_2 + \lambda_1 \partial g_1/\partial u_2 + \lambda_2 \partial g_2/\partial u_2) h_2] \, dt = 0 \tag{11}$$

along $(t, x_1^*, x_2^*, u_1^*, u_2^*)$. Equation (11) must hold for any modification functions h_1 and h_2. In particular, taking

$$h_i = \partial f/\partial u_i + \lambda_1 \partial g_1/\partial u_i + \lambda_2 \partial g_2/\partial u_i, \qquad i = 1, 2,$$

leads to the conclusion that the optimal solution obeys

$$\partial f/\partial u_1 + \lambda_1 \partial g_1/\partial u_1 + \lambda_2 \partial g_2/\partial u_1 = 0,$$
$$\partial f/\partial u_2 + \lambda_1 \partial g_1/\partial u_2 + \lambda_2 \partial g_2/\partial u_2 = 0. \tag{12}$$

We have shown that if $(x_1^*, x_2^*, u_1^*, u_2^*)$ maximize (1) subject to (2), then there must be continuously differentiable functions λ_1, λ_2 such that $(x_1^*, x_2^*, u_1^*, u_2^*, \lambda_1, \lambda_2)$ satisfy not only (2), but also (9), (10), and (12). These necessary conditions (exclusive of boundary requirements) can be stated in terms of the Hamiltonian,

$$H(t, x_1, x_2, u_1, u_2, \lambda_1, \lambda_2) = f + \lambda_1 g_1 + \lambda_2 g_2.$$

We have the state equations,

$$x_i' = \partial H/\partial \lambda_i = g_i, \qquad i = 1, 2; \tag{13a}$$

the multiplier equations,

$$\lambda_i' = -\partial H/\partial x_i = -(\partial f/\partial x_1 + \lambda_1 \partial g_1/\partial x_i + \lambda_2 \partial g_2/\partial x_i), \quad i = 1, 2; \tag{13b}$$

and the optimality conditions,

$$\partial H/\partial u_j = \partial f/\partial u_j + \lambda_1 \partial g_1/\partial u_j + \lambda_2 \partial g_2/\partial u_j = 0, \qquad j = 1, 2, \tag{13c}$$

as necessary conditions to be obeyed by $x_1^*, x_2^*, u_1^*, u_2^*, \lambda_1, \lambda_2$ in addition to the boundary conditions

$$x_1(t_0), x_2(t_0) \qquad \text{given}, \quad \lambda_1(t_1) = \lambda_2(t_2) = 0. \tag{14}$$

Further, $u_1^*(t)$, $u_2^*(t)$ must maximize $H(t, x_1^*(t), x_2^*(t), u_1, u_2, \lambda_1(t), \lambda_2(t))$ with respect to u_1, u_2 at each t.

The extension to n state variables and m control variables is straightforward. To save writing, compact vector notation is employed. If $y = [y_1, \ldots, y_p]$ and $z = [z_1, \ldots, z_p]$, then

$$y \cdot z = \sum_{i=1}^{p} y_i z_i.$$

Suppose that there are n state variables, m control variables, and a salvage term:

$$\max \int_{t_0}^{t_1} f(t, x_1(t), \ldots, x_n(t), u_1(t), \ldots, u_m(t))\, dt + \phi(x_1(t_1), \ldots, x_n(t_1)) \tag{15}$$

$$\text{subject to} \quad x_i'(t) = g_i(t, x_1(t), \ldots, x_n(t), u_1(t), \ldots, u_m(t)), \tag{16}$$

$$x_i(t_0) \text{ fixed}, \; x_i(t_1) \text{ free}, \; i = 1, \ldots, n, \tag{17}$$

where the functions f, g_i are continuously differentiable in all arguments. Note that n need not equal m; indeed $n \gtreqless m$. Let $\mathbf{x} = [x_1, \ldots, x_n]$, $\mathbf{u} = [u_1, \ldots, u_m]$, $\mathbf{g} = [g_1, \ldots, g_n]$ denote the vectors indicated. Let $\mathbf{u}^*$ be the vector of optimal control functions. Solving simultaneously the system (16), (17) with $\mathbf{u} = \mathbf{u}^*$ generates the corresponding optimal state variables $\mathbf{x}^*$. The necessary conditions can be developed as above.

If the vector control function $\mathbf{u}^*$ and corresponding vector state $\mathbf{x}^*$ solve (15)–(17), then there are continuously differentiable functions $\boldsymbol{\lambda}(t) = [\lambda_1(t), \ldots, \lambda_n(t)]$ such that, defining

$$H(t, \mathbf{x}, \mathbf{u}, \boldsymbol{\lambda}) = f(t, \mathbf{x}, \mathbf{u}) + \sum_{i=1}^{n} \lambda_i g_i(t, \mathbf{x}, \mathbf{u}), \tag{18}$$

$\mathbf{u}^*$, $\mathbf{x}^*$, $\boldsymbol{\lambda}$ together simultaneously satisfy

$$x_i' = \partial H/\partial \lambda_i = g_i(t, \mathbf{x}, \mathbf{u}), \quad x_i(t_0) \text{ fixed}, \quad i = 1, \ldots, n; \tag{19}$$

$$\lambda_k' = -\partial H/\partial x_k = -\left(\partial f/\partial x_k + \sum_{i=1}^{n} \lambda_i \partial g_i/\partial x_k\right), \quad k = 1, \ldots, n; \tag{20}$$

$$\lambda_k(t_1) = \partial \phi(x(t_1))/\partial x_k, \quad k = 1, \ldots, n; \tag{21}$$

$$0 = \partial H/\partial u_j = \partial f/\partial u_j + \sum_{i=1}^{n} \lambda_i \partial g_i/\partial u_j, \quad j = 1, \ldots, m; \tag{22}$$

respectively. In addition $\mathbf{u}^*(t)$ maximizes $H(t, \mathbf{x}^*(t), \mathbf{u}, \boldsymbol{\lambda}(t))$ with respect to $\mathbf{u}$ at each t.

EXERCISES

1. Show that (19)–(22) are necessary conditions for an optimal solution to (15)–(17).

2. Compare and verify the equivalence of the necessary conditions for solution of the calculus of variations problem

$$\max \int_{t_0}^{t_1} F(t, x_1, \ldots, x_n, x_1', \ldots, x_n')\, dt + \phi(x_1, \ldots, x_n)$$

$$\text{subject to} \quad x_i(t_0) \text{ fixed}, \quad x_i(t_1) \text{ free}, \quad i = 1, \ldots, n,$$

and the necessary conditions for the equivalent optimal control problem.

3. Show that $\lambda_i(t)$ can be interpreted as the marginal valuation of the ith state variable at t in problem (15)–(17).

FURTHER READING

Seater developed a model in which an individual divides time among leisure, work, and search for a higher wage. The state variables are the current wage and wealth (assets). Davis and Elzinga analyze a model in which a firm chooses the rate at which earnings are retained and new equity capital raised in order to maximize the present value of owners' holdings. The state variables are the market price of a share of stock and the equity per share. Harris studied household purchase and consumption rates to maximize utility; state variables are the stock of goods and the stock of money. Sampson chooses consumption and search for a resource and improved technology; the state variables are the resource stock and the technology level.

Section 6

Fixed Endpoint Problems

Suppose both the initial and terminal values of the state variable are fixed:

$$\max \int_{t_0}^{t_1} f(t, x, u)\, dt \tag{1}$$

$$\text{subject to} \quad x'(t) = g(t, x, u), \tag{2}$$

$$x(t_0) = x_0, \quad x(t_1) = x_1, \quad t_0, t_1, \text{ fixed}. \tag{3}$$

Because of the requirement that the comparison control guide the state to the fixed position x_1 at time t_1, we modify our approach to finding necessary conditions obeyed by optimal solutions. We assume that a feasible solution exists.

Let u^* denote an optimal control function and let x^* be the corresponding state function, generated by putting $u = u^*$ into (2) and solving (2) and (3). Denote the maximum value in (1) as J^*. Let u denote another feasible control function and let x be the corresponding state function obtained by solving (2) and (3) with this control. Denote the value of the integrals in (1) following x, u by J. Then using (2.3) and (3), we write

$$J - J^* = \Delta J = \int_{t_0}^{t_1} [f(t, x, u) + \lambda g(t, x, u) + x\lambda'$$
$$-f(t, x^*, u^*) - \lambda g(t, x^*, u^*) - x^*\lambda']\, dt. \tag{4}$$

Substitute for the integrand in (4) its Taylor series around (t, x^*, u^*):

$$\Delta J = \int_{t_0}^{t_1} [(f_x + \lambda g_x + \lambda')(x - x^*)$$
$$+(f_u + \lambda g_u)(u - u^*)]\, dt + \int_{t_0}^{t_1} h.o.t. \tag{5}$$

The partial derivatives of f and g are evaluated along (t, x^*, u^*). Define

$$\delta x = x - x^*, \qquad \delta u = u - u^*. \tag{6}$$

The part of ΔJ that is linear in δx, δu is called the *first variation* of J and is written

$$\delta J = \int_{t_0}^{t_1} \left[(f_x + \lambda g_x + \lambda') \,\delta x + (f_u + \lambda g_u) \,\delta u \right] dt. \tag{7}$$

The variation δJ looks familiar; it was derived earlier by a slightly different approach. The current method is more flexible. It provides the first variation (or the linear part of the difference in functional values) even when a one-parameter family of feasible functions may be difficult to construct.

If x^*, u^* are optimal for (1)–(3), then no modification of that policy (say to x, u) can improve the value of J. As before, we choose λ to satisfy

$$\lambda'(t) = -\left[f_x(t, x^*, u^*) + \lambda(t) g_x(t, x^*, u^*) \right], \tag{8}$$

so the coefficient of δx in (7) will be zero. Then we need

$$\delta J = \int_{t_0}^{t_1} \left[f_u(t, x^*, u^*) + \lambda g_u(t, x^*, u^*) \right] \delta u \, dt \le 0 \tag{9}$$

for any arbitrary feasible modification of the control δu. (Recall that δJ is the linear part of $J - J^*$ and J^* is the maximum.) Note that feasibility now includes the requirement that the corresponding modified state variable terminate at x_1.

It is shown in the appendix to this section that, if there is a feasible control u that drives the state from $x(t_0) = x_0$ to $x(t_1) = x_1$, then the coefficient of δu in (9) must be zero.

$$f_u(t, x^*, u^*) + \lambda g_u(t, x^*, u^*) = 0 \tag{10}$$

when λ obeys (8). *In sum*, if x^*, u^* are optimal for (1)–(3), then there is a function λ such that x^*, u^*, λ simultaneously satisfy (2), (3), (8), and (10). The Hamiltonian is maximized with respect to u at each t. Note that there is no transversality condition; the fixed value x_1 provides the needed condition.

The necessity of (10) is no longer obvious, since δu cannot be chosen completely arbitrarily. The modified control must be feasible; it must drive the state variable to x_1 at t_1. The necessity of (10) in the present case can be verified by a somewhat lengthy construction, as shown in the appendix to this section. We first suppose x^*, u^* is optimal but does not satisfy (10). Then we construct a modification of the control δu that is feasible and that improves J. This contradicts the assumed optimality and completes the demonstration of necessity of (10).

We have implicitly assumed that certain regularity conditions hold; otherwise the foregoing must be modified. A full statement of necessary conditions includes a multiplier λ_0 associated with f that may be either 0 or 1. We have implicitly assumed that λ_0 can always be chosen equal to 1; yet without regularity, it may be necessary to choose $\lambda_0 = 0$. As an example, consider

$$\max \int_0^T u\,dt$$

$$\text{subject to} \qquad x' = u^2, \quad x(0) = x(T) = 0.$$

In this problem, $u = 0$, $0 \le t \le T$, is the *only* feasible control. Writing

$$H = u + \lambda u^2,$$

we have

$$H_u = 1 + 2\lambda u = 0.$$

which is not satisfied by $u = 0$. The correct version is

$$H = \lambda_0 u + \lambda u^2,$$

so

$$H_u = \lambda_0 + 2\lambda u = 0.$$

A choice of $\lambda_0 = 0$ and $u = 0$ does satisfy this condition. We shall implicitly assume in the following that we can optimally choose $\lambda_0 = 1$. See also Section 14 for a more complete treatment.

Example 1. We solve our production planning problem by optimal control. Let $u(t)$ be the production rate and $x(t)$, the inventory level:

$$\min \int_0^T (c_1 u^2 + c_2 x)\,dt \tag{11}$$

$$\text{subject to} \quad x'(t) = u(t), \quad x(0) = 0, \quad x(T) = B, \quad u(t) \ge 0. \tag{12}$$

The initial state (inventory level) is zero and is supposed to reach B by time T. Inventory holding cost accrues at c_2 per unit and production cost increases with the square of the production rate. We form the Hamiltonian:

$$H = c_1 u^2 + c_2 x + \lambda u.$$

Then

$$\partial H/\partial u = 2c_1 u + \lambda = 0, \tag{13}$$

$$\lambda' = -\partial H/\partial x = -c_2. \tag{14}$$

To find x, u, λ that satisfy (12)–(14), integrate (14) to get λ and substitute into (13). Then put u into (12) and integrate. The two constants of integration are found using the boundary conditions, yielding

$$x(t) = c_2 t(t - T)/4c_1 + Bt/T,$$

$$u(t) = c_2(2t - T)/4c_1 + B/T,$$

$$\lambda(t) = c_2 T/2 - 2c_1 B/T - c_2 t,$$

so long as $u \ge 0$ for $0 \le t \le T$. This is the solution obtained by the calculus of variations, as it should be.

Example 2. Two factors, capital $K(t)$ and an extractive resource $R(t)$ are used to produce a good Q according to the production function $Q = AK^{1-a}R^a$ (Cobb-Douglas), where $0 < a < 1$. The product may be consumed, yielding utility $U(C) = lnC$, where $C(t)$ is the consumption, or it may be invested. The total amount of extractive resource available is X_0. We wish to maximize utility over a fixed horizon T:

$$\max \int_0^T \ln C(t)\, dt$$

$$\begin{aligned} \text{subject to} \quad & X' = -R, \quad X(0) = X_0, \quad X(T) = 0, \\ & K' = AK^{1-a}R^a - C, \quad K(0) = K_0, \quad K(T) = 0, \\ & C \geq 0, \quad R \geq 0. \end{aligned}$$

The remaining stock of extractive resource is $X(t)$. Terminal stocks of extractive resource and capital equal zero since they are of no value at T. (A better formulation involves nonnegativity conditions on terminal stocks; see Section 7.) This problem has two state variables K and X and two control variables C and R.

It will be convenient to define $y(t) = R/K$, the ratio of resource to capital. Replacing R with Ky, the problem becomes

$$\max \int_0^T \ln C\, dt \tag{15}$$

$$\begin{aligned} \text{subject to} \quad & X' = -Ky, \quad X(0) = X_0, \quad X(T) = 0, \\ & K' = AKy^a - C, \quad K(0) = K_0, \quad K(T) = 0, \quad (16) \\ & C \geq 0, \quad y \geq 0. \end{aligned}$$

Since the marginal utility of C becomes arbitrarily large as $C \to 0$, an optimal solution has $C > 0$. Likewise, since the marginal productivity of the resource becomes infinitely large as its use shrinks toward zero, an optimal solution has $y > 0$. Thus, we are assured that the nonnegativity conditions on the controls will be met.

Let λ_1 and λ_2 be the multipliers associated with X and K, respectively. They are the marginal values of extractive resource reserves and of capital. The Hamiltonian is

$$H = lnC - \lambda_1 Ky + \lambda_2 (AKy^a - C). \tag{17}$$

An optimal solution $C, y, X, K, \lambda_1, \lambda_2$ obeys (16), (17) and

$$H_C = 1/C - \lambda_2 = 0, \tag{18}$$

$$H_y = -\lambda_1 K + a\lambda_2 AKy^{a-1} = 0, \tag{19}$$

$$\lambda_1' = -H_x = 0, \tag{20}$$

$$\lambda_2' = -H_K = \lambda_1 y - \lambda_2 Ay^a. \tag{21}$$

We shall show that the optimal capital/output ratio grows linearly and that the optimal consumption grows faster or slower than linearly according to whether $a < 1/2$ or $a > 1/2$. The way to find all the variables will be explained later. From (19), so long as $K > 0$, we have

$$\lambda_1 = \lambda_2 aAy^{a-1}. \tag{22}$$

Since λ_1 is constant, from (20), this implies that

$$\lambda_2'/\lambda_2 = (1-a)y'/y. \tag{23}$$

But substituting from (22) for λ_1, into (21) gives

$$\lambda_2'/\lambda_2 = -(1-a)Ay^a. \tag{24}$$

Combining (23) and (24) gives $y^{-(a+1)}y' = -A$. Integrating yields

$$Ay^a = 1/(k_1 + at). \tag{25}$$

The optimal resource/capital ratio declines, according to (25). And since $K/Q = K/AKy^a = k_1 + at$, the optimal capital/output ratio grows linearly at the rate a.

Substituting from (25) into (24) gives

$$\lambda_2'/\lambda_2 = -(1-a)/(k_1 + at).$$

Integrating produces

$$\lambda_2(t) = (k_1 + at)^{-(1-a)/a}/k_2$$

which, when combined with (18) yields

$$C(t) = k_2(k_1 + at)^{(1-a)/a}. \tag{26}$$

Optimal consumption grows over time according to (16). It grows at an increasing rate if $a < 1/2$, at a constant rate if $a = 1/2$, and at a decreasing rate if $a > 1/2$.

One can substitute for C and y from (26) and (25) into (17) to obtain the differential equation obeyed by K. It is a linear first order equation with integrating factor $(k_1 + at)^{-1/a}$. Finally, putting the result in (16) and integrating gives X. The two constants of integration generated here plus the two, k_1 and k_2, generated above can then be found with the aid of the four boundary conditions in (16) and (17).

APPENDIX TO SECTION 6

In this appendix we prove the necessity of (10). First, δu must be feasible, thus, it must render $\delta x_1 = 0$. Second, it must be improving, and hence, render $\delta J > 0$. Let us compute each of these variations for an arbitrary δu. For an arbitrary multiplier function γ

$$J = \int_{t_0}^{t_1} f(t, x, u)\, dt = \int_{t_0}^{t_1} (f + \gamma g - \gamma x')\, dt$$

$$= \int_{t_0}^{t_1} (f + \gamma g + x\gamma')\, dt - \gamma(t_1)x(t_1) + \gamma(t_0)x(t_0). \tag{27}$$

Compute

$$\delta J = \int_{t_0}^{t_1} \left[(f_x + \gamma g_x + \gamma')\, \delta x + (f_u + \gamma g_u)\, \delta u\right] dt - \gamma(t_1)\, \delta x(t_1)$$
$$+ \gamma(t_0)\, \delta x(t_0). \tag{28}$$

If we let

$$\gamma'(t) = -(f_x + \gamma g_x), \qquad \gamma(t_1) = 0, \tag{29}$$

then

$$\delta J = \int_{t_0}^{t_1} (f_u + \gamma g_u)\, \delta u\, dt. \tag{30}$$

Also, by (2), we get

$$\int_{t_0}^{t_1} \left[m(t) g(t, x, u) - m(t) x'(t)\right] dt = 0 \tag{31}$$

for any multiplier function $m(t)$. Integrating the last term by parts,

$$\int_{t_0}^{t_1} (mg + xm')\, dt - m(t_1)x(t_1) + m(t_0)x(t_0) = 0. \tag{32}$$

Now compute the variation in (32) (taking the linear part of the Taylor expansion as was done in (4) and (5)).

$$\int_{t_0}^{t_1} \left[(mg_x + m')\, \delta x + mg_u\, \delta u\right] dt - m(t_1)\, \delta x_1 = 0. \tag{33}$$

Choose m to satisfy

$$m'(t) = -mg_x, \qquad m(t_1) = 1. \tag{34}$$

Then (33) becomes

$$\delta x_1 = \int_{t_0}^{t_1} mg_u\, \delta u\, dt, \tag{35}$$

and therefore, for any constant k, from (30) to (35),

$$\delta J + k\, \delta x_1 = \int_{t_0}^{t_1} (f_u + \gamma g_u + kmg_u)\, \delta u\, dt, \tag{36}$$

where γ and m are assumed to follow (28) and (34), respectively. We have at our disposal the constant k and the function $\delta u(t)$. If we let

$$\delta u = [f_u + (\gamma + km) g_u], \tag{37}$$

and choose k so that $\delta x_1 = 0$ (it is shown below that k can be chosen to achieve this), then

$$\delta J + k\,\delta x_1 = \delta J = \int_{t_0}^{t_1} [f_u + (\gamma + km) g_u]^2 \, dt \geq 0. \tag{38}$$

The last relation of (38) holds with strict inequality unless

$$f_u + (\gamma + km) g_u = 0, \quad t_0 \leq t \leq t_1. \tag{39}$$

Thus, the choice of (37) improves the objective value unless (39) holds; (39) is therefore necessary if no improvement is to be possible.

It remains to verify that k can be selected to assure feasibility. Substitute from (37) into (35) to get

$$\delta x_1 = \int_{t_0}^{t_1} m g_u [f_u + (\gamma + km) g_u] \, dt = 0 \tag{40}$$

or

$$\int_{t_0}^{t_1} m g_u (f_u + \gamma g_u) \, dt + k \int_{t_0}^{t_1} m^2 g_u^2 \, dt = 0. \tag{41}$$

Provided that $m^2 g_u^2 \neq 0$, we can solve (41) for

$$k = - \int_{t_0}^{t_1} m g_u (f_u + \gamma g_u) \, dt \bigg/ \int_{t_0}^{t_1} m^2 g_u^2 \, dt. \tag{42}$$

Note, as claimed, putting k according to (42) into (40) renders $\delta x_1 = 0$. In sum, a modification δu given in (37) with k, $\gamma(t)$, and $m(t)$ specified in (42), (28), and (34), respectively, will improve J (see 38)) *unless* (39) is satisfied. Thus, (39) is necessary for optimality of x^*, u^*, as claimed earlier. But write

$$\lambda = \gamma + km. \tag{43}$$

Then, from (28), (34) and (43),

$$\lambda' = \gamma' + km' = -[f_x + (\gamma + km) g_x] = -(f_x + \lambda g_x) \tag{44}$$

and

$$\lambda(t_1) = \gamma(t_1) + km(t_1) = k. \tag{45}$$

Equation (45) relates the parameters and indicates that $\lambda(t_1)$ will be determined within the problem. Now (44) and (45) specify exactly the function λ of (7)–(10). Hence (39) is equivalent to (10). This completes the demonstration. In future cases of fixed values of the state variable at terminal points, no such proofs are provided. Results are given under the supposition that there is a feasible control to insure that the given constraints of the problem are satisfied.

EXERCISES

1.

$$\text{minimize} \quad \int_0^1 u^2(t)\, dt$$

$$\text{subject to} \quad x'(t) = x(t) + u(t), \quad x(0) = 1, \quad x(1) = 0.$$

2. Solve Exercise I4.5 using optimal control.

3. In Example 1, explain why the multiplier declines over time at rate c_2.

4. Suppose that future utilities in Example 2 are discounted at rate r. Show that

a. the optimal capital/output ratio grows linearly at rate a:

$$K/Q = k_1 + at.$$

b. the optimal consumption path is a single peaked function of time that eventually declines if T is sufficiently large:

$$C(t) = k_2 e^{-rt}(k_1 + at)^{(1-a)/a}$$

5. Show there is no solution to

$$\max \quad \int_0^1 u\, dt$$

$$\text{subject to} \quad x' = x + u^2, \quad x(0) = 1, \quad x(1) = 0.$$

FURTHER READING

The appendix is based on a discussion of Bryson and Ho. Example 2 is based on Dasgupta and Heal. They use more general functional forms and include a random exogenous occurrence of a technical advance that releases the dependence of the economy on the exhaustible resource (e.g., solar energy replaces oil). See also Kamien and Schwartz (1978a, 1979) for extensions involving endogenous technical advance and pollution, respectively.

Section 7

Various Endpoint Conditions

Results for a fixed value of the state variable at the endpoint, a free value of the state variable at the endpoint, a free value of the upper limit of integration, several functions, and a salvage term emerge from a single problem incorporating all these features. The possibility of terminal nonnegativity requirement on some state variables, separately and together, will also be considered.

We seek to maximize

$$\int_{t_0}^{t_1} f(t, x, u)\, dt + \phi(t_1, x(t_1)) \tag{1}$$

subject to

$$x_i'(t) = g_i(t, x, u), \qquad i = 1, \ldots, n; \tag{2}$$

$$x_i(t_0) = x_{i0} \text{ fixed}, \qquad i = 1, \ldots, n; \tag{3}$$

$$x_i(t_1) = x_{i1} \text{ fixed}, \qquad i = 1, \ldots, q; \tag{4}$$

$$x_i(t_1) \text{ free}, \qquad i = q + 1, \ldots, r; \tag{5}$$

$$x_i(t_1) \geq 0, \qquad i = r + 1, \ldots, s; \tag{6}$$

$$K(x_{n+1}, \ldots x_n, t_1) \geq 0 \quad \text{at} \quad t_1; \tag{7}$$

where

$$1 \leq q \leq r \leq s \leq n,$$

$$x(t) = [x_1(t), \ldots, x_n(t)], \qquad u(t) = [u_1(t), \ldots, u_m(t)],$$

and K is assumed to be a continuously differentiable function. Provision has been made for some components of the state vector to be fixed at terminal time, for others to be free, for still others to be nonnegative, and lastly, for some to obey a certain inequality condition.

To generate the necessary conditions obeyed by an optimal solution, associate a continuously differentiable multiplier function $\lambda_i(t)$ with the ith equation of (2), let $\lambda = (\lambda_1, \ldots, \lambda_n)$ and write

$$J = \int_{t_0}^{t_1} [f + \lambda \cdot (\mathbf{g} - \mathbf{x}')]\, dt + \varphi(t_1, \mathbf{x}(t_1)). \tag{8}$$

Familiar integration by parts, term by term, yields

$$J = \int_{t_0}^{t_1} (f + \lambda \cdot \mathbf{g} + \lambda' \cdot \mathbf{x})\, dt + \varphi(t_1, \mathbf{x}(t_1))$$
$$+ \lambda(t_0) \cdot \mathbf{x}(t_0) - \lambda(t_1) \cdot \mathbf{x}(t_1). \tag{9}$$

Let $\mathbf{x}^*, \mathbf{u}^*$ be optimal and let $J^*, f^*, \mathbf{g}^*, \varphi^*$ denote values attained through evaluation along $(t, \mathbf{x}^*, \mathbf{u}^*)$. Let $\mathbf{x}, \mathbf{u}$ be a nearby feasible path satisfying (2)–(7) on $t_0 \le t \le t_1 + \delta t_1$ and let $J, f, \mathbf{g}, \varphi$ denote values attained when evaluated along the feasible path $(t, \mathbf{x}, \mathbf{u})$. Using (9) and the same function λ in each case, compute the difference in value achieved using the two programs:

$$J - J^* = \int_{t_0}^{t_1} (f + \lambda \cdot \mathbf{g} + \lambda' \cdot \mathbf{x} - f^* - \lambda \cdot \mathbf{g}^* - \lambda' \cdot \mathbf{x}^*)\, dt + \varphi - \varphi^*$$
$$- \lambda(t_1) \cdot [\mathbf{x}(t_1) - \mathbf{x}^*(t_1)] + \int_{t_1}^{t_1 + \delta t_1} f(t, \mathbf{x}, \mathbf{u})\, dt. \tag{10}$$

Substitute for the integrand in (10) and also $\phi - \phi^*$ their Taylor series around $(\mathbf{x}^*, \mathbf{u}^*)$. The first variation of J consists of the linear terms in that series. It is the first order approximation to the change in J resulting from a feasible modification of the control. We have

$$\delta J = \int_{t_0}^{t_1} [(f_{\mathbf{x}} + \lambda \mathbf{g}_{\mathbf{x}} + \lambda') \cdot \mathbf{h} + (f_{\mathbf{u}} + \lambda \mathbf{g}_{\mathbf{u}}) \cdot \delta \mathbf{u}]\, dt$$
$$+ \varphi_{\mathbf{x}} \cdot \delta \mathbf{x}_1 + \varphi_t \delta t_1 - \lambda(t_1) \cdot \mathbf{h}(t_1) + f(t_1) \delta t_1, \tag{11}$$

where

$$\mathbf{h} = [h_1, \ldots, h_n], \qquad h_i(t) = x_i(t) - x_i^*(t),$$
$$\delta \mathbf{u} = (\delta u_1, \ldots, \delta u_n), \qquad \delta u_j(t) = u_j(t) - u_j^*(t),$$
$$\delta \mathbf{x}_1 = \mathbf{x}(t_1 + \delta t_1) - \mathbf{x}^*(t_1). \tag{12}$$

All the functions in (11) are to be evaluated along $(\mathbf{x}^*(t), \mathbf{u}^*(t))$. Using the construction in (I9.9), we have approximately

$$\mathbf{h}(t_1) = \mathbf{x}(t_1) - \mathbf{x}^*(t_1) \approx \delta \mathbf{x}_1 - \mathbf{x}^{*\prime}(t_1) \delta t_1 = \delta \mathbf{x}_1 - \mathbf{g}^*(t_1) \delta t_1. \tag{13}$$

Substituting (13) into (11) and collecting terms,

$$\delta J = \int_{t_0}^{t_1} [(f_x + \lambda g_x + \lambda') \cdot h + (f_u + \lambda g_u) \cdot \delta u] \, dt$$
$$+ [\phi_x - \lambda(t_1)] \cdot \delta x_1 + (f + \lambda \cdot g + \phi_t) |_{t_1} \delta t_1 \leq 0. \quad (14)$$

Now (14), with definitions (12), is the desired form of the variation. It must be nonnegative for feasible modifications of the path. The multipliers λ are chosen to satisfy

$$\lambda' = -(f_x + \lambda g_x) \quad (15)$$

along (x^*, u^*); that is,

$$\lambda_i' = -\left(\partial f / \partial x_i + \sum_{j=1}^{n} \lambda_j \partial g_j / \partial x_i \right), \qquad i = 1, \ldots, n,$$

where the partial derivatives are all evaluated along the optimal path (x^*, u^*). Now, since the comparison path could end at the same point in time and some values of the state variables as the optimal path does, we need

$$\int_{t_0}^{t_1} (f_u^* + \lambda g_u^*) \cdot \delta u \, dt \leq 0 \quad (16)$$

for all feasible δu. It can be shown as in Section II.6 that if there is a feasible solution, (16) will be assured only if

$$f_u + \lambda g_u = 0 \quad (17)$$

along (x^*, u^*); that is

$$\partial f / \partial u_i + \sum_{k=1}^{n} \lambda_k \partial g_k / \partial u_j = 0, \qquad j = 1, \ldots, m.$$

Furthermore, the familiar condition that the Hamiltonian be maximized with respect to the controls is obtained. With (15) and (17), (14) reduces to

$$(\phi_x - \lambda) \cdot \delta x_1 + (f + \lambda \cdot g + \phi_t) \, \delta t_1 \leq 0 \quad \text{at} \quad t_1 \quad (18)$$

or

$$\sum_{i=1}^{n} (\partial \phi / \partial x_i - \lambda_i) \, \delta x_{i1} + \left(f + \sum_{j=1}^{n} \lambda_j g_j + \partial \phi / \partial t \right) \delta t_1 \leq 0$$

for all feasible $\delta x_{i1} = \delta x_i(t_1), \delta t_1$. Owing to (4), $\delta x_i(t_1) = 0$ for $i = 1, \ldots, q$. Owing to (5), $\delta x_i(t_1)$ may have any sign for $i = q + 1, \ldots, r$; therefore we choose

$$\lambda_i(t_1) = \partial \phi / \partial x_j, \qquad i = q + 1, \ldots r. \quad (19)$$

If the value of a state variable is freely chosen at terminal time, then the value of its multiplier at terminal time is its marginal contribution to the salvage term.

For the components of x in (6), we have $x_i(t_1) > 0$ or $x_i(t_1) = 0$. In the former case, $\delta x_i(t_1)$ may have any sign, so let

$$\lambda_i(t_1) = \partial\phi/\partial x_i \quad \text{if } x_i(t_1) > 0, \qquad i = r+1, \ldots, s. \tag{20}$$

If the nonnegativity condition is not binding, the case is similar to that of the terminal value being free. In the latter case, feasible modifications must not decrease $x_i(t_1)$, so $\delta x_i \geq 0$ only. But then $[\partial\phi/\partial x_i - \lambda_i(t_1)]\,\delta x_i(t_1) \leq 0$ for all feasible $\delta x_i(t_1)$ only if

$$\lambda_i(t_1) \geq \partial\phi/\partial x_i \qquad \text{when} \quad x_i(t_1) = 0, \qquad i = r+1, \ldots, s. \tag{21}$$

If the nonnegativity condition is active, the terminal marginal value of the state variable is no less than its contribution to the salvage term. The statements in (2) and (21) can be combined into

$$x_i(t_1) \geq 0, \qquad \lambda_i(t_1) \geq \partial\phi/\partial x_i,$$
$$x_i(t_1)\left[\lambda_i(t_1) - \partial\phi/\partial x_i\right] = 0, \qquad i = r+1, \ldots, s. \tag{22}$$

In particular, if x_i does not enter the salvage term, then (22) tells us that the terminal values of state and its multiplier are nonnegative and at least one is zero. If $x_i > 0$, its valuation λ_i is zero. If $x_i = 0$, its valuations may be positive at t_1.

Now

$$\sum_{i=s+1}^{n} [\partial\phi/\partial x_i - \lambda_i(t_1)]\,\delta x_i(t_1) + \left(f + \sum_{i=1}^{n} \lambda_i g_i + \phi_t\right)\delta t_1 \leq 0 \tag{23}$$

must hold for all feasible modifications in the last $n - s$ coordinates of the state vector and in the terminal time. If (7) is satisfied with strict inequality in an optimal solution, then the modifications may have any sign and, if $K > 0$,

$$\lambda_i(t_1) = \partial\phi/\partial x_i, \qquad i = s+1, \ldots, n,$$
$$f + \sum_{i=1}^{n} \lambda_i g_i + \phi_t = 0. \tag{24}$$

But if $K = 0$ in an optimal solution, then the values that $\delta x_{s+1}, \ldots, \delta x_n, \delta t_1$ may assume at t_1 must leave the value of K unchanged or else increase it. Thus, we require

$$dK = \sum_{t=1}^{n} (\partial K/\partial x_i)\,\delta x_i + (\partial K/\partial t_1)\,\delta t_1 \geq 0. \tag{25}$$

Apply Farkas' lemma (Section A6) to the statement that (23) must hold for all $\delta x_{s+1}, \ldots, \delta x_n, \delta t_1$ satisfying (25). By Farkas' lemma, since $(\partial K/\partial x_{x+1}, \ldots, \partial K/\partial t_1)$ is a $1 \times (n - s + 1)$ matrix, there must be a number $p \geq 0$ such that

$$\lambda_i(t_1) = \partial\phi/\partial x_i + p\,\partial K/\partial x_i, \qquad i = s+1, \ldots, n,$$

$$f + \sum_{i=1}^{n} \lambda_i g_i + \phi_t + p\,\partial K/\partial t_1 = 0 \tag{26}$$

at t_1 whenever $K = 0$. Conditions (24) and (26) can be combined into

$$\lambda_i(t_1) = \partial\phi/\partial x_i + p\,\partial K/\partial x_i, \qquad i = s+1, \ldots, n,$$

$$f + \sum_{i=1}^{n} \lambda_i g_i + \phi_t + p\,\partial K/\partial t_1 = 0,$$

$$p \geq 0, \qquad K \geq 0, \qquad pK = 0 \quad \text{at} \quad t_1. \tag{27}$$

If $K > 0$, then (27) assures that $p = 0$ so (27) yields (24). If $K = 0$, then $p \geq 0$ and (27) gives (26).

The transversality condition for a variety of special cases may be found from (27). For instance, if K does not depend on t_1 and t_1 may be freely chosen, then we have the condition that

$$f + \sum_{i=1}^{n} \lambda_i g_i + \phi_t = 0 \quad \text{at} \quad t_1 \qquad \text{if} \quad t_1 \text{ is free}. \tag{28}$$

If (7) is the condition for terminal time to be bounded above,

$$K \equiv T - t_1 \geq 0, \tag{29}$$

then the second line of (27) implies the requirement that

$$f + \sum_{i=1}^{n} \lambda_i g_i + \phi_t \geq 0 \tag{30}$$

with strict equality if $t_1 < T$.

If there are several terminal conditions $K_k \geq 0$ to be satisfied, the necessary conditions can be found, as above, using Farkas' lemma. There will be a number $p_k \geq 0$ associated with each such constraint, along with the requirement that $p_k K_k = 0$. Necessary conditions corresponding to a terminal equality relation $K = 0$ is equivalent to $K \geq 0$ and $-K \geq 0$. Conditions (27) are obtained, *except* that the associated multiplier p is unrestricted in sign.

Necessary conditions obeyed by an optimal solution x^*, u^*, include existence of continuous functions $\boldsymbol{\lambda} = [\lambda_1(t), \ldots (\lambda_n(t)]$ and numbers p as specified below such that the following are obeyed.

Necessary Conditions

a. State equations:

$$x_i' = g_i(t, \mathbf{x}, \mathbf{u}), \qquad i = 1, \ldots, n.$$

b. Multiplier (costate, auxiliary, adjoint) equations:

$$\lambda_i' = -\left(\partial f/\partial x_i + \sum_{j=1}^{n} \lambda_j \partial g_j/\partial x_i\right), \qquad i = 1, \ldots, n.$$

c. Optimality conditions:

(i) $\partial f/\partial u_j + \sum_{k=1}^{n} \lambda_k \partial g_k/\partial u_j = 0$, $j = 1, \ldots, M$;

(ii) $H(t, \mathbf{x}^*, \mathbf{u}, \boldsymbol{\lambda})$ is maximized by $\mathbf{u} = \mathbf{u}^*$.

d. Transversality conditions

(i) $\lambda_i(t_1) = \partial\phi/\partial x_i$ if $x_i(t_1)$ is free;

(ii) $x_i(t_1) \geq 0$, $\lambda_i(t_1) \geq \partial\phi/\partial x_i$, $x_i(t_1)[\lambda_i(t_1) - \partial\phi/\partial x_i] = 0$;

(iii) $\lambda_i(t_1) = \partial\phi/\partial x_i + p\,\partial K/\partial x_i$,
$i = q, \ldots, n$, $p \geq 0$, $pK = 0$, if
$K(x_q(t_1), \ldots, x_n(t_1)) \geq 0$ is required;

(iv) $\lambda_i(t_1) = \partial\phi/\partial x_i + p\,\partial K/\partial x_i$, $i = q, \ldots, n$,
if $K(x_q(t_1), \ldots, x_n(t_1)) = 0$ is required;

(v) $f + \sum_{i=1}^{n} \lambda_i g_i + \phi_t = 0$ at t_1 if t_1 is free;

(vi) $f + \sum_{i=1}^{n} \lambda_i g_i + \phi_t \geq 0$ at t_1, with strict equality in case $t_1 < T$,
if $T - t_1 \geq$ is required;

(vii) $\lambda_i(t_1) = \partial\,\partial\phi/\partial x_i + p\,\partial K/\partial x_i$, $i = q, \ldots, n$,
$f + \sum_{i=1}^{n} \lambda_i g_i + \phi_t + p\,\partial K/\partial t_1 = 0$,
$p \geq 0$, $K \geq 0$, $pK = 0$ at t_1 if
$K(x_q(t_1), \ldots, x_n(t_1), t_1) \geq 0$ is required.

Note that in (18) the coefficient of δt_1 is the Hamiltonian plus the partial derivative of the salvage term with respect to time. If the upper limit of the integration can be freely chosen then this coefficient must equal zero (recall (28)). In particular, in the absence of a salvage term the Hamiltonian must be zero if t_1 can be chosen freely. Moreover, if t_1 is restricted by an upper bound then the Hamiltonian (in the absence of a salvage term) must be nonnegative, according to (29) and (30). Now, using arguments similar to the ones in Section I.3 and I.12, it follows that the Hamiltonian is just the shadow price of extending the upper limit of integration by a small amount.

Also, by recalling the deviation of the canonical form of the Euler equation in Section I.3, expressions (13) and (15), it follows that the necessary condition $\partial H/\partial u = 0$ in the optimal control formulation is just the counterpart of the definition of p in (I.3.13). That is, $\partial H/\partial u = f_u + \lambda g_u = 0$ implies that $\lambda = -f_u/g_u$, $g_u \neq 0$. But if $x' = u$, then $g_u = 1$ and $\lambda = -p$, in (I.3.13). In other words, $\lambda = -p/g_u$. (Note that the p here is not the same as in (26) above.) From this it follows that the Hamiltonion defined in optimal control theory is just the negative of the Hamiltonion defined in the calculus of variations.

Example 1. In Section I10 we showed that for problems of the form

$$\text{optimize} \quad \int_{t_0}^{t_1} f(t, x)(1 + u^2)^{1/2}\, dt$$

$$\text{subject to} \qquad x' = u, \quad x(t_0) = x_0, \quad R(t_1) = x_1$$

the optimal path will be perpendicular to the terminal curve $R(t_1) = x_1$ at t_1. In the optimal control format, let

$$H = f(t, x)(1 + u^2)^{1/2} + \lambda u,$$

$$K = x_1 - R(t_1).$$

Then x, u and λ also satisfy

$$H_u = f(t, x)u/(1 + u^2)^{1/2} + \lambda = 0, \tag{31}$$

$$\lambda' = -H_x = -f_x(t, x)(1 + u^2)^{1/2}, \tag{32}$$

$$\lambda(t_1) = p, \tag{33}$$

$$f(t, x)(1 + u^2)^{1/2} + \lambda u - pR'(t_1) = 0 \quad \text{at} \quad t_1, \tag{34}$$

where transversality conditions (33) and (34) are adapted from (d.iv) with p unrestricted in sign since the terminal condition is an equality.

Substituting from (31) and (33) into (34) and collecting terms gives

$$uR' = -1 \qquad \text{at} \quad t_1;$$

that is

$$u(t_1) = x'(t_1) = -1/R'(t_1),$$

and so the optimal path is orthogonal to the terminal curve, as claimed.

Example 2. Reconsider the problem of choosing consumption over a known finite horizon T to maximize the value of the discounted utility stream:

$$\max \quad \int_0^T e^{-rt} U(C(t))\, dt$$

$$\text{subject to} \qquad K' = iK - C, \quad C \geq 0, \quad K(0) = K_0 > 0, \quad K(T) \geq 0,$$

where K is capital stock, i is its earnings rate, $\lim_{C \to 0} U'(C) = \infty$, and $U'' < 0$. The condition on the utility function for small rates of consumption assures that $C > 0$. The terminal capital stock must be nonnegative but need not be zero.

The Hamiltonian for the problem is

$$H = e^{-rt} U(C) + \lambda(iK - C).$$

The functions K, C, and λ must obey the constraints and also

$$H_C = e^{-rt}U'(C) - \lambda = 0, \qquad \text{so} \quad U'(C(t)) = e^{rt}\lambda(t),$$

$$\lambda' = -i\lambda, \qquad \text{so} \quad \lambda(t) = \lambda_0 e^{-it},$$

$$\lambda(T) \geq 0 \qquad \text{and} \qquad \lambda(T)K(T) = 0.$$

Combining the first two lines gives

$$U'(C(t)) = e^{(r-i)t}\lambda_0.$$

There are now two possibilities: either $K(T) = 0$ or $\lambda(T) = 0$. If the latter holds, then $\lambda_0 = 0$ so that $U'(C(t)) = 0$ for all $0 \leq t \leq T$. This is feasible only if the utility function has a "bliss point" C^* at which utility is maximal ($U'(C^*) = 0$; recall $U'' < 0$ by hypothesis) and if iK_0 is large enough to support consumption at the rate C^* throughout the period (i.e., if $iK_0 \geq C^*(1 - e^{-iT})$). Under these circumstances, the marginal valuation of capital is zero. The bliss point is achieved throughout and additional capital would add no utility. If the conditions just described do not all hold, then $\lambda_0 > 0$, $\lambda(T) > 0$, and $K(T) = 0$. Capital has a positive marginal valuation, reflecting the increment in utility that could be attained. Capital is exhausted at the end of the period.

EXERCISES

1.

$$\max \int_0^1 (-u^2/2)\, dt$$

subject to $\quad x' = y, \quad y' = u, \quad x(0) = 0, \quad y(0) = 0, \quad x(1) + y(1) \geq 2.$

2. Compare the necessary conditions of calculus of variations with the necessary conditions of optimal control for the problem

$$\max \int_{t_0}^{t_1} f(t, x, x')\, dt + \phi(t_1, x(t_1))$$

subject to $\quad x(t_0) = x_0$ fixed

if it is also required (alternatively) that

a. $x(t_1) \geq 0$,
b. $x(t_1) = R(t_1)$,
c. $x(t_1) \geq R(t_1)$,
d. $t_1 \leq T$.

3. Solve Example 2 by the calculus of variations.

4. Write a minimization problem analogous to (1)–(7) and find necessary conditions for its solution.

5. Find necessary conditions for the solution to problem (1)–(3) with the added constraints

$$K_1(x_1(t_1),\ldots,x_n(t_1)) \geq 0$$
$$K_2(x_1(t_1),\ldots,x_n(t_1)) = 0.$$

6. Reconsider problem (1)–(7) with t_1 fixed. Suppose that the functions $f, g_1, \ldots, g_n$ are all concave in $(x_1, \ldots, x_n, u_1, \ldots, u_m)$ and that both ϕ and K are concave in $(x_1, \ldots, x_n)$. State and prove a sufficiency theorem. (Review Section 3.)

7. Find the shortest path from the circle $x^2 + t^2 = 1$ to the straight line $t = 2$.

8. Find the shortest path from the circle $x^2 + t^2 = 1$ to the straight line $x = t - 2$.

Section 8

Discounting, Current Values, Comparative Dynamics

For many problems of economic interest, future values of rewards and expenditures are discounted, say, at rate r:

$$\max \int_0^T e^{-rt} f(t, x, u)\, dt \tag{1}$$

$$\text{subject to} \qquad x' = g(t, x, u), \quad x(0) = x_0. \tag{2}$$

In terms of the Hamiltonian

$$H = e^{-rt} f(t, x, u) + \lambda g(t, x, u), \tag{3}$$

we require (x, u, λ) to satisfy

$$H_u = e^{-rt} f_u + \lambda g_u = 0, \tag{4}$$

$$\lambda' = -H_x = e^{-rt} f_x - \lambda g_x, \qquad \lambda(T) = 0. \tag{5}$$

All values are discounted back to time 0; in particular, the multiplier $\lambda(t)$ gives a marginal valuation of the state variable at t discounted back to time zero.

It is often convenient to conduct the discussion in terms of current values, that is, values at t rather than their equivalent at time zero. Further, if t is not an explicit argument of f or g, then the differential equations describing an optimum solution will be autonomous when the multiplier is given in its current value form. These points will now be illustrated.

Write (3) in the form

$$H = e^{-rt}\left[f(t, x, u) + e^{rt} \lambda g(t, x, u) \right] \tag{6}$$

and define

$$m(t) \equiv e^{rt} \lambda(t) \tag{7}$$

as the *current value multiplier* associated with (2). Whereas $\lambda(t)$ gives the marginal value of the state variable at t, discounted back to time zero (when the whole problem is being solved), the new current value multiplier $m(t)$ gives the marginal value of the state variable at time t in terms of values at t. Also let

$$\mathcal{H} \equiv e^{rt}H = f(t, x, u) + mg(t, x, u). \tag{8}$$

We call $\mathcal{H}$ the *current value Hamiltonian*. Differentiating (7) with respect to time gives

$$\begin{aligned} m' &= re^{rt}\lambda + \lambda' e^{rt} \\ &= rm - e^{rt}H_x \end{aligned} \tag{9}$$

on substituting from (7) and (5). In view of (8), $H = e^{-rt}\mathcal{H}$, so (9) becomes

$$\begin{aligned} m' &= rm - e^{rt}\partial(e^{-rt}\mathcal{H})/\partial x \\ &= rm - e^{rt}e^{-rt}\mathcal{H}_x \\ &= rm - f_x - mg_x. \end{aligned} \tag{10}$$

In addition, (4) can be written as

$$H_u = \partial(e^{-rt}\mathcal{H})/\partial u = e^{-rt}\partial\mathcal{H}/\partial u = 0,$$

which implies

$$\partial\mathcal{H}/\partial u = 0. \tag{11}$$

Finally, (2) may be recovered in terms of the current value Hamiltonian:

$$x' = \partial\mathcal{H}/\partial m = g. \tag{12}$$

In sum, (3)–(5) may be equivalently stated as

$$\mathcal{H} = f(t, x, u) + mg(t, x, u), \tag{13}$$

$$\partial\mathcal{H}/\partial u = f_u + mg_u = 0, \tag{14}$$

$$m' = rm - \partial\mathcal{H}/\partial x = rm - f_x - mg_x. \tag{15}$$

Terminal conditions may be stated in terms of the current value multipliers by using conditions already derived and definition (7). For example, if $x(T)$ is free, then $\lambda(T) = e^{-rT}m(T) = 0$ is required. If $x(T) \geq 0$ is needed, then $e^{-rt}m(T) \geq 0$ and $e^{-rT}m(T)x(T) = 0$.

Notice first that (14) and (15) do not contain any discount terms. Second, note that if t is not an explicit argument of f or g, then Equations (2), (14), and (15) reduce to

$$\begin{aligned} x' &= g(x, u), \\ f_u(x, u) + mg_u(x, u) &= 0, \\ m' &= rm - f_x(x, u) - mg_x(x, u), \end{aligned}$$

which is an *autonomous* set of equations; that is, they do not depend on time explicitly. Solving the second equation for $u = u(m, x)$ in terms of m and x and substituting into the equations for x', m' results in a pair of autonomous differential equations. In general, autonomous differential equations are easier to solve than nonautonomous ones. Furthermore, even if an explicit solution is not possible, phase diagram analysis of the qualitative properties of the solution may be possible when there are autonomous equations.

The following example uses the current value multiplier in an infinite horizon problem. Diagrammatic analysis and the use of the approximating linear differential equation system in the neighborhood of the steady state are shown. Finally, comparative dynamic analysis is illustrated.

Example. Reconsider the example in Section 4, but let $T = \infty$:

$$\max \quad \int_0^\infty e^{-rt}[P(x) - C(u)]\,dt \tag{16}$$

$$\text{subject to} \quad x' = u - bx, \quad x(0) = x_0 \geq 0, \tag{17}$$

$$u \geq 0. \tag{18}$$

The *current value Hamiltonian* is

$$H = P(x) - C(u) + m(u - bx).$$

If the optimal investment rate is positive, it satisfies

$$C'(u) = m, \tag{19}$$

where the current value multiplier obeys

$$m' = (r + b)m - P'(x). \tag{20}$$

The solution x, u, m must satisfy the foregoing conditions (if it involves $u > 0$). (Compare this with the conditions developed in Section 4.) We cannot find the solution explicitly without specification of P and C. Nonetheless, we can qualitatively characterize the solution by sketching paths compatible with the conditions for either the x-m plane or the x-u plane.

We eliminate u and consider the x-m plane. Since $C'' > 0$, C' is a monotone function and may be inverted. Thus, from (19)

$$u = C'^{-1}(m) = g(m), \tag{19'}$$

where $g \equiv C'^{-1}$. The properties of g are readily obtained from the properties of C'. Since $C'(0) = 0$, it follows that $g(0) = 0$. Also, differentiating (19) gives

$$C''(u)\,du = dm,$$

so that

$$du/dm = 1/C'' = g' > 0.$$

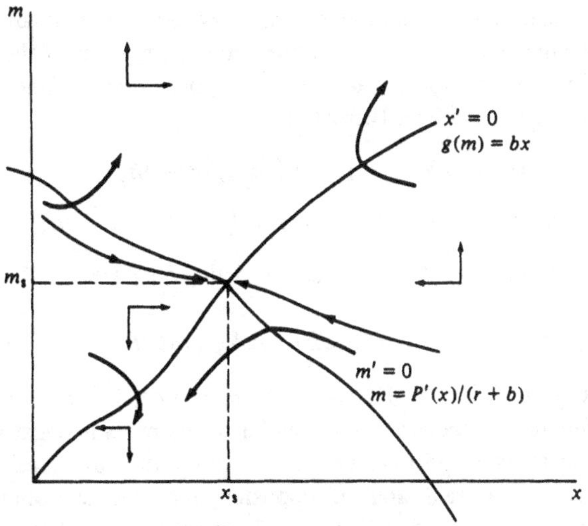

Figure 8.1

Putting (19′) in (17) gives

$$x' = g(m) - bx. \tag{21}$$

Now (20) and (21) form a pair of differential equations in x and m. To sketch the directions of movement compatible with these two equations, we first consider the $x' = 0$ locus, namely, points satisfying

$$g(m) = bx. \tag{22}$$

Since $g(0) = 0$ and $g' > 0$, this locus passes through the origin and is increasing. At a point (x_a, m_a) on the locus, (22) is satisfied. At a point $(x_a, m_a + k)$, $k > 0$, above the locus, we have

$$g(m_a + k) - bx_a > 0$$

since g is an increasing function and so $x' > 0$ at such a point. Similarly, one can show that x is decreasing at points below the $x' = 0$ locus.

Next, consider the points for which $m' = 0$, namely,

$$(r + b)m = P'(x). \tag{23}$$

Since $P'' < 0$, this is a downward sloping curve. Above the curve, m is increasing, whereas m decreases below the $m' = 0$ locus. Figure 8.1 reflects the analysis and shows typical paths consistent with the differential equations.

Since the problem is infinite horizon and autonomous, we inquire about a steady state. The steady state is defined by $x' = m' = 0$, or equivalently by the solution x_s, m_s of (22) and (23). There is a unique path from x_0 to x_s. To verify that this steady state is a saddlepoint, we show that the characteristic roots of the linearized differential equation system are real and of opposite sign. Take the linear terms of the Taylor series expansion of the right side of (20) and (21) around x_s, m_s to get the approximating linear differential equation system (see (B5.21)–(B5.24)):

$$x' = -b(x - x_s) + g'(m_s)(m - m_s),$$
$$m' = -P''(x_s)(x - x_s) + (r + b)(m - m_s),$$

where $x = x(t)$, $m = m(t)$. The characteristic roots are

$$k_1, k_2 = r/2 \pm \left[(r + 2b)^2 - 4g'(m_s)P''(x_s)\right]^{1/2}/2.$$

Since $g' > 0$ and $P'' < 0$, the roots are real. Because the argument of the square root function exceeds r^2, the smaller of the roots is negative. And since the sum of the roots is positive (r), the larger root must surely be positive. Hence, the roots are real and of opposite sign; the stationary point is a saddlepoint as illustrated. (See Section B5, especially (B5.18) and the discussion thereof.)

If $x_0 < x_s$, the steady state will be approached monotonically, with x growing steadily and m falling. Since u is an increasing function of m, (recall (19)), it follows that the investment rate is also decreasing monotonically in this case. Other patterns of behavior, arising from a finite horizon and/or alternative terminal conditions are apparent from the diagram.

Comparative statics analysis of the steady state can be made. An increase in the discount rate r lowers the $m' = 0$ locus and leaves the $x' = 0$ locus unaffected. The new intersection involves lower values of m_s and x_s (see Figure 8.2). Thus, an increase in the discount rate reduces the equilibrium capital stock, its marginal valuation, and (from (19)) the equilibrium investment rate:

$$\partial x_s/\partial r < 0, \qquad \partial m_s/\partial r < 0, \qquad \partial u_s/\partial r < 0.$$

An increase in the decay rate b shifts both the $m' = 0$ and the $x' = 0$ loci down, so the new intersection has a lower value of x_s. While the equilibrium capital stock falls, the influence on the marginal valuation of that stock and on the equilibrium investment rate is unclear. These comparative statics results can also be found by differentiating the system of equations (22) and (23). Similar analyses can be made with respect to shifts in the marginal profit function $P'(x)$ or the marginal cost $C'(u)$ (i.e., $g(m)$).

Comparative dynamics, which involve analysis of changes in the entire optimal path with respect to a change in a parameter, not just the steady state, is more difficult, but sometimes possible. To determine the effect of an increase in the discount rate on the optimal path, let $r_2 > r_1$. The impact on the steady state was just discussed. Suppose $x_0 < x_{s2}$. We show that the optimal path corresponding to r_2 lies below the optimal path corresponding to r_1 by showing that the paths cannot cross. The optimal paths *cannot* be sketched as in Figure 8.2. Clearly, there is no intersection between x_{s2} and

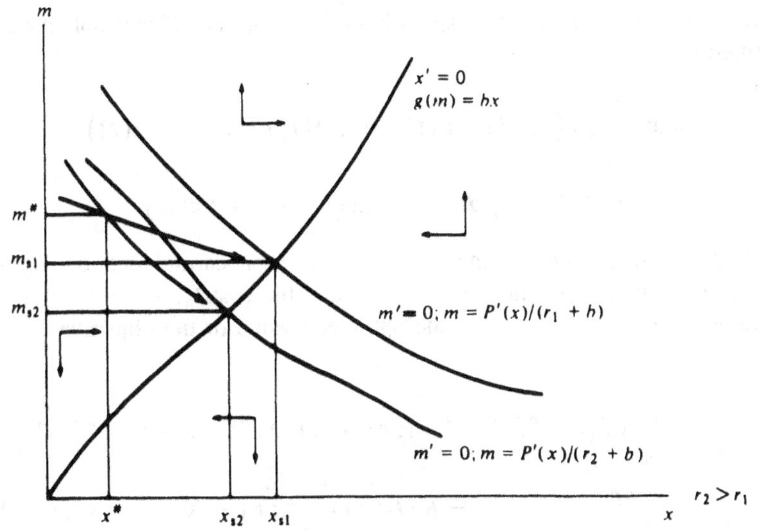

Figure 8.2

x_{s1}. Suppose $x^{\#}$ is the x coordinate of the intersection point closest to x_{s2}. From Figure 8.2 the slope of the optimal path associated with r_2 must be smaller than that associated with r_1 at $(x^{\#}, m^{\#})$. The slope of the optimal path is

$$dm/dx = m'/x' = \left[(r+b)m - P'(x)\right]/\left[g(m) - bx\right].$$

At the intersection, b, $x^{\#}$, and $m^{\#}$ will be the same for both paths. Therefore the required inequality holds if $r_2 < r_1$, which is a contradiction and the optimal paths corresponding to r_1 and r_2 cannot cross. If the discount rate rises, the optimal path of m, and hence of investment, is shifted down and the stationary level of x is decreased. Similar analyses would reveal that a downward shift in the marginal profit function or an upward shift in the marginal cost function results in a downward shift of the optimal path of advertising corresponding to each given level of the stock x.

Another approach to comparative dynamics is to determine how the maximized value of the objective changes with respect to a parameter of the problem. For instance, it might be asked how a change in the discount rate r changes the value of the maximized objective (16). This can be done in general by the following procedure.

Consider the simplest optimal control problem (see (1.1), (1.2), (1.3)):

$$\max \quad \int_{t_0}^{t_1} f(t, x(t), u(t); r)\, dt$$

$$\text{subject to} \quad x'(t) = g(t, x(t), u(t)),$$

$$t_0, t_1, x(t_0) = x_0, \text{ fixed}; \quad x(t_1) \text{ free},$$

where the parameter r appears explicitly in the integrand. This problem can be rewritten as

$$\max \int_{t_0}^{t_1} [f(t, x(t), u(t); r) + \lambda(t) g(t, x(t), u(t)) + x(t)\lambda'(t)]\, dt - \lambda(t_1)x(t_1) + \lambda(t_0)x(t_0)$$

(recall 2.3). Now, once the maximization has been carried out, the optimal control, $u^* = u^*(r)$, the state, $x^* = x^*(r)$, and the costate, $\lambda^* = \lambda^*(r)$ are all functions of the parameter r. So the optimized value of the objective

$$V^*(r) = \int_{t_0}^{t_1} [f(t, x^*(r), u^*(r); r) + \lambda^*(r) g^*(t, x^*(r), u^*(r)) + x^*(r)\lambda^{*\prime}(r)]\, dt - \lambda^*(t_1; r)x^*(t_1; r) + \lambda^*(t_0; r)x^*(t_0; r). \tag{24}$$

Then,

$$\begin{aligned} \partial V^*(r)/\partial r = \int_{t_0}^{t_1} [& f_r^* + f_x^*\, \partial x^*/\partial r + f_u^*\, \partial u^*/\partial r \\ &+ \lambda^*(g_x^*\, \partial x^*/\partial r + g_u^*\, \partial u^*/\partial r) \\ &+ g^* \partial\lambda^*/\partial r + x^* \partial\lambda^{*\prime}/\partial r + \lambda^{*\prime}\, \partial x^*/\partial r]\, dt \\ &- \lambda^*(t_1; r)\, \partial x^*(t_1; r)/\partial r \\ &- x^*(t_1; r) \partial\lambda^*(t_1; r)/\partial r + \lambda^*(t_0; r)\, \partial x^*(t_0; r)/\partial r \\ &+ x^*(t_0; r)\, \partial\lambda^*(t_0; r)/\partial r, \end{aligned} \tag{25}$$

where $f_r^*, f_x^*, f_u^*, g_x^*, g_u^*$ refer to the partial derivatives of f and g with respect to r, x, and u, respectively, evaluated along u^*, x^*. Note that f is affected by a change in r directly, and indirectly through the changes in x^* and u^*. However, by regrouping the terms inside the integral we get

$$\begin{aligned} &\int_{t_0}^{t_1} [f_r^* + (f_x^* + \lambda^* g_x^* + \lambda^{*\prime})\, \partial x^*/\partial r + (f_u^* + \lambda^* g_u^*)\, \partial u^*/\partial r \\ &\qquad + g^* \partial\lambda^*/\partial r + x^* \partial\lambda^{*\prime}/\partial r]\, dt \\ &\quad = \int_{t_0}^{t_1} [f_r^* + x^{*\prime} \partial\lambda^*/\partial r + x^* \partial\lambda^{*\prime}/\partial r]\, dt, \end{aligned} \tag{26}$$

because the coefficients of $\partial x^*/\partial r$ and $\partial u^*/\partial r$ are just the necessary conditions that must equal zero along the optimal path and $g^* = x^{*\prime}$. Thus,

$$\begin{aligned}\partial V^*/\partial r = &\int_{t_0}^{t_1} \left[f_r^* + x^{*\prime} \partial \lambda^* / \partial r + x^* \partial \lambda^{*\prime} / \partial r \right] dt \\ &- \lambda^*(t_1; r)\, \partial x^*(t_1; r)/\partial r - x^*(t_1; r)\, \partial \lambda^*(t_1; r)/\partial r \\ &+ \lambda^*(t_0; r)\, \partial x^*(t_0; r)/\partial r + x^*(t_0; r)\, \partial \lambda^*(t_0; r)/\partial r. \quad (27)\end{aligned}$$

Now note that

$$x^{*\prime} \partial \lambda^* / \partial r + x^* \partial \lambda^{*\prime} / \partial r = d(x^* \partial \lambda^* / \partial r)/dt$$

so that

$$\int_{t_0}^{t_1} [d(x^* \partial \lambda^* / \partial r)/dt] dt = x^*(t_1; r) \partial \lambda^*(t_1; r)/\partial r - x^*(t_0; r) \partial \lambda^*(t_0; r)/\partial r. \quad (28)$$

Substituting (28) back into (27) and cancelling terms gives

$$\partial V^*(r)/\partial r = \int_{t_0}^{t_1} f_r^* dt - \lambda^*(t_1; r) \partial x^*(t_1; r)/\partial r + \lambda^*(t_0; r) \partial x^*(t_0; r)/\partial r. \quad (29)$$

But because $x(t_1)$ is free, $\lambda^*(t_1; r) = 0$ by the transversality condition and because $x(t_0)$ is fixed, $\partial x^*(t_0; r)/\partial r \equiv 0$. So,

$$\partial V^*(r)/\partial r = \int_{t_0}^{t_1} f_r^* \, dt. \quad (30)$$

In terms of the above example

$$\partial V^*(r)/\partial r = -\int_{t_0}^{\infty} t e^{-rt} [\, \mathrm{p}(x^*) - C(u^*)] \, dt < 0. \quad (31)$$

Expression (30) is an optimal control equivalent of the "envelope theorem" that arises in connection with static optimization problems. In the static version of the envelope theorem the partial derivative of the optimized objective function with respect to r would just be f_r^*. Here it is the integral of f_r^*. If one wanted to determine, say, $\partial u^*(t)/\partial r$ then as in static optimization problems, one would have to appeal to the second order conditions.

EXERCISES

1. In (16)–(18), let $P(x) = ax - x^2/2$ and $C(u) = cu^2$. Solve explicitly and relate your findings to the analysis of the general case (16)–(18).

2. Provide a diagrammatic analysis of the solution to (16)–(18) in the x-u plane. (Differentiate (19) totally and use (19) and (20) to eliminate m and m' from the system.)

3. In (16)–(18), suppose that the profit function is linear, $P(x) = px$, and solve the problem. Show that m and u reach their respective steady state levels immediately, but x approaches its steady state only asymptotically.

4. If the discount rate $r(t)$ is time dependent, transformation to current values is still possible, although it cannot lead to an autonomous problem. Consider

$$\int_0^T e^{-\rho(t)} f(t, x(t), u(t))\, dt$$

$$\text{subject to} \quad x' = g(t, x, u), \quad x(0) = x_0,$$

where

$$\rho(t) = \int_0^t r(s)\, ds.$$

If $r(t) = r$ constant, the $\rho(t) = rt$, so the generalization reduces to the familiar form for a constant discount rate. Following the outline above, find the necessary conditions for optimality using a current value multiplier.

5. Utility $U(C, P)$ increases with the consumption rate C and decreases with the stock of pollution, P. For $C > 0,\ P > 0$,

$$U_C > 0, \qquad U_{CC} < 0, \qquad \lim_{C \to 0} U_C = \infty;$$

$$U_P < 0, \qquad U_{PP} < 0, \qquad \lim_{P \to 0} U_P = 0, \qquad U_{CP} = 0.$$

The constant rate of output $\bar{C}$ is to be divided between consumption and pollution control. Consumption contributes to pollution, while pollution control reduces it; $Z(C)$ is the net contribution to the pollution flow, with $Z' > 0$, $Z'' > 0$. For small C, little pollution is created and much abated; thus net pollution declines: $Z(C) < 0$. But for large C, considerable pollution is created and few resources remain for pollution control, therefore on net, pollution increases; $Z(C) > 0$. Let C^* be the consumption rate that satisfies $Z(C^*) = 0$. In addition, the environment absorbs pollution at a constant proportionate rate b. Characterize the consumption path $C(t)$ that maximizes the discounted utility stream:

$$\int_0^\infty e^{-rt} U(C, P)\, dt$$

subject to

$$P' = Z(C) - bP, \qquad P(0) = P_0, \qquad 0 \le C \le \bar{C}, \qquad 0 \le P.$$

Also characterize the corresponding optimal pollution path and the steady state.

FURTHER READING

The discussion of the example and Exercise 3 have been drawn from Gould. Several capital stock-investment problems lead to a similar formulation. For instance, x might be the stock of goodwill possessed by the firm; then u is the rate of advertising to increase that goodwill (see Gould). Or if x is human capital, then u is related to education and training (see Ben-Porath, Blinder and Weiss, Haley, Heckman). If x is the stock of a durable good whose services are rented out for profit (for example, autos, heavy machinery, computers, and copying machines), then u is the rate of acquisition or manufacture of new machines (see Kamien and Schwartz 1974c). On applications to health capital, see Cropper (1977) and Grossman. And see Kotowitz and Mathewson for recent extensions to optimal advertising.

See Gould, Oniki, and Epstein for further discussion of comparative dynamics. Caputo in three articles presents a derivation of the dynamic envelope theorem and an example of its application.

Exercise 5 is analyzed by Forster. See Cropper (1976) for extensions.

Section 9

Equilibria in Infinite Horizon Autonomous Problems

The example in the last section was an infinite horizon "autonomous" problem. It is said to be autonomous because time enters only through the discount term. In infinite horizon problems, a transversality condition needed to provide a boundary condition is typically replaced by the assumption that the optimal solution approaches a steady state. This assumption is plausible since one might expect that, in the long run, the optimal solution would tend to "settle down" as the environment is stationary by hypothesis. However, some autonomous problems have no stable equilibrium.

In the example of Section 8, the optimal path of the state variable x was monotonic over time. In infinite horizon autonomous problems with just one state variable, if there is an optimal path to a steady state, then the state variable is monotonic. Further, in such cases, that steady state must be a "saddlepoint." There may not be an optimal path to a steady state; it may be unstable, with paths leading away from it. In general, there may be no steady state, a unique steady state, or multiple steady states. This section is devoted to demonstrating these results, and generally, to cataloging the kinds of equilibria that may exist. The results depend heavily upon the assumptions that the problem has a single state variable, a single control, and an infinite horizon and is autonomous.

Consider

$$\max \int_{0}^{\infty} e^{-rt} f(x, u)\, dt \tag{1}$$

$$\text{subject to} \qquad x' = g(x, u), \quad x(0) = x_0. \tag{2}$$

The current value Hamiltonian for (1) is

$$H(x, u, m) = f(x, u) + mg(x, u). \tag{3}$$

Since u maximizes (2), we have

$$H_u = f_u(x, u) + mg_u(x, u) = 0, \tag{4}$$

$$H_{uu} < 0. \tag{5}$$

Only weak inequality is required in (5), but we assume that strong inequality holds. (This is used in (9).) Further,

$$m' = rm - H_x = rm - f_x - mg_x. \tag{6}$$

Since H_u is strictly monotone along an optimal path, by (5), (4) implicitly gives u as a function of x and m. Write

$$u = U(x, m). \tag{7}$$

We assume that U is a continuously differentiable function. The properties of U may be found by differentiating (4) and (7) totally:

$$dH_u = H_{uu}\,du + H_{ux}\,dx + g_u\,dm = 0. \tag{8}$$

From (8) we have

$$du = -(H_{ux}/H_{uu})\,dx - (g_u/H_{uu})\,dm. \tag{9}$$

while from (7) $du = U_x\,dx + U_m d_m$. Hence, from (9)

$$U_x = -H_{ux}/H_{uu}, \qquad U_m = -g_u/H_{uu}. \tag{10}$$

Substituting from (7) into (2) and (6) gives a pair of differential equations for x and m:

$$x' = g(x, U(x, m)), \qquad x(0) = x_0, \tag{11}$$

$$m' = rm - H_x(x, U(x, m), m). \tag{12}$$

A steady state (x_s, m_s), assumed to exist, satisfies

$$g(x, U(x, m)) = 0, \tag{13}$$

$$rm - H_x(x, U(x, m), m) = 0. \tag{14}$$

Let $u_s = U(x_s, m_s)$.

To determine the nature of any steady state, we study the linear differential equation system that approximates (11) and (12) at x_s, m_s. Take the linear terms of the Taylor series expansion of the right sides of (11) and (12) around x_s, m_s:

$$x' = (g_x + g_u U_x)(x - x_s) + g_u U_m(m - m_s), \tag{15}$$

$$m' = -(H_{xx} + H_{xu}U_x)(x - x_s) + (r - g_x - H_{xu}U_m)(m - m_s), \tag{16}$$

where the partial derivatives are evaluated at (x_s, m_s) and $x = x(t)$, $m = m(t)$. Using (10) to eliminate U_x and U_m gives

$$x' = a(x - x_s) + b(m - m_s), \tag{17}$$

$$m' = c(x - x_s) + (r - a)(m - m_s). \tag{18}$$

where the coefficients are

$$a = g_x - g_u H_{ux}/H_{uu}, \qquad b = -(g_u^2/H_{uu}),$$
$$c = (H_{xu}^2 - H_{uu}H_{xx})/H_{uu} \tag{19}$$

and all the partial derivatives are evaluated at (x_s, m_s). In studying (17) and (18), it will save writing if we consider the variables to be $x - x_s$ and $m - m_s$. The steady state for the transformed variables is the origin (0, 0). Rather than introduce a new notation for the transformed variables, we shall continue to use the symbols x and m. Context will generally indicate whether $x - x_s$ or x is meant. As an exercise, it may be useful to repeat the analysis in terms of the original variables.

The characteristic equation associated with (17) and (18) is

$$k^2 - rk + a(r - a) - bc = 0 \tag{20}$$

with roots

$$k_1, k_2 = r/2 \pm \left[(r - 2a)^2 + 4bc\right]^{1/2}/2. \tag{21}$$

The solution has the form

$$x(t) = Ae^{k_1 t} + Be^{k_2 t} \tag{22}$$

for k_1, k_2 real and distinct, or

$$x(t) = (A + Bt)e^{rt/2} \tag{23}$$

if $k_1 = k_2 = r/2$, or

$$x(t) = e^{rt/2}(A \cos pt + B \sin pt), \qquad p = \left[-(r - 2a)^2 - 4bc\right]^{1/2} \tag{24}$$

if p is real. (See also Section B3.) If the roots are real, then the larger root is necessarily positive (recall (21)). The smaller root may be either positive or negative. It will be negative if $r < [(r - 2a)^2 + 4bc]^{1/2}$; that is, if

$$bc > a(r - a), \tag{25}$$

the roots are real and of opposite signs. Let $k_1 > 0 > k_2$ if (25) holds. Then (22) will converge to 0 provided we take $A = 0$. The assumption of convergence to the steady state has provided a condition with which to evaluate a constant of integration. Thus, if the inequality of (25) holds, then the roots will certainly be real and the steady state will satisfy the conditions to be a saddlepoint.

On the other hand, the roots will be real and nonnegative if $(r - 2a)^2 + 4bc \geq 0$ and (25) fails. These two conditions may be written

$$a(r - a) \geq bc \geq a(r - a) - (r/2)^2 \quad \text{implies roots are real and nonnegative.} \tag{26}$$

As long as the roots are both nonnegative, the path ((22) or (23)) cannot converge to the steady state. It will move away from it, unless the initial position happens to be the steady state.

Finally, the roots will be complex if $(r-2a)^2 + 4bc < 0$; that is,

$$a(r-a) - (r/2)^2 > bc \qquad \text{implies roots are complex.} \tag{27}$$

Then the path follows (24). But note that since the real part of the complex roots is positive ($= r/2$), the path moves away from the steady state. The steady state is an unstable equilibrium.

In sum, a solution to (17) and (18) can converge to the origin only if (25) holds. In all other cases, all paths diverge from the origin. The solution to an *arbitrary* pair of linear differential equations with constant coefficients may exhibit still other patterns. The roots could be both real and negative or complex with negative real parts. In either case, the equilibrium would be stable with all paths converging to it. (See Section B3). But a pair of linear differential equations (17) and (18) arising from solution to the problem given by (1) and (2) *cannot* have both roots with negative real parts; the solution pattern cannot be totally stable with all paths converging to a steady state.

Figure 9.1 illustrates a phase diagram for (17) and (18) in the neighborhood of the steady state. Since (17) and (18) are just (15) and (16) and approximate (11) and (12) in that neighborhood, Figure 9.1 also reflects the behavior of (11)–(12) near the steady state. The diagrams shown depend on the signs and magnitudes of the various parameters.

From (17) and (18), the $x' = 0$ locus is given by

$$m = -ax/b, \tag{28}$$

whereas the $m' = 0$ locus is

$$m = -cx/(r-a). \tag{29}$$

In view of (5) and definition (19), $b > 0$. However, a and c could have any sign. In the usual manner, we find that the x coordinate will be rising above the $x' = 0$ locus if $a > 0$, will be falling if $a < 0$, and so on. Also note that the $x' = 0$ locus will be steeper than the $m' = 0$ locus if and only if $-a/b > -c/(r-a)$. With these considerations in mind, one may develop nine possible diagrams, illustrated in Figure 9.1, based on the signs and relative magnitudes of parameters. We distinguish between equilibria that are saddlepoints and those that are not; no indication is given as to whether roots of unstable equilibria are real or complex.

Figure 9.1 not only catalogs the various possibilities according to the signs and relative magnitudes of the expressions in (19) but also illustrates the need to check whether (25) holds. One can show that

$$bc - a(r-a) = (g_u^2 H_{xx} - 2g_u g_x H_{ux} + g_x^2 H_{uu})/H_{uu} - r(g_x - g_u H_{xu}/H_{uu}). \tag{25'}$$

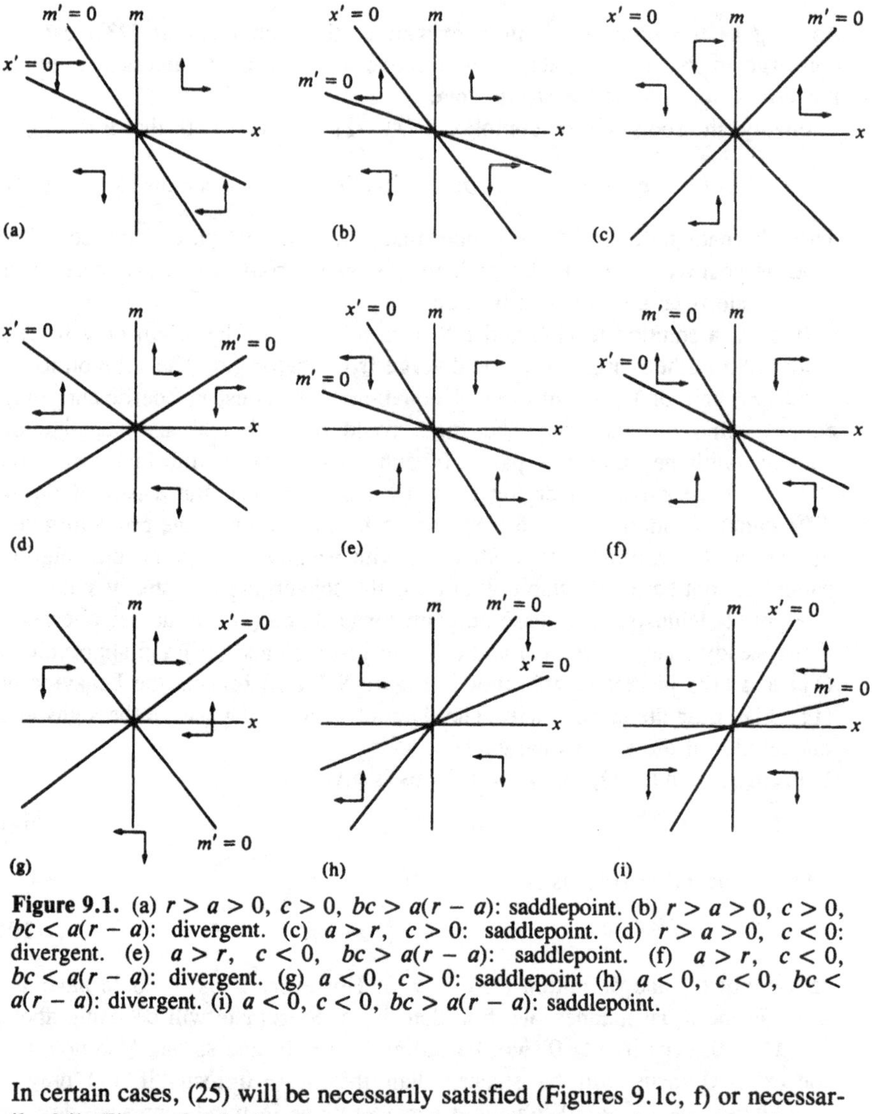

Figure 9.1. (a) $r > a > 0$, $c > 0$, $bc > a(r - a)$: saddlepoint. (b) $r > a > 0$, $c > 0$, $bc < a(r - a)$: divergent. (c) $a > r$, $c > 0$: saddlepoint. (d) $r > a > 0$, $c < 0$: divergent. (e) $a > r$, $c < 0$, $bc > a(r - a)$: saddlepoint. (f) $a > r$, $c < 0$, $bc < a(r - a)$: divergent. (g) $a < 0$, $c > 0$: saddlepoint (h) $a < 0$, $c < 0$, $bc < a(r - a)$: divergent. (i) $a < 0$, $c < 0$, $bc > a(r - a)$: saddlepoint.

In certain cases, (25) will be necessarily satisfied (Figures 9.1c, f) or necessarily fail (Figure 9.1d). In other cases, the condition must be checked. To see this, compare figures 9.1h and 9.1i. Their general appearance is similar and it may seem that there would be a path to the origin in either case; yet (25) is satisfied in Case i but violated in Case h; the former equilibrium is a saddlepoint but the latter is not!

If the Hamiltonian is concave in x, y then $c > 0$ (review 19)). A check of Figures 9.1e and 9.1b indicates that concavity of H is neither necessary nor sufficient for an equilibrium to be a saddlepoint (and hence a convergent path). However, if H is concave and the discount rate is small, the equilibrium will be a saddlepoint.

The analysis above has been conducted for the neighborhood of a single equilibrium. Of course (13) and (14) may have several solutions; (11) and (12) may have multiple equilibria. The analysis may be applied separately to each equilibrium. If there are multiple equilibria, the desired steady state and the qualitative nature of the optimal solution may depend crucially on the initial conditions.

We can now show that the optimal path to a steady state (if one exists) will always be monotonic in the state variable for the problems of (1) and (2). Suppose we follow the optimal solution to (1) and (2) up to a time t_1, arriving at $x_1 = x(t_1)$, and then stop to reconsider the problem at that time:

$$\max \int_{t_1}^{\infty} e^{-r(t-t_1)} f(x, u)\, dt \qquad \text{subject to} \quad x' = g(x, u), \quad x(t_1) = x_1. \tag{30}$$

Problem (30) has been written with values discounted back to t_1. (Of course, multiplication by e^{-rt_1} will discount future values back to time 0 if desired.)

Problems (1) and (30) are structurally the same, differing only in the initial value of the state variable. To see this more clearly, change the variable of integration in (30) from t to $s = t - t_1$. If we define $V(x_0)$ as the maximum that can be achieved in (1), starting in state x_0, the maximum value in (30) is $V(x_1)$. The fact that $V(x)$ is a function of a single variable (time does not enter) rests on the assumptions that the problem has an infinite horizon and that it is autonomous.

Following an argument similar to the one given at the beginning of Section 4, it can be shown that in an optimal solution to (1), the current value multiplier m associated with (1) is related to the current value function $V(x)$ by

$$m(t) = V'(x(t)) \tag{31}$$

wherever the derivative exists. Substituting from (31) into (3) gives

$$H = f(x, u) + V'(x) g(x, u). \tag{32}$$

The control u is chosen to maximize (32). Since (32) depends only on x and u, the maximizing value of $u = u(x)$ depends only on x. If $g(x, u(x)) = 0$, then $x' = 0$. If x is stationary, then $m = V'(x)$ and $u = u(x)$ are also stationary and the value of x cannot change further. Since x' cannot change sign, the state x must be a monotone function of time. Note however that the optimal control function u need *not* be monotone. Recall that the steady state need not be unique and that there does not need to be a path converging to a particular steady state. We have established only that from any given initial point, if the optimal path for (1) goes to some steady state (and $V'(x)$ exists along that path), then that path will be monotone in x.

In example of Section 8, the optimal path is monotone in u as well as in x. While this need *not* always be the case, there are sets of conditions under which it will be so. In particular, suppose that the current value Hamiltonian is concave, with

$$H_{uu} < 0, \qquad H_{uu}H_{xx} - H_{ux}^2 \geq 0 \tag{33}$$

and further that

$$g_u > 0, \qquad g_x < 0, \qquad H_{xu} < 0, \qquad m(t) \geq 0 \qquad \text{for all} \quad t \geq 0. \tag{34}$$

Then the necessary conditions are sufficient for optimality. In addition, the approach of (x, u, m) to unique steady state values (x_s, u_s, m_s) will be monotonic, as will now be shown.

Under assumptions (33) and (34), (13) describes an upward sloping curve in the m-x plane, while (14) describes a downward sloping curve. Thus there is at most one intersection, and therefore at most one steady state. Further, under assumptions (33) and (34), we have, by definition (19), $a < 0$, $c \geq 0$. The present case therefore corresponds to Figure 9.1g. It is clear from the sketch (extending it to permit the $x' = 0$ and $m' = 0$ loci to be nonlinear while retaining their positive and negative slopes, respectively, as required by the assumptions) that both x and m must approach the equilibrium monotonically. If $x_0 < x_s$, then x increases and m decreases. Further, since $du/dt = U_x\,dx/dt + U_m\,dm/dt$, u decreases in this case, as by (10), $U_x < 0$ and $U_m > 0$. If, on the other hand, $x_0 > x_s$, then x decreases while m and u increase throughout.

In conclusion, this section has been devoted to cataloging the nature of equilibria and the approach to those equilibria that can arise for problems of the form (1) and (2). The following example illustrates the analysis that may be performed in a particular instance.

Example. The market for a product is n people, of whom $x(t)$ know of the good. Their purchases generate profits of $P(x)$, where $P(0) = 0$, $P' > 0$, and $P'' < 0$. People learn of the good by discussion with others who are knowledgeable. At a cost $C(u)$ (where $C(0) = 0$, $C' > 0$, $C'' > 0$), the firm influences the contact rate $u(t)$, the rate at which people discuss the good. The $x(t)$ knowledgeable people inform $x(t)u(t)$ people, of whom only a fraction $1 - x/n$ will be newly informed. Knowledgeable people forget at a constant proportionate rate b. Characterize the program that maximizes the present value of the firm's profit stream:

$$\max \int_0^\infty e^{-rt}[P(x) - C(u)]\,dt \tag{35}$$

$$\text{subject to} \qquad x' = -bx + xu(1 - x/n), \quad 0 < x_0 < n. \tag{36}$$

From (36), $x < n$ throughout since the positive term tends toward zero as x approaches n. The current value Hamiltonian is

$$H = P(x) - C(u) + m(-bx + xu - x^2u/n).$$

An optimal solution satisfies (36) and

$$C'(u) = mx(1 - x/n), \tag{37}$$

$$m' = (r + b - u)m + 2mxu/n - P'(x). \tag{38}$$

For the phase diagram analysis, differentiate (37) totally and use (35) to eliminate m and (37) and (38) to eliminate m' from the result. The resulting system is (36) and

$$u' = [r + bx/(n - x)]C'(u)/C''(u) - P'(x)x(1 - x/n)/C''(u). \tag{39}$$

We have $x' = 0$ along $x = 0$ and for

$$u = bn/(n - x). \tag{40}$$

This is an increasing convex function that grows without bound as x approaches n. Above this locus, x is increasing; x decreases below the locus.

The points (x, u) such that $u' = 0$ satisfy

$$C'(u) = P'(x)x(1 - x/n)/[r + bx/(n - x)] \equiv h(x). \tag{41}$$

The left side is an increasing function of u. The right side depends only on x. It is zero at both $x = 0$ and $x = n$; it is increasing for small x and decreasing for large x. Figure 9.2 illustrates the identification of pairs (x, u) satisfying (41). As illustrated, two values of x correspond to each value of u, $0 \le u < u_3$. They are closer together as u gets larger. The phase diagram in the case that the curves intersect can now be drawn (Figure 9.3).

There are two equilibria in Figure 9.3. To characterize each, we linearize the system of (36) and (39) in the neighborhood of an equilibrium (x_s, u_s), getting

$$x' = a_1(x - x_s) + b_1(u - u_s),$$
$$u' = a_2(x - x_s) + b_2(u - u_s), \tag{42}$$

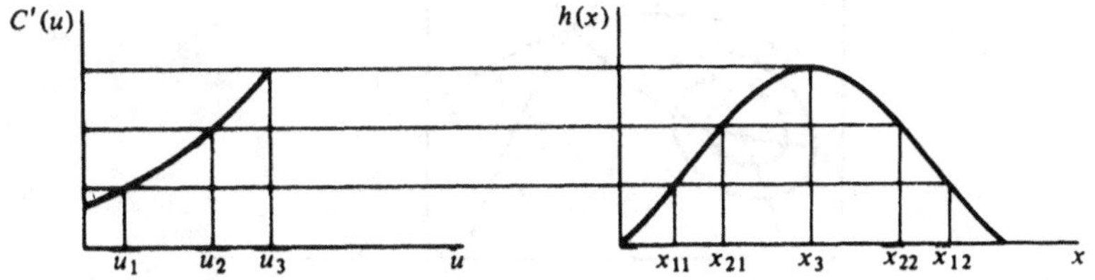

Figure 9.2

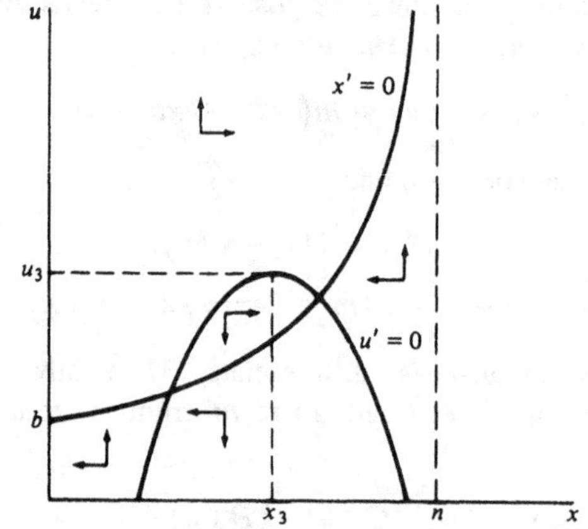

Figure 9.3

where

$$a_1 = -x_s u_s / n < 0,$$
$$b_1 = x_s(1 - x_s/n) > 0,$$
$$b_2 = r + bx_s/(n - x_s) > 0. \tag{43}$$

The sign of a_2 is ambiguous. The roots of the characteristic equation associated with (42) will be real and of opposite sign if and only if

$$a_1 b_2 - a_2 b_1 < 0 \tag{44}$$

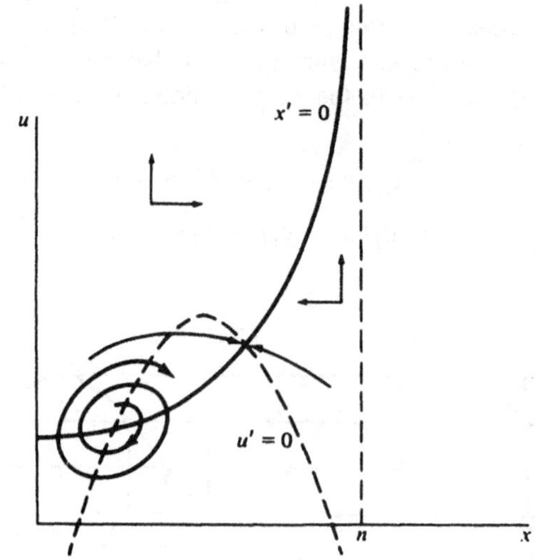

Figure 9.4

(review (B5.12) and (B5.13)). Thus an equilibrium at which (44) holds will be a saddlepoint; the solution diverges from an equilibrium at which (44) does not hold.

The linearized $x' = 0$ locus has slope $du/dx = -a_1/b_1 > 0$. The linearized $u' = 0$ locus has slope $du/dx = -a_2/b_2$. In view of the signs given in (43), it follows that (44) will hold if and only if $-a_1/b_1 > -a_2/b_2$; that is, the linearized $x' = 0$ locus is *steeper* than the linearized $u' = 0$ locus. Consulting Figure 9.4, one sees that the second equilibrium, the one with the larger coordinates, is characterized by the $x' = 0$ locus being steeper than the $u' = 0$ locus; it is a saddlepoint. The equilibrium with the smaller coordinates is totally unstable since the $x' = 0$ locus is less steep than the $u' = 0$ locus there. If x_0 is sufficiently large, the firm can get on the unique path to the saddlepoint equilibrium (sketched as the heavy curve). Note that while x is monotonic on this path, u may not be.

EXERCISES

1. Discuss the optimal solution to the Neoclassical Growth Model in Exercise I17.4.

2. Show that if, in the Neoclassical Growth Model of Exercise 1, utility $U(c, k)$ depends on capital stock (wealth), as well as on consumption,

$$U_c > 0, \qquad U_{cc} < 0, \qquad U_{ck} = 0, \qquad U_k > 0, \qquad U_{kk} < 0,$$

then there may be multiple equilibria. Show further that these equilibria alternate in character between being saddlepoints and being totally unstable.

3. A population of N fish in a certain lake grow at rate

$$N'(t) = aN(t) - bN^2(t)$$

if undisturbed by people. Fish can be withdrawn from the lake and consumed at rate $c(t)$, yielding utility $u(c(t))$ to the consuming community and reducing the fish growth rate accordingly:

$$N'(t) = aN(t) - bN^2(t) - c(t),$$

Assume future utilities to the community are discounted at constant rate r. Characterize a fishing (consumption) plan to maximize the present value of the discounted stream of utilities. Assume that $N(0) = a/b$ (why?), and that $u' > 0$, $u'' < 0$.

4. Verify (31).

5. An example in which H is concave and there is no finite saddlepoint equilibria is

$$J = \max \int_0^\infty e^{-rt}(-u^2/2 + xu - x^2/2)\, dt$$

$$\text{subject to} \qquad x' = u, \quad x(0) = x_0 > 0$$

where $r > 1$. Show the following.

a. Optimal functions $x(t)$ and $m(t)$ satisfy

$$x' = x + m, \qquad m' = (r - 1)m.$$

b. The only steady state solution to the system in a. is (0, 0) and it is totally unstable.

c. The optimal solution is

$$x(t) = x_0 e^t, \qquad u(t) = x_0 e^t, \qquad m(t) = 0, \qquad J = 0.$$

FURTHER READING

The advertising example was drawn from Gould. Exercise 2 is discussed in detail by Kurz (1968b). Exercise 3 is drawn from Colin Clark, whereas Exercise 5 is discussed by Kurz (1968a) and Samuelson (1972).

The analysis of local behavior of the optimal solution in the neighborhood of a steady state is considerably more difficult if there is more than one state variable. The class of problems

$$\max \int_0^\infty e^{-rt} F(x, u)\, dt$$

$$\text{subject to} \qquad x' = u, \quad x(0) = x_0,$$

where $x = [x_1, \ldots, x_n]$, $u = [u_1, \ldots, u_n]$ and F is a twice differentiable concave function of its $2n$ arguments, has been studied by means of a linear approximation in the neighborhood of a steady state. Levhari and Liviatan have shown that if F is *strictly concave* in its $2n$ arguments, then if m_i is a root of the characteristic equation, so is $r - m_i$. This means that complete stability is impossible; if the real part of m_i is negative, then the real part of $r - m_i$ cannot be negative. Further, they have shown that if F is strictly concave and $r = 0$, then there are no purely imaginary roots. See also Kurz (1968a) and Samuelson (1972).

A related question is *global stability* of an equilibrium. Under what conditions will the optimal solution converge to a particular equilibrium regardless of the initial conditions? For example, see Ryder and Heal for an extensive analysis of a particular problem with two state variables. See Brock (1977) for a survey of results on global asymptotic stability. See also Rockafellar, Magill (1977a, b), Cass and Shell, Brock and Scheinkman (1976, 1977), and Scheinkman (1978) for representative papers on stability.

The questions of appropriate solution concept and of the existence and properties of solutions to infinite horizon problems have been addressed by Halkin (1974), Haurie (1976), Brock and Haurie (1976). Halkin showed that in general there are no necessary transversality conditions for an infinite horizon problem. See also Seierstad and Sydsaeter (1977, 1987), and Michel.

Section 10

Bounded Controls

The control may be bounded, as in

$$\max \int_{t_0}^{t_1} f(t, x, u)\, dt \tag{1}$$

$$\text{subject to} \qquad x' = g(t, x, u), \quad x(t_0) = x_0, \tag{2}$$

$$a \le u \le b. \tag{3}$$

Absence of a bound is a special case with either $a \to -\infty$ or $b \to \infty$, as appropriate. For instance, gross investment may be required to be nonnegative.

Let J denote the value of the integral in (1). After appending (2) with a multiplier and integrating by parts, one can compute the variation δJ, the linear part of $J - J^*$.

$$\delta J = \int_{t_0}^{t_1} [(f_x + \lambda g_x + \lambda')\, \delta x + (f_u + \lambda g_u)\, \delta u]\, dt - \lambda(t_1)\, \delta x(t_1). \tag{4}$$

Choose λ to satisfy

$$\lambda' = -(f_x + \lambda g_x), \qquad \lambda(t_1) = 0, \tag{5}$$

so that (4) reduces to

$$\delta J = \int_{t_0}^{t_1} (f_u + \lambda g_u)\, \delta u\, dt. \tag{4'}$$

In order for x, u, λ to provide an optimal solution, no comparison path can yield a larger value to the objective. Thus,

$$\delta J = \int_{t_0}^{t_1} (f_u + \lambda g_u)\, \delta u\, dt \le 0 \tag{6}$$

is required for all feasible modifications δu. Feasible modifications are those that maintain (3). If the optimal control is at its lower bound a at some t, then the modified control $a + \delta u$ can be no less than a for feasibility, so $\delta u \geq 0$ is required. Similarly, if the optimal control is at its upper bound b, then any feasible modification satisfies $\delta u \leq 0$. Summarizing,

$$\begin{aligned}
&\delta u \geq 0 \quad \text{whenever} \quad u = a, \\
&\delta u \leq 0 \quad \text{whenever} \quad u = b, \\
&\delta u = \text{unrestricted} \quad \text{whenever} \quad a < u < b.
\end{aligned} \tag{7}$$

We need (6) to be satisfied for all δu consistent with (7). Therefore, u will be chosen so that

$$\begin{aligned}
u(t) &= a & &\text{only if} \quad f_u + \lambda g_u \leq 0 \quad \text{at} \quad t, \\
a < u(t) &< b & &\text{only if} \quad f_u + \lambda g_u = 0 \quad \text{at} \quad t, \\
u(t) &= b & &\text{only if} \quad f_u + \lambda g_u \geq 0 \quad \text{at} \quad t.
\end{aligned} \tag{8}$$

For instance, if $u^*(t) = a$, then (from (7)) $\delta u \geq 0$ is required, and thus $(f_u + \lambda g_u)\,\delta u \leq 0$ only if $f_u + \lambda g_u \leq 0$. Similarly, if $u^*(t) = b$, then $\delta u \leq 0$ is required for a feasible modification and thus $(f_u + \lambda g_u)\,\delta u \leq 0$ only if $f_u + \lambda g_u \geq 0$. And, as usual, if $a < u^*(t) < b$, then δu may have any sign so that $(f_u + \lambda g_u)\,\delta u \leq 0$ can be assured only if $f_u + \lambda g_u = 0$ at t. A statement equivalent to (8) is

$$\begin{aligned}
f_u + \lambda g_u &< 0 \quad \text{implies} \quad u(t) = a, \\
f_u + \lambda g_u &= 0 \quad \text{implies} \quad a \leq u(t) \leq b, \\
f_u + \lambda g_u &> 0 \quad \text{implies} \quad u(t) = b.
\end{aligned} \tag{8$'$}$$

Thus, if x^*, u^* solve (1)–(3), then there must be a function λ such that x^*, u^*, λ satisfy (2), (3), (5), and (8). These necessary conditions can be generated by means of the Hamiltonian

$$H = f(t, x, u) + \lambda g(t, x, u).$$

Then (2) and (5) result from

$$x' = \partial H/\partial \lambda, \qquad \lambda' = -\partial H/\partial x.$$

Conditions (8) can be generated by maximizing H subject to (3); this is an ordinary nonlinear programming problem in u.

Solve

$$\begin{aligned}
&\max \quad H = f + \lambda g \\
&\text{subject to} \quad a \leq u \leq b
\end{aligned} \tag{9}$$

by appending the constraints to the objective with multipliers w_1, w_2. The Lagrangian for (9) is (see Section A6)

$$L = f(t, x, u) + \lambda g(t, x, u) + w_1(b - u) + w_2(u - a), \tag{10}$$

from which we obtain the necessary conditions for a constrained maximum with respect to u:

$$\partial L/\partial u = f_u + \lambda g_u - w_1 + w_2 = 0, \tag{11}$$

$$w_1 \geq 0, \qquad w_1(b-u) = 0, \tag{12}$$

$$w_2 \geq 0, \qquad w_2(u-a) = 0. \tag{13}$$

Conditions (11)–(13) are equivalent to conditions (8) and constitute an alternative statement of the requirement, as will be shown in Exercise 6. (If $u^*(t) = a$, then $b - u^* > 0$, so (12) requires $w_1 = 0$; hence, from (11), $f_u + \lambda g_u + w_2 = 0$. Since $w_2 \geq 0$, we have $f_u + \lambda g_u \leq 0$ if $u^*(t) = a$. This is the first instance in (8). One continues similarly for the other two possibilities.)

Example 1. We solved our production planning problem

$$\min \quad \int_0^T (c_1 u^2 + c_2 x)\, dt$$

$$\text{subject to} \qquad x' = u, \quad x(0) = 0, \quad x(T) = B, \quad u(t) \geq 0$$

in Section 6 and elsewhere in the case of $B \geq c_2 T^2/4c_1$. If $B < c_2 T^2/4c_1$, that plan is not feasible and explicit account must be taken of the nonnegativity constraint $u \geq 0$. We now discuss this case.

This control $u(t)$ is to be chosen at each t to minimize the Hamiltonian

$$H = c_1 u^2 + c_2 x + \lambda u, \qquad \text{subject to} \quad u \geq 0.$$

The Lagrangian, with multiplier function w, is

$$L = c_1 u^2 + c_2 x + \lambda u - wu.$$

Necessary conditions for u to be minimizing (see Exercise 1) are

$$\partial L/\partial u = 2c_1 u + \lambda - w = 0, \tag{14}$$

$$w \geq 0, \qquad u \geq 0, \qquad wu = 0. \tag{15}$$

Further,

$$\lambda' = -\partial H/\partial x = -c_2,$$

so that

$$\lambda(t) = k_0 - c_2 t \tag{16}$$

for some constant k_0. Substituting from (16) into (14) and rearranging gives

$$u(t) = (w - \lambda)/2c_1 = (c_2 t - k_0 + w)/2c_1. \tag{17}$$

To solve, we make a conjecture about the structure of the solution and then seek a path with this structure satisfying the conditions. Since the time span T

is long relative to the amount B to be produced, we guess that there is an initial period, say $0 \le t \le t^*$ (for some t^* to be determined), with no production or inventory. Production begins at t^*. Thus our hypothesis is

$$\begin{aligned} u(t) &= 0, \qquad 0 \le t < t^*, \\ u(t) &> 0, \qquad t^* \le t \le T, \end{aligned} \tag{18}$$

for some t^* to be determined.

When $u(t) = 0$, we have from (17)

$$w(t) = k_0 - c_2 t \ge 0, \qquad 0 \le t < t^*. \tag{19}$$

Nonnegativity in (19) is required by (15). From (19), $w(t)$ decreases on $0 \le t < t^*$, so nonnegativity is assured provided

$$k_0 - c_2 t^* \ge 0. \tag{20}$$

When $u(t) > 0$, (15) implies $w(t) = 0$. Then from (17)

$$u(t) = (c_2 t - k_0)/2c_1 \ge 0, \qquad t^* \le t \le T. \tag{21}$$

Since $u(t)$ increases after t^*, $u(t) \ge 0$ for $t^* \le t \le T$ provided that $u(t^*) = (c_2 t^* - k_0)/2c_1 \ge 0$. This requirement and (20) together imply that

$$k_0 = c_2 t^*. \tag{22}$$

Hypothesis (18) now takes the more concrete form

$$\begin{aligned} u(t) &= 0, \qquad 0 \le t < t^*, \\ u(t) &= c_2(t - t^*)/2c_1, \qquad t^* \le t \le T. \end{aligned} \tag{23}$$

Recalling that $u = x'$ and integrating yields

$$\begin{aligned} x(t) &= 0, \qquad 0 \le t < t^*, \\ x(t) &= c_2(t - t^*)^2/4c_1, \qquad t^* \le t \le T. \end{aligned} \tag{24}$$

The constants of integration were evaluated using the final condition $x(0) = 0$ and the required continuity of x (so $x(t^*) = 0$). Finally, combining (24) with the terminal condition $x(T) = B$ gives

$$t^* = T - 2(c_1 B/c_2)^{1/2}. \tag{25}$$

With a distant delivery date T, the duration of the production period $T - t^*$ varies directly with $c_1 B/c_2$; it increases with the amount to be produced and the production cost coefficient c_1 and decreases with the unit holding cost c_2. It is precisely the period obtained under the supposition that production had to begin immediately but that the delivery date T could be chosen optimally; see Example I9.1.

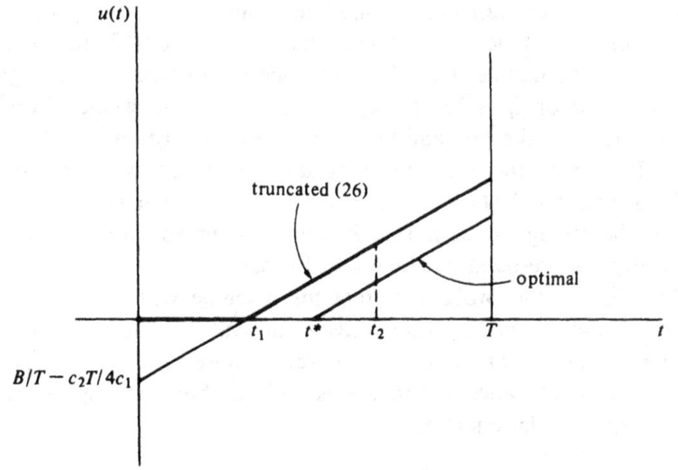

Figure 10.1

In sum, the solution is given in the accompanying table, where t^* is given by (25). Extending our sufficiency theorem to cover a constrained control region shows the solution tabulated to be optimal (see Section 15).

	$0 \le t < t^*$	$t^* \le t \le T$
$u(t)$	0	$c_2(t - t^*)/2c_1$
$x(t)$	0	$c_2(t - t^*)^2/4c_1$
$\lambda(t)$	$c_2(t^* - t)$	$c_2(t^* - t)$
$w(t)$	$c_2(t^* - t)$	0

Observe that the solution could *not* be obtained by taking the solution to the unconstrained problem

$$u(t) = c_2(2t - T)/4c_1 + B/T \tag{26}$$

and deleting the nonfeasible portion! Taking this solution, appropriate only if $B \ge c_2T^2/4c_1$, and setting $u = 0$ whenever it dictates $u < 0$ does not provide the optimal solution. This is algebraically clear and is also illustrated graphically in Figure 10.1 where $t_1 = T/2 - 2c_1B/c_2T$, $t_2 = 2t_1 = T - 4c_1B/c_2t$, and t^* is given in (25).

In Figure 10.1, the "solution" (26) begins with $u < 0$ and reaches $u = 0$ at t_1. Since output is negative for $0 \le t < t_1$, inventory is likewise negative. Production from t_1 until t_2 is devoted to driving inventory back up to zero! Production from t_2 until T fulfills the total delivery requirement B. If production were to begin at t_2 and follow the truncated path, a total of B would be produced but costs would be needlessly high. The cost minimizing plan is reflected in the optimal path.

Example 2. Machine Maintenance and Sale Date.

A functioning machine yields a constant revenue $R > 0$, net of all costs except for machine maintenance. This revenue flow ceases when the machine stops functioning or is sold. Its age at failure is stochastic. Preventative maintenance reduces the probability of failure by any age. A failed machine is worthless. If the machine is still working at age t, its salvage or resale value is $S(t)$. We assume that $S'(t) \leq 0$ and that $S(t) < R/r$, where r is the discount rate. Thus, the salvage value is less than the discounted revenue stream that a hypothetically infinite-lived machine could produce.

Let $F(t)$ denote the probability that the machine will fail by age t. The probability of failure at any time, given survival to that time, depends on current maintenance $u(t)$ and on a "natural" failure rate $h(t)$ that governs in the absence of maintenance and increases with machine age. Specifically $F(t)$ satisfies the differential equation

$$F'(t) = [1 - u(t)]h(t)[1 - F(t)], \qquad h(0) = 0, \tag{27}$$

where $h(t)$ is a known function satisfying

$$h(t) \geq 0, \qquad h'(t) \geq 0.$$

This means that the instantaneous probability density of failure at t, given survival to t, does not depend on the maintenance history, but only on the current maintenance effort. (Of course the probability of survival to t does depend on the maintenance history.)

$$F(t) = 1 - \exp\left(-\int_0^t [1 - u(s)]h(s)\,ds\right).$$

The maintenance rate $u(t)$ is bounded:

$$0 \leq u(t) \leq 1. \tag{28}$$

If $u = 0$ is selected, then the failure rate $F'/(1 - F)$ assumes its natural value $h(t)$. If $u = 1$, the failure rate will be zero. Maintenance cost is $M(u)h$ where $M(u)$ is a known function satisfying

$$M(0) = 0, \qquad M'(u) > 0, \qquad M''(u) > 0 \qquad \text{for} \quad 0 \leq u \leq 1. \tag{29}$$

The discounted expected profit from the machine is

$$\int_0^T e^{-rt}[R - M(u(t))h(t)][1 - F(t)]\,dt + e^{-rT}S(T)[1 - F(T)]. \tag{30}$$

We seek a maintenance policy u and sale date T to maximize (30) subject to (27) and (28). The control is u; the state is F.

The current value Hamiltonian is

$$H = \left[R - M(u)h\right](1 - F) + \lambda(1 - u)h(1 - F) + w_1 u + w_2(1 - u),$$

where λ is the current value multiplier associated with (27). An optimal maintenance policy u^* must maximize H at each age t, $0 \le t \le T$, where λ satisfies

$$\lambda' = r\lambda - H_F = r\lambda + R - M(u)h + \lambda(1 - u)h. \tag{31}$$

Since $F(T)$ and T are to be chosen optimally, transversality conditions (7.d.i) and (7.d.v), respectively, apply. The salvage term corresponding to ϕ is

$$\phi(F(T), T) = e^{-rT}S(T)\left[1 - F(T)\right]$$

where the state variable is F and terminal time is T. Since

$$\phi_F(F, T) = -e^{-rT}S(T),$$
$$\phi_T(F, T) = e^{-rT}\left[1 - F(T)\right]\left[S'(T) - rS(T)\right],$$

application of (7.d.i) to this problem gives

$$\lambda(T) = -S(T), \tag{32}$$

and (7.d.v) yields

$$H(T) = \left[rS(T) - S'(T)\right]\left[1 - F(T)\right]. \tag{33}$$

We have

$$H_u = -h(1 - F)\left[M'(u) + \lambda\right] + w_1 - w_2 = 0,$$
$$w_1 \ge 0, \qquad w_2 \ge 0, \qquad w_1 u = 0, \qquad w_2(1 - u) = 0, \tag{34}$$

so from (34) and (29),

$$\begin{aligned}
&\text{if } M'(0) + \lambda(t) > 0,\\
&\qquad \text{then select } u^*(t) = 0, \quad w_2 = 0;\\
&\text{if } M'(1) + \lambda(t) < 0,\\
&\qquad \text{then select } u^*(t) = 1, \quad w_1 = 0;\\
&\text{otherwise select } u^*(t)\\
&\qquad \text{to satisfy } M'(u) + \lambda(t) = 0, \quad w_1 = w_2 = 0.
\end{aligned} \tag{35}$$

Since the functional forms of M and h have not been specified, we can only qualitatively characterize the optimal solution. Define

$$Q(t) \equiv M'(u(t)) + \lambda(t).$$

On an interval of time during which $0 < u^* < 1$, we have $Q(t) = 0$. Since Q is constant, $Q'(t) = 0$. Therefore

$$\begin{aligned} Q'(t) &= M''(u)u' + \lambda' \\ &= M''(u)u' + [r + (1-u)h]\lambda + R - M(u)h \\ &= M''(u)u' - M'(u)[r + (1-u)h] + R - M(u)h = 0, \end{aligned}$$

where we have used (31) to eliminate λ' and then used $Q(t) = 0$ to eliminate λ. The last equation may be rearranged to find that the optimal maintenance policy satisfies

$$M''(u)u' = M'(u)[r + (1-u)h] - R + M(u)h \tag{36}$$

whenever the bounds of (28) are not tight. Since $M'' > 0$ by assumption, the sign u' is the same as the sign of the right side of (36).

It will facilitate our study of u' to consider first the locus of points (h, u) such that $u' = 0$ would be obtained. Such points satisfy

$$M'(u)[r + (1-u)h] - R + M(u)h = 0. \tag{37}$$

The shape of the locus may be discerned by differentiating (37) implicitly to get

$$du/dh = -[M + (1-u)M']/[r + (1-u)h]M'' < 0.$$

The negativity follows by inspection, using the assumed properties of M. The intercepts of the locus (h, u) for which $u' = 0$ are also obtainable from (37). This is illustrated in Figure 10.2.

Next, we seek the direction of movement of a trajectory (h, u) through time where h is a given function and u obeys the differential equation (36). We know that h is nondecreasing; hence, the rightward pointing arrow in Figure 10.2. Further, since the right side of (36) is an increasing function of u, a path at a point above the $u' = 0$ locus will rise and a path at a point below it will fall. The arrows in Figure 10.2 illustrate these facts.

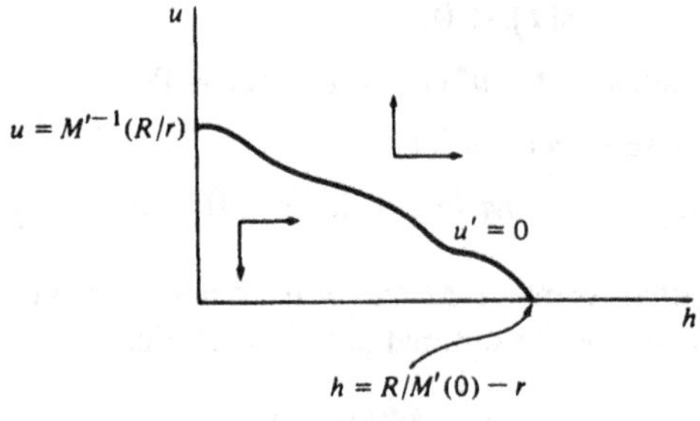

Figure 10.2

Figure 10.2 differs from those drawn earlier. Since the state variable is absent from (36), we were able to include the nonautonomous exogenous function $h(t)$ in the diagram.

Substitute from (32) into (33) using the definition of H to get

$$R - M(u)h - [r + (1 - u)h]S = -S' \geq 0 \quad \text{at} \quad T. \tag{38}$$

At the time of sale, the excess of the value obtained by keeping the machine a little longer over the value of selling it should be no less than the loss in resale value from postponing sale slightly.

We can now show that the optimal maintenance policy involves $u'(t) \leq 0$, $0 \leq t \leq T$, in every case. (But the actual expenditure $M(u)h$ may either increase or decrease through time since $h' \geq 0$.) If $0 < u(T) < 1$, then $Q(T) = 0$, so $\lambda(T) = -M'(u(T)) = -S(T)$. Substituting this into (38) and recalling (36), we obtain $u'(T) \leq 0$. From Figure 10.2, if $u'(T) \leq 0$, then u must be nonincreasing throughout $(0, T)$. It is likewise clear from the figure that u is nonincreasing throughout in case $u(T) = 0$.

Finally, suppose $u(T) = 1$. Let t_1 be the first moment that $u = 1$. From (31) and (32), we get

$$\lambda(t) = -\int_t^T e^{-r(s-t)}[R - M(1)h(s)]\,ds - e^{-r(T-t)}S(T), \qquad t_1 \leq t \leq T.$$

Replace $h(t)$ by its upper bound $h(T)$ for $0 \leq t \leq T$, so that

$$\lambda(t) \leq -(1 - e^{-r(T-t)})[R - M(1)h(T)]/r - e^{-r(T-t)}S(T).$$

Then, since $Q = M' + \lambda$,

$$Q(t) \leq M'(1) - (1 - e^{-r(T-t)})[R - M(1)h(T)]/r - e^{-r(T-t)}S(T).$$

From (35) and (32),

$$M'(1) + \lambda(T) = M'(1) - S(T) < 0;$$

therefore

$$M'(1) < S(T).$$

Replacing $M'(1)$ by its upper bound finally gives

$$Q(t) \leq (1 - e^{-r(T-t)})[rS(T) - R + M(1)h(T)]/r < 0, \qquad t_1 \leq t \leq T,$$

where the second inequality follows from (38). In particular, $Q(t_1) < 0$. Therefore, there can be no period of $u < 1$ ending at t_1 (since $Q(t_1) = 0$ in that case). This means that $u = 1$ for $0 \leq t \leq T$ in case $u(T) = 1$. Thus u is a nonincreasing function of machine age in every case.

EXERCISES

1. Find necessary conditions for solution of the problem

$$\min \int_{t_0}^{t_1} f(t, x, u)\, dt$$

subject to $\quad x' = g(t, x, u), \quad x(t_0) = x_0, \quad a(t) \le u(t) \le b(t).$

2. Find the control function $u(t)$ that solves

$$\max \int_0^2 (2x - 3u - u^2)\, dt$$

subject to $\quad x' = x + u, \quad 0 \le u \le 2, \quad x(0) = 5, \quad x(2)$ free.

3. Provide an interpretation for the result that $\lambda(t^*) = 0$ in the solution of Example 1.

4. An individual's earnings are proportional to the product of his "human capital" $K(t)$ and the fraction of his time, $1 - s(t)$, spent working. Human capital decays at a constant proportionate rate b and grows with investment, an increasing concave function of capital and the proportion of time $s(t)$ devoted to education.

   Discuss the optimal study-work program to maximize the discounted earnings over a known remaining lifetime T:

$$\max \int_0^T e^{-rt}(1 - s)K\, dt$$

subject to $\quad K' = A(sK)^a - bK, \quad K(0) = K_0 > 0, \quad 0 \le s \le 1,$

where $A > 0$, $0 < a < 1$, $b \ge 0$.

5. Reconsider Example 2 in the case in which a failed machine is worth a positive constant junk value.

6. Show that conditions (11)–(13) are equivalent to conditions (8).

FURTHER READING

Example 2 is discussed by Kamien and Schwartz (1971b). See Ben-Porath and Haley for analysis of Exercise 4, as well as Blinder and Weiss; Heckman; Ryder, Stafford and Stephan; Southwick and Zionts. Another paper with bounded controls is by Aarrestad.

Section 11

Further Control Constraint

The effects of constraints on the controls, depending on t and x, will now be examined in a problem with several states and controls. Feasible controls may be interrelated. For example, if one wishes to divide a single resource among three uses, one can let u_i, $i = 1, 2$, be the share in the ith use, and $1 - u_1 - u_2$ will then be the share in the third use. For each share to be nonnegative, we stipulate that

$$u_1 \geq 0, \qquad u_2 \geq 0, \qquad 1 - u_1 - u_2 \geq 0.$$

In general, control constraints that do not depend on the current state can be written

$$h_j(t, u_1(t), \ldots, u_m(t)) \geq 0, \qquad j = 1, \ldots, p, \tag{1}$$

where the functions h_j are assumed continuously differentiable. Note that (1) encompasses equality constraints since

$$h_j \geq 0 \quad \text{and} \quad -h_j \geq 0 \quad \text{imply} \quad h_j = 0.$$

Consider an optimal control problem with a single constraint of the form in (1):

$$\max \int_{t_0}^{t_1} f(t, x, u)\, dt \tag{2}$$

$$\text{subject to} \quad x_i' = g_i(t, x, u), \quad i = 1, \ldots, n \tag{3}$$

$$h(t, u) \geq 0, \tag{4}$$

$$x(t_0) = x_0, \tag{5}$$

where $x = [x_1, \ldots, x_n]$ and $u = [u_1, \ldots, u_m]$. Without incorporating (4) at this point, append (3) with a vector of continuous functions $\lambda = [\lambda_1, \ldots, \lambda_n]$ to the objective J:

$$J = \int_{t_0}^{t_1} \left\{ f(t, x, u) + \sum_{i=1}^{n} \lambda_i [g_i(t, x, u) - x_i' \right\} dt.$$

Integrate by parts and compute the variation, taking into account (5):

$$\delta J = \int_{t_0}^{t_1} [(f_x + \lambda g_x + \lambda') \cdot \delta x + (f_u + \lambda g_u)]\, dt - \lambda \cdot \delta x_1 \le 0. \quad (6)$$

Let x and u solve (2)–(5); then $\delta J \le 0$ at x, u, as indicated. Choose $\lambda_k(t)$ to satisfy

$$\lambda_k' = -\left(\partial f / \partial x_k + \sum_{i=1}^{n} \lambda_i \, \partial g_i / \partial x_k \right), \qquad k = 1, \ldots, n, \quad (7)$$

$$\lambda_k(t_1) = 0, \qquad k = 1, \ldots, n, \quad (8)$$

where x and u are optimal. In view of (7) and (8), (6) reduces to

$$\delta J = \int_{t_0}^{t_1} \sum_{j=1}^{m} \left(\partial f / \partial u_j + \sum_{i=1}^{n} \lambda_i \, \partial g_i / \partial u_j \right) \delta u_j \, dt \le 0, \quad (9)$$

so at each t, the optimal x, u must satisfy

$$\sum_{j=1}^{m} \left(\partial f / \partial u_j + \sum_{i=1}^{n} \lambda_i \, \partial g_i / \partial u_j \right) \delta u_j \le 0 \quad (10)$$

for all feasible modifications $[\delta u_1, \ldots, \delta u_m]$. When $h(t, u_1, \ldots, u_m) > 0$ in an optimal solution, then feasible modifications δu_j, $j = 1, \ldots, m$, may have any sign, so that

$$\partial f / \partial u_j + \sum_{i=1}^{n} \lambda_i \, \partial g_i / \partial u_j = 0 \quad \text{at } t \quad \text{if} \quad h(t, u_1, \ldots, u_m) > 0 \quad (11)$$

is necessary. If, however, $h(t, u_1, \ldots, u_m) = 0$, then feasible modifications in the controls are those that do not reduce the value of h. This means that

$$dh = \sum_{j=1}^{m} (\partial h / \partial u_j) \, \delta u_j \ge 0 \quad (12)$$

is required. Hence (10) must be satisfied at the optimal x, u for all $[\delta u_1, \ldots, \delta u_m]$ that obey (12). We assume h satisfies a regularity condition so that only feasible modifications satisfy (12). Applying Farkas' lemma (see Section A6) to the statement that "(10) holds for all $[\delta u_1, \ldots, \delta u_m]$ that obey (12)," indicates that an equivalent statement is "there is a function $w(t) \geq 0$ such that

$$\partial f/\partial u_j + \sum_{i=1}^{n} \lambda_i \, \partial g_i/\partial u_j + w \, \partial h/\partial u_j = 0, \qquad j = 1, \ldots, m, \quad (13)$$

whenever $h(t, u_1, \ldots, u_m) = 0$." Cases (11) and (13) can be combined into

$$\partial f/\partial u_j + \sum_{i=1}^{n} \lambda_i \, \partial g_i/\partial u_j + w \, \partial h/\partial u_j = 0, \qquad j = 1, \ldots, m, \quad (14)$$

$$w \geq 0, \qquad wh = 0, \quad (15)$$

for $t_0 \leq t \leq t_1$. When $h > 0$, (15) assures that $w = 0$ and then (14) yields the requirement of (11). Otherwise (14)–(15) imply satisfaction of (13).

In sum, necessary conditions for x, u to solve (2)–(5) are that x, u satisfy (3)–(5) and that there are continuous functions $\lambda(t) = [\lambda_1, \ldots, \lambda_n]$ and a function $w(t)$ such that x, u, λ, satisfy (7), (8), (14), and (15). Conditions (14) and (15) are obtained by maximizing $H = f + \sum_{i=1}^{n} \lambda_i g_i$ subject to (4); the Lagrange multiplier is w. Form the Lagrangian

$$L = f + \sum_{i=1}^{n} \lambda_i g_i + wh$$

and compute

$$\lambda'_k = -\partial L/\partial x_k = -\left(\partial f/\partial x_k + \sum_{i=1}^{n} \lambda_i \, \partial g_i/\partial x_k\right),$$

$$\lambda_k(t_1) = 0, \qquad k = 1, \ldots, n$$

$$\partial L/\partial u_j = \partial f/\partial u_j + \sum_{i=1}^{n} \lambda_i \, \partial g_i/\partial u_j + w \, \partial h/\partial u_j = 0, \qquad j = 1, \ldots, m.$$

A regularity condition or constraint qualification that must hold is discussed at the end of this section.

If constraint (4) were replaced by (1), then there would be a multiplier $w_j \geq 0$ associated with each, along with the requirement that $w_j h_j = 0$ at each t, $t_0 \leq t \leq t_1$. The necessary conditions can be found as above, considering the admissible modifications $[\delta u_1, \ldots, \delta u_m]$ that preserve adherence to (1) (using Farkas' lemma). Alternatively, each constraint of (1) may be appended to J by a multiplier $w_j \geq 0$ and then computing $\delta J \leq 0$, and so on. The restriction $w_j h_j = 0$ is also needed.

The admissible controls may depend on the state. For instance, the amount of resource to be allocated among several uses (controls) may be bounded by the amount available (state). In general, the admissible control region at t may be described by

$$h_j(t, x_1, \ldots, x_n, u_1, \ldots, u_m) \geq 0, \qquad j = 1, \ldots, p, \tag{16}$$

where the functions h_j are assumed to be continuously differentiable. In this case, (16) *must* be incorporated into J for the impact of modifications in the state variables on the feasible control region to be reflected in the multipliers $\lambda_1, \ldots, \lambda_n$.

We now develop necessary conditions obeyed by a solution to

$$\max \quad \int_{t_0}^{t_1} f(t, \boldsymbol{x}(t), \boldsymbol{u}(t))\, dt \tag{17}$$

$$\text{subject to} \quad x_i' = g_i(t, \boldsymbol{x}(t), \boldsymbol{u}(t)), \quad i = 1, \ldots, n, \tag{18}$$

$$h_j(t, \boldsymbol{x}(t), \boldsymbol{u}(t)) \geq 0, \quad j = 1, \ldots, p, \tag{16}$$

$$x_i(t_0) = x_{i0}, \quad i = 1, \ldots, n, \tag{19}$$

where $\boldsymbol{x} = [x_1, \ldots, x_n]$ and $\boldsymbol{u} = [u_1, \ldots, u_m]$. It is assumed that for every attainable (t, x), there exists a control function $\boldsymbol{u}$ that satisfies (16). It is also assumed that the constraints (16) satisfy a regularity condition or constraint qualification along the optimal path, to be discussed later.

Let J denote (17); append (18) with differentiable multipliers $\lambda_i(t)$, $i = 1, \ldots, n$, and (16) with multipliers $w_j(t)$, $j = 1, \ldots, p$, that satisfy

$$w_j \geq 0, \qquad w_j h_j = 0, \qquad t_0 \leq t \leq t_1. \tag{20}$$

Then

$$J = \int_{t_0}^{t_1} \left(f + \sum_{i=1}^{n} \lambda_i g_i - \sum_{i=1}^{n} \lambda_i x_i' + \sum_{j=1}^{p} w_j h_j \right) dt.$$

Integration by parts gives

$$J = \int_{t_0}^{t_1} (f + \boldsymbol{\lambda} \cdot \boldsymbol{g} + \boldsymbol{\lambda}' \cdot \boldsymbol{x} + \boldsymbol{w} \cdot \boldsymbol{h})\, dt - \boldsymbol{\lambda} \cdot \boldsymbol{x}|_{t_1} + \boldsymbol{\lambda} \cdot \boldsymbol{x}|_{t_1},$$

where

$$\boldsymbol{\lambda} = [\lambda_1, \ldots, \lambda_n], \qquad \boldsymbol{g} = [g_1, \ldots, g_n]$$
$$\boldsymbol{w} = [w_1, \ldots, w_p], \qquad \boldsymbol{h} = [h_1, \ldots, h_p].$$

Compute the variation, taking notice of (19):

$$\delta J = \int_{t_0}^{t_1} [(f_x + \boldsymbol{\lambda} \boldsymbol{g}_x + \boldsymbol{\lambda}' + \boldsymbol{w}\boldsymbol{h}_x) \cdot \delta \boldsymbol{x} + (f_u + \boldsymbol{\lambda} \boldsymbol{g}_u + \boldsymbol{w}\boldsymbol{h}_u) \cdot \delta \boldsymbol{u}]\, dt$$
$$- \boldsymbol{\lambda} \cdot \delta \boldsymbol{x}_1 \leq 0. \tag{21}$$

The variation (21) must be nonnegative at the optimal x, u. Choose the multipliers λ so that the coefficients of each δx_k will be zero:

$$\lambda'_k = -\left(\partial f/\partial x_k + \sum_{i=1}^{n} \lambda_i \partial g_i/\partial x_k + \sum_{j=1}^{p} w_j \partial h_j/\partial x_k\right), \quad (22)$$

$$\lambda_k(t_1) = 0. \quad (23)$$

Now (21) reduces to

$$\delta J = \int_{t_0}^{t_1} \sum_{k=1}^{m} \left(\partial f/\partial u_k + \sum_{i=1}^{n} \lambda_i \partial g_i/\partial u_k + \sum_{j=1}^{p} w_j \partial h_j/\partial u_k\right) \delta u_k \, dt, \quad (24)$$

which must be nonnegative for all feasible modifications $\delta u_1, \ldots, \delta u_m$. Since all the constraints are incorporated into J, this implies

$$\partial f/\partial u_k + \sum_{i=1}^{n} \lambda_i \partial g_i/\partial u_k + \sum_{j=1}^{p} w_j \partial h_j/\partial u_k = 0, \qquad k = 1, \ldots, m. \quad (25)$$

The necessary conditions for $x_1, \ldots, x_n, u_1, \ldots, u_n$ to solve (16)–(19) are that there are continuous functions $\lambda_1, \ldots, \lambda_n$ and functions $w_1, \ldots, w_p$ such that (16), (18)–(20), (22), (23), and (25) are obeyed. At each t, the Hamiltonian is maximized with respect to u, subject to (16). Two of these conditions can be generated from the Lagrangian

$$L = f + \sum_{i=1}^{n} \lambda_i g_i + \sum_{j=1}^{p} w_j h_j.$$

Then

$$\partial L/\partial u_j = 0, \qquad j = 1, \ldots, m, \qquad \text{yields (25);}$$

$$\lambda'_k = -\partial L/\partial x_k, \qquad k = 1, \ldots, n, \qquad \text{yields (22).}$$

Note that whenever $h_j > 0$, we have $w_j = 0$ and the terms involving partial derivatives of h_j have no impact. However, whenever the jth constraint of (16) is tight, choice of control variables is restricted to maintain feasibility. The multipliers $\lambda_i(t)$ give the marginal valuation of the corresponding state variable x_i at t. Note that (22) reflect not only the direct effect of changes in x_i on the current reward f and on the state changes through $g_1, \ldots, g_n$, but also the effect of changes in x_i on the control region through $h_1, \ldots, h_p$.

Since the Kuhn-Tucker theorem has been used in maximizing the Hamiltonian, a constraint qualification of Kuhn-Tucker theory is needed (review Section A6.) Any of the Kuhn-Tucker constraint qualifications will do. For instance, it will be satisfied if all the functions h are concave in u and the set of u_i satisfying (16) has a nonempty interior. It is satisfied if the matrix of partial derivatives of *active* constraints in (16) with respect to the controls has rank equal to the number of active constraints (which in turn does not exceed m).

Constraints on the state space alone (without any component of u) *cannot* be studied in the way just suggested. At each time t, there must be a way of insuring feasibility, which means choice of u. State space constraints will be discussed in Section 17.

EXERCISES

1. Follow the procedure suggested at the beginning of this section and use Farkas' lemma to find necessary conditions for an optimal solution of (1)–(3) and (5).

2. Find necessary conditions for an optimal solution to (2), subject to (3), (5), and

$$h_j(t, u) \geq 0, \qquad j = 1, \ldots, p_1 - 1,$$
$$h_j(t, u) = 0, \qquad j = p_1, p_1 + 1, \ldots, p_2.$$

Show that w_j, $j = p_1, \ldots, p_2$, may be of any sign. [Hint: Recall that an equality can be written as two inequalities.]

3. Find necessary conditions for an optimal solution to

$$\min \int_{t_0}^{t_1} f(t, x, u)\, dt$$
$$\text{subject to} \quad x' = g(t, x, u), \quad h(t, u) \leq 0, \quad x(t_0) = 0.$$

4. A firm's investment must be self-financed, that is, paid from current revenues. Let $R(K)$ be the net revenue earned from a stock of capital K. Let $C(I)$ be the cost of investing at rate I. Assume $R' > 0$, $R'' < 0$, $C' > 0$, $C'' > 0$. Discuss the optimal solution to

$$\max \int_0^\infty e^{-rt}[R(K) - C(I)]\, dt$$
$$\text{subject to} \quad K' = I - bK, \quad K(0) = K_0, \quad 0 \leq I \leq R(K).$$

5. Show that the first variation for

$$\max_{x, u, t_1} \int_{t_0}^{t_1} f(t, x, u)\, dt + \phi(x(t_1), t_1)$$
$$\text{subject to} \quad x_i' = g_i(t, x, u), \quad i = 1, \ldots, n,$$
$$h_j(t, x, u) \geq 0, \quad j = 1, \ldots, r,$$
$$x(t_0) = x_0,$$
$$x_i(t_1) = x_{i1}, \quad i = 1, \ldots, q \leq n,$$
$$K_k(x(t_1), t_1) \geq 0, \quad k = 1, \ldots, s,$$

where $\boldsymbol{x} = [x_1, \ldots, x_n]$, $\boldsymbol{u} = [u_1, \ldots, u_m]$, is

$$\begin{aligned}
\delta J = {} & \left(f + \sum_{i=1}^{n} \lambda_i g_i \right)\Bigg|_{t_1} \delta t_1 + \left(\partial\phi/\partial t_1 + \sum_{k=1}^{s} p_k \, \partial K_k/\partial t_1 \right) \delta t_1 \\
& + \sum_{i=q+1}^{n} \left(\partial\phi/\partial x_i + \sum_{k=1}^{s} p_k \, \partial K_k/\partial x_i - \lambda_i \right) \delta x_i \\
& + \int_{t_0}^{t_1} \left[\sum_{k=1}^{n} \left(\partial f/\partial x_k + \sum_{i=1}^{n} \lambda_i \, \partial g_i/\partial x_k + \lambda_k' + \sum_{j=1}^{r} w_j \, \partial h_j/\partial x_k \right) \delta x_k \right. \\
& \left. + \sum_{k=1}^{m} \left(\partial f/\partial u_k + \sum_{i=1}^{n} \lambda_i \, \partial g_i/\partial u_k + \sum_{j=1}^{r} w_j \, \partial h_j/\partial u_k \right) \delta u_k \right] dt.
\end{aligned}$$

Section 12

Discontinuous and Bang-Bang Control

The continuity requirement on the control can be relaxed. Admissible controls are now *piecewise continuous* functions of time. This means that the control will be continuous, except possibly at a finite number of points of time. Any discontinuity involves a finite jump. The state variable x, the multiplier function λ, and the Hamiltonian H must (still) be continuous, regardless of discontinuity in u.

Discontinuities in u correspond to corners in x in the calculus of variations. Recall from Section I13 that if the solution $x(t)$ to

$$\text{optimize} \quad \int_{t_0}^{t_1} F(t, x, x')\, dt$$

has corners, the functions

$$F_{x'} \qquad \text{and} \qquad F - x'F_{x'}$$

must nonetheless be continuous. To see what this means in the optimal control format, rewrite the problem

$$\text{optimize} \quad \int_{t_0}^{t_1} F(t, x, u)\, dt$$

$$\text{subject to} \qquad x' = u.$$

The Hamiltonian is

$$H = F + \lambda u.$$

Then

$$H_u = F_u + \lambda = 0, \qquad \text{so} \quad \lambda = -F_{x'}.$$

Thus continuity of λ throughout corresponds to continuity of $F_{x'}$. Furthermore, substituting this expression for λ into H gives, since $x' = u$,

$$H = F - x'F_{x'}.$$

Thus, continuity of H corresponds to the second corner condition.

The solution to a problem that is linear in u frequently involves discontinuities in the control. Consider

$$\max \int_{t_0}^{t_1} [F(t, x) + uf(t, x)]\, dt \tag{1}$$

$$\text{subject to} \quad x' = G(t, x) + ug(t, x), \tag{2}$$

$$x(0) = x_0, \tag{3}$$

$$a \le u \le b. \tag{4}$$

The Hamiltonian is

$$\begin{aligned} H &= F(t, x) + uf(t, x) + \lambda G(t, x) + \lambda ug(t, x) \\ &= F + \lambda G + (f + \lambda g)u. \end{aligned} \tag{5}$$

The necessary conditions include

$$\lambda' = -\partial H/\partial x, \tag{6}$$

$$u = \begin{Bmatrix} a \\ ? \\ b \end{Bmatrix} \quad \text{whenever} \quad f + \lambda g \begin{Bmatrix} < \\ = \\ > \end{Bmatrix} 0. \tag{7}$$

If $f + \lambda g = 0$ cannot be sustained over an interval of time, then the control is "bang bang"; it is at its minimum level while its coefficient in H is negative and is at its maximum level when its coefficient in H is positive. Alternatively, one could write the Lagrangian

$$L = H + w_1(u - a) + w_2(b - u)$$

with necessary conditions (6) and

$$L_u = f + \lambda g + w_1 - w_2 = 0, \qquad w_1 \ge 0, \qquad w_1(u - a) = 0,$$

$$w_2 \ge 0, \qquad w_2(b - u) = 0. \tag{7'}$$

Of course, (7) and (7′) are equivalent conditions.

Example 1. A particle starts at a given point x_0 and can move along a straight line at an acceleration that can be controlled within limits. Let $x(t)$ denote its position at t. Then $x''(t) = u$, $x(0) = x_0$ where u is constrained to obey $-1 \le u \le 1$. The problem is to choose u so the particle comes to rest ($x' = 0$) at the origin ($x = 0$) as quickly as possible. Let $x_1 = x$, $x_2 = x'$.

Then the problem can be stated

$$\min_{u,T} \int_0^T dt$$

subject to $x_1' = x_2$, $x_1(0) = x_0$, $x_1(T) = 0$,

$x_2' = u$, $x_2(0) = 0$, $x_2(T) = 0$,

$-1 \le u \le 1$ (i.e., $u - 1 \le 0$ and $-u - 1 \le 0$).

Associate λ_i with x_i. The Lagrangian is

$$L = 1 + \lambda_1 x_2 + \lambda_2 u + w_1(u-1) - w_2(u+1).$$

An optimal solution satisfies

$$\partial L/\partial u = \lambda_2 + w_1 - w_2 = 0 \tag{8}$$

where

$$w_1 \ge 0, \qquad w_1(u-1) = 0,$$

$$w_2 \ge 0, \qquad w_2(u+1) = 0;$$

$$\lambda_1' = -\partial L/\partial x_1 = 0, \tag{9}$$

$$\lambda_2' = -\partial L/\partial x_2 = -\lambda_1. \tag{10}$$

Also, since the terminal time is free,

$$L(T) = 0. \tag{11}$$

Indeed, since the problem is autonomous, $L = 0$ throughout $[0, T]$.

Conditions (8) imply either $\lambda_2 > 0$, in which case $w_2 > 0$ and $u = -1$, or $\lambda_2 < 0$, in which case $w_1 > 0$ and $u = 1$. Thus, $u = -1$ when $\lambda_2 > 0$ and $u = 1$ when $\lambda_2 < 0$. From (9), λ_1 is constant. Integrating (10) then gives

$$\lambda_2(t) = -\lambda_1 t + c. \tag{12}$$

Since λ_2 is a linear function of t, it changes sign at most once. But since the value u depends on the sign of λ_2, this means u changes value (switches) at most once.

The control u is not determined from (8) if $\lambda_2 = 0$, but $\lambda_2 = 0$ for, at most, an instant. To see this, suppose $\lambda_2 = 0$ for an interval of time. Then, from (12), $\lambda_1 = 0$ and therefore $H = 1 + 0 + 0 \ne 0$, which contradicts (11). Thus, (8) determines u for all t, except possibly a single moment of switching.

During an interval of time in which $u = 1$, $x_2' = u = 1$, so that $x_2 = t + c_0$. Hence $x_1' = x_2 = t + c_0$, so that $x_1 = t^2/2 + c_0 t + c_1$. Substituting $t = x_2 - c_0$, we obtain

$$x_1 = (x_2 - c_0)^2/2 + c_0(x_2 - c_0) + c_1 = x_2^2/2 + c_2$$

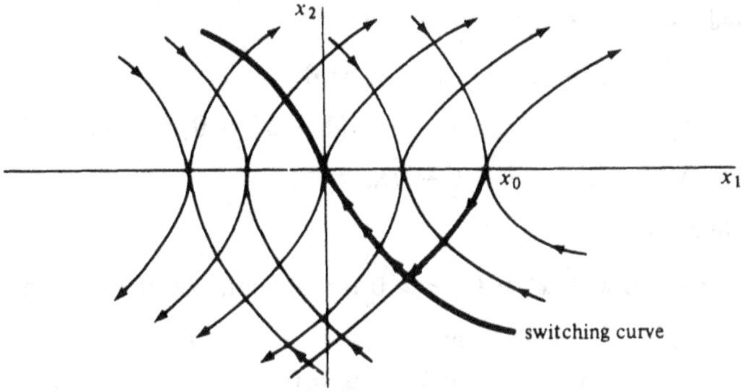

Figure 12.1

where $c_2 = -c_0^2/2 + c_1$. The optimal path is a parabola in the x_1-x_2 plane when $u = 1$. When $u = -1$, $x_2' = -1$, which implies $x_2 = -t + c_3$, and $x_1' = x_2 = -t + c_3$, which gives $x_1 = -(t - c_3)^2/2 + c_4 = -x_2^2/2 + c_4$.

The two parabolas passing through $x_1 = x_2 = 0$ are $x_1 = x_2^2/2$ and $x_1 = -x_2^2/2$. The final portion of the path must be along one of these two curves. The solution consists of at most two portions since u changes value at most once. In Figure 12.1, follow the parabola from the initial point to one of the two parabolas passing through the origin; then follow the parabola to the origin.

The parabolas with x_2 increasing correspond to solutions with $x_2' = u = 1 > 0$, while the parabolas with downward-pointing arrows correspond to $x_2' = u = -1 < 0$ (recall the problem formulation). The heavily marked path indicates an optimal solution from initial position $(x_0, 0)$. The optimal solution involves maximum acceleration for half the distance to the origin and then maximum deceleration for the remaining distance to bring the particle to rest at the origin.

Example 2. A good produced at rate $x(t)$ can either be reinvested to expand productive capacity or sold. Productive capacity grows at the reinvestment rate. What fraction $u(t)$ of the output at time t should be reinvested to maximize total sales over the fixed period $[0, T]$? The initial capacity is c.

With the definitions provided, we seek to

$$\max \int_0^T [1 - u(t)]\, x(t)\, dt \tag{13}$$

$$\text{subject to} \quad x'(t) = u(t)\, x(t), \quad x(0) = c > 0, \tag{14}$$
$$0 \le u(t) \le 1.$$

The Lagrangian is

$$L = (1 - u)x + \lambda u x + w_1(1 - u) + w_2 u.$$

An optimal solution satisfies

$$L_u = x(\lambda - 1) + w_2 - w_1 = 0, \tag{15}$$

$$w_1 \ge 0, \qquad w_2 \ge 0, \qquad w_1(1-u) = 0, \qquad w_2 u = 0, \tag{16}$$

$$\lambda' = -L_x = u - 1 - u\lambda, \tag{17}$$

$$\lambda(T) = 0. \tag{18}$$

Since $x(0) > 0$ and $x' \ge 0$, $x > 0$ throughout; therefore, from (15)–(17)

$$\begin{aligned} &\text{Either} \quad \lambda > 1 \quad \text{and} \quad u = 1, \quad \text{so} \quad \lambda' = -\lambda, \\ &\text{or} \quad \lambda < 1 \quad \text{and} \quad u = 0, \quad \text{so} \quad \lambda' = -1. \end{aligned} \tag{19}$$

This means λ is decreasing over $0 \le t \le T$. Furthermore, by (18), it is zero at T. Thus, there is a final interval $t^* \le t \le T$ during which $\lambda < 1$, $u = 0$, $\lambda' = -1$, and then $x' = 0$. Hence

$$\left.\begin{aligned} u(t) &= 0, \\ \lambda(t) &= T - t, \\ x(t) &= x(t^*), \end{aligned}\right\} \qquad t^* \le t \le T. \tag{20}$$

The time t^* is such that $\lambda(t^*) = 1$; that is,

$$t^* = T - 1 \tag{21}$$

provided $T \ge 1$. If $T \le 1$, then the solution is given by (20) with $t^* = 0$.

If $T > 1$, then there is an initial interval $0 \le t \le T - 1$ during which $\lambda > 1$, $u = 1$, $\lambda' = -\lambda$, $x' = x$. Using $x(0) = c$ and the required continuity of x and λ at $t^* = T - 1$ gives

$$\left.\begin{aligned} u(t) &= 1, \\ \lambda(t) &= \exp(T - t - 1), \\ x(t) &= ce^t, \end{aligned}\right\} \qquad 0 \le t < T - 1. \tag{22}$$

Values of w_1 and w_2 are found from (15) and (16), using (20)–(22). If $T \le 1$, then

$$w_1(t) = 0, \qquad w_2(t) = c(1 - T + t), \qquad 0 \le t \le T - 1. \tag{23}$$

If $T > 1$, then

$$\begin{aligned} w_1(t) &= \begin{cases} ce^t[\exp(T - t - 1) - 1], & 0 \le t \le T - 1, \\ 0, & T - 1 \le t \le T; \end{cases} \\ w_2(t) &= \begin{cases} 0, & 0 \le t < T - 1, \\ c(1 - T + t)\exp(T - 1), & T - 1 \le t \le T. \end{cases} \end{aligned} \tag{24}$$

These functions satisfy the nonnegativity conditions. In this problem, they happen to be continuous but continuity of the w_1 is not required.

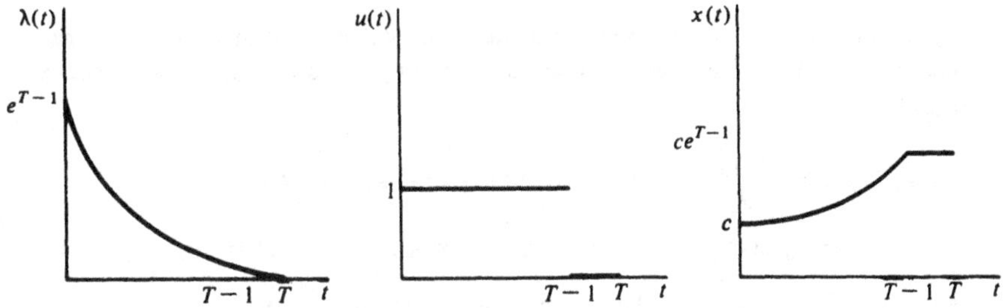

Figure 12.2

If the planning horizon is short, then it is optimal to sell all the output. Otherwise, it is optimal to reinvest all the output until $t = T - 1$, building capacity at the maximal rate, and thereafter sell all the output. (See Figure 12.2.)

EXERCISES

1.

$$\max \int_0^T e^{-rt}(1-u)x\,dt$$

$$\text{subject to} \quad x' = ux, \quad x(0) = x_0 > 0, \quad 0 \le u \le 1.$$

2.

$$\max \int_0^2 (2x - 3u)\,dt$$

$$\text{subject to} \quad x' = x + u, \quad 0 \le u \le 2, \quad x(0) = 5, \quad x(2) \text{ free}.$$

3. Repeat Exercise 2, replacing "max" by "min."

4.

$$\max \int_0^T (x - u)\,dt$$

$$\text{subject to} \quad x' = u, \quad 0 \le u \le x,$$
$$x(0) = x_0 \text{ given}, \quad x(T) \text{ free}, \quad T \text{ given}.$$

5.

$$\min \int_0^1 (2 - 5t)u\,dt$$

$$\text{subject to} \quad x' = 2x + 4te^{2t}u, \quad x(0) = 0, \quad x(t) = e^2, \quad -1 \le u \le 1.$$

6. The revenue $px(t)$ that a machine earns at any time t is proportional to its quality $x(t)$. The quality decays at a constant proportionate rate b but can be enhanced by

expenditure $u(t)$ on maintenance. The machine will be sold at a predetermined time T; the sales price is proportional to its quality. Find the maintenance policy $u(t)$, $0 \le t \le T$, to

$$\max \quad \int_0^T e^{-rt}[px(t) - u(t)]\,dt + e^{-rT}sx(T)$$

$$\text{subject to} \qquad x' = u - bx, \quad 0 \le u \le \bar{u}, \quad x(0) = x_0,$$

where $s < 1 < p/(r + b)$. Also interpret the stated inequalities.

7. Reconsider Exercise 10.4 in case $a = 1$ and $b = 0$.

FURTHER READING

Example 1 is drawn from Pontryagin, who also gives several related examples. These examples are given as well in Bryson and Ho and in Lee and Markus. Example 2 is from Berkovitz. See Thompson and Bensoussan, Hurst, Naslund for maintenance problems related to the one in Exercise 6. See Ijiri and Thompson for another application of bang-bang control. Exercise 5 is from Hestenes. Arvan and Moses provide a more recent, novel application of a bang-bang control.

Section 13

Singular Solutions and Most Rapid Approach Paths

SINGULAR SOLUTIONS

In the examples of Section 12, the Hamiltonian was linear in the control u. The coefficient u in the Hamiltonian could be equal to zero only for isolated instants. In other problems, the coefficient of u in H is equal to zero over some period of time. During such periods, the control does not affect the Hamiltonian, and therefore, the choice of u is not determined in the usual way. In these cases, the value of u is said to be *singular*. One can usually manipulate the various other conditions to determine the value of the control.

In the notation of (12.1)–(12.7), if $f + \lambda g = 0$, then the value of u is not apparent. It clearly cannot be chosen to maximize H, since in this case the value of H is independent of the value of u. However, if $f + \lambda g = 0$ over a period of time, this fact may often be used to advantage.

Example 1. The altitude of the land is given by the differentiable function $y(t)$, $0 \le t \le T$. Find the altitude $x(t)$ at which to build a road over $0 \le t \le T$ so that the grade never exceeds $|a| \ge 0$ and the total cost

$$\int_0^T [x(t) - y(t)]^2 \, dt \tag{1}$$

is minimized. The cost at each t accumulates with the square of the amount of filling or excavation needed.

The constraint can be written

$$x' = u, \qquad -a \le u \le a. \tag{2}$$

The Lagrangian is

$$L = -(x - y)^2 + \lambda u + w_1(a - u) + w_2(u + a),$$

so the optimal solution satisfies

$$L_u = \lambda - w_1 + w_2 = 0, \qquad w_1 \ge 0, \qquad w_1(a - u) = 0,$$

$$w_2 \ge 0, \qquad w_2(u + a) = 0; \tag{3}$$

$$\lambda' = -H_x = 2(x - y); \tag{4}$$

$$\lambda(0) = \lambda(T) = 0. \tag{5}$$

Transversality conditions (5) hold since the height of the road at the endpoints is not specified. Combining (4) and (5) gives

$$\lambda(t) = 2\int_0^t [x(s) - y(s)]\, ds, \tag{6}$$

$$\int_0^T [x(s) - y(s)]\, ds = 0. \tag{7}$$

Combining (3) and (6) indicates that there are three types of intervals:

a. $u(t) = -a$, $\int_0^t [x(s) - y(s)]\, ds < 0$;
b. $x(t) = y(t)$, $u(t) = y'(t)$, $\int_0^t [x(s) - y(s)]\, ds = 0$;
c. $u(t) = a$, $\int_0^t [x(s) - y(s)]\, ds > 0$.

The optimal road consists of sections built along the terrain exactly and of sections of maximum allowable grade.

These specifications are sufficient to determine the optimal road construction uniquely. For instance, in Figure 13.1, t_1 and $x(t_1)$ are determined by

$$\int_0^{t_1} [x(s) - y(s)]\, ds = 0, \qquad \int_{t_1}^T [x(s) - y(s)]\, ds = 0,$$

while in Figure 13.2, t_1 and t_2 are specified by

$$\int_0^{t_1} [x(s) - y(s)]\, ds = 0, \qquad \int_{t_2}^T [x(s) - y(s)]\, ds = 0.$$

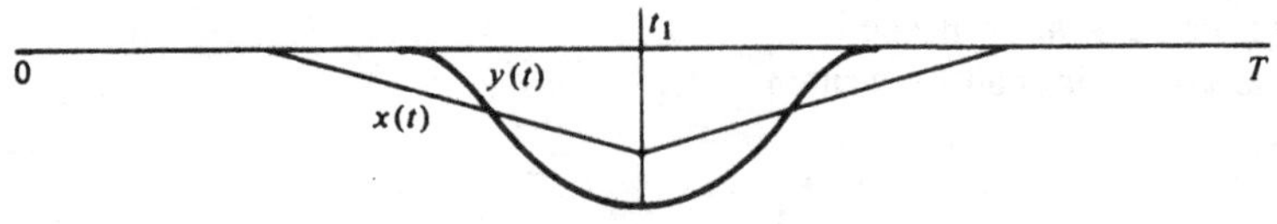

Figure 13.1

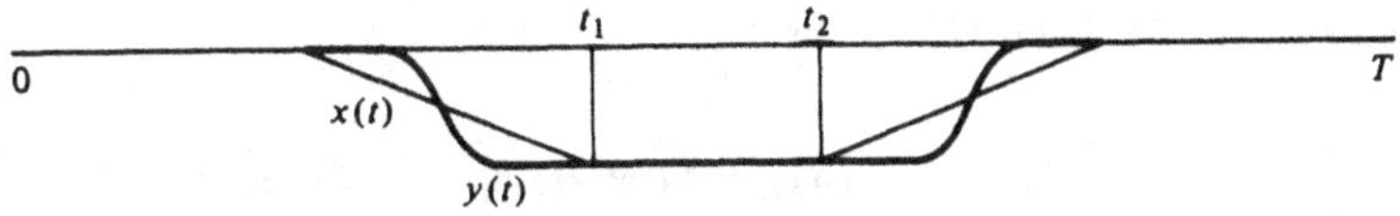

Figure 13.2

Example 2. The firm's problem of selecting investment I to

$$\max \int_0^\infty e^{-rt}[p(t)f(K(t)) - c(t)I(t)]\,dt$$

$$\text{subject to} \quad K' = I - bK, \quad K(0) = K_0, \quad I \geq 0,$$

where $p(t)$ and $c(t)$ are given functions of time indicating the unit price of output $f(K)$ and of investment I, has current value Hamiltonian

$$H = pf(K) - cI + m(I - bK).$$

Necessary conditions obeyed by a solution include

$$m(t) \leq c(t), \qquad I(t)[c(t) - m(t)] = 0, \tag{8}$$

$$m' = (r + b)m - pf'(K). \tag{9}$$

The marginal value m of a unit of capital never exceeds its marginal cost c. If the marginal value is less than the marginal cost, then no investment occurs. On any interval that $I > 0$, we have $m = c$ so $m' = c'$; making these substitutions into (9) gives the familiar rule for the optimal capital stock:

$$pf'(K) = (r + b)c - c' \quad \text{while} \quad I > 0 \tag{10}$$

(review I5.2)). I is selected to maintain (10), as long as that is feasible. (Differentiate (10) totally with respect to t to get an explicit equation involving I.) This is the singular solution.

To see what conditions hold while $I = 0$, collect terms in m in (9) and integrate:

$$e^{-(r+b)t}m(t) = \int_t^\infty e^{-(r+b)s}p(s)f'(K(s))\,ds, \tag{11}$$

where we used the assumption that $\lim_{t\to\infty} e^{-(r+b)t}m(t) = 0$ to evaluate the constant of integration. Also, by the fundamental theorem of integral calculus,

$$e^{-(r+b)t}c(t) = -\int_t^\infty [d(e^{-(r+b)s}c(s))/ds]\,ds$$

$$= \int_t^\infty e^{-(r+b)s}[c(s)(r + b) - c'(s)]\,ds. \tag{12}$$

Combining (8), (11), and (12) gives

$$\int_t^\infty e^{-(r+b)s}[p(s)f'(K(s)) - (r+b)c(s) + c'(s)]\,ds \le 0 \quad (13)$$

with equality in case $I > 0$. Therefore, capital is not acquired at any time that the discounted stream of revenue produced by a marginal unit of capital is insufficient to cover the corresponding discounted stream of "user cost." Now suppose that $I(t) = 0$ on $t_1 \le t \le t_2$ with $I(t) > 0$ just prior to t_1 and immediately after t_2. Thus,

$$\int_t^\infty e^{-(r+b)s}[pf' - (r+b)c + c']\,ds = 0 \quad (14)$$

for $t = t_1$ and for $t = t_2$. It follows that

$$\int_t^{t_2} e^{-(r+b)s}[pf' - (r+b)c + c']\,ds \le 0, \qquad t_1 \le t \le t_2, \quad (15)$$

with *equality* for $t = t_1$.

Therefore the myopic rule (10), marginal cost = marginal benefit holds *at each moment* t of an interval on which $I > 0$. Furthermore, it holds *on average* (taking $t = t_1$ in (15)) over an interval $t_1 \le t \le t_2$ that $I = 0$. Over an interval between investment periods, the integral of discounted cost of capital employment equals the discounted marginal value of its product.

Example 3. The Vidale-Wolfe Advertising Model.

Sethi's formulation of the advertising model proposed by Vidale and Wolfe is as follows. The total profit, excluding advertising cost, from industry sales per unit time is P. The firm supplies a fraction x of industry sales and thereby collects a gross profit of Px. Advertising expenditure $u(t)$ affects its market share:

$$x'(t) = au(t)[1 - x(t)] - bx(t), \qquad x(0) = x_0, \quad (16)$$

where

$$0 \le u(t) \le \bar{u}. \quad (17)$$

Parameter a reflects the efficacy of advertising in attracting sales. Sales are lost to others at the constant proportionate rate b. The firm can affect the rate of gain of new sales, but not the rate of loss of repeat sales. The profit, to be maximized by choice of $u(t)$ subject to (16) and (17), is

$$\int_0^\infty e^{-rt}[Px(t) - u(t)]\,dt. \quad (18)$$

Let m be the current value multiplier associated with (16) and write the current value Lagrangian

$$\begin{aligned} L &= Px - u + m[au(1-x) - bx] + w_1 u + w_2(\bar{u} - u) \\ &= Px - mbx + u[ma(1-x) - 1] + w_1 u + w_2(\bar{u} - u). \end{aligned}$$

Necessary conditions for solution include

$$m' = (r + b + au)m - P; \tag{19}$$

$$\partial L/\partial u = ma(1-x) - 1 + w_1 - w_2 = 0,$$

$$w_1 \geq 0, \quad w_1 u = 0, \quad w_2 \geq 0, \quad w_2(\bar{u} - u) = 0; \tag{20}$$

$$u = \begin{Bmatrix} \bar{u} \\ ? \\ 0 \end{Bmatrix} \quad \text{when} \quad ma(1-x)\begin{Bmatrix} > \\ = \\ < \end{Bmatrix} 1. \tag{21}$$

Sethi gives the following solution:

I. If $Pa \leq r + b$, then $u = 0$ for all t.

II. If $Pa > r + b$ and $Pab < (r + b + a\bar{u})(b + a\bar{u})$, then the steady state,

$$\begin{aligned} x_s &= 1 - \left[r + (r^2 + 4abP)^{1/2}\right]/2aP, \\ u_s &= bx_s/a(1 - x_s) \\ m_s &= 1/a(1 - x_s), \end{aligned}$$

can be attained and it should be approached as rapidly as possible by selecting

$$u = \begin{Bmatrix} \bar{u} \\ u_s \\ o \end{Bmatrix} \quad \text{whenever} \quad x(t)\begin{Bmatrix} < \\ = \\ > \end{Bmatrix} x_s.$$

III. If $Pa > r + b$ and $Pab \geq (r + b + a\bar{u})(b + a\bar{u})$ then

$$u(t) = \begin{cases} \bar{u} & \text{if} \quad x(t) < 1 - (r + b + a\bar{u})/aP, \\ 0 & \text{otherwise}. \end{cases}$$

In Case I, the profit available is too low relative to the efficacy of advertising, the rate of discounting, and the loss rate of repeat sales for any advertising to be worthwhile. In Case II, advertising is worthwhile, and an optimal state x is attained as rapidly as possible and maintained thereafter. In Case III, advertising is worthwhile but the ceiling on the advertising rate prevents maintenance of the optimal share x_s.

It is left to the reader to verify the solution by checking that the conditions (16), (17), and (19)–(21) are satisfied in each case. How this solution was developed in the first place is of more interest. We find the singular solution of Case II.

According to (21), advertising will be at either its lower or its upper bound except when $ma(1 - x) = 1$. Suppose this equation were to hold for some interval of time. Since $m(1 - x)$ would be constant,

$$0 = dm(1 - x)/dt = m'(1 - x) - x'm.$$

Substitute for x' from (16), for m' from (19), and $m = 1/a(1 - x)$:

$$aP(1 - x)^2 - r(1 - x) - b = 0.$$

This quadratic equation in $1 - x$ has two solutions, but for it to be meaningful only the positive root can be relevant:

$$1 - x_s = \left[r + (r^2 + 4abP)^{1/2}\right]/2aP. \tag{22}$$

Since $x_s \geq 0$ is also required for sense, we must have

$$Pa \geq r + b. \tag{23}$$

Since x is constant over the interval being discussed,

$$m_s = 1/a(1 - x_s) \tag{24}$$

and, from (16), u satisfies

$$u_s = bx_s/a(1 - x_s) \tag{25}$$

provided

$$u_s \leq \bar{u}. \tag{26}$$

It can be shown that (26) is equivalent to

$$(a\bar{u} + b + r)(a\bar{u} + b) \geq abP. \tag{27}$$

Thus, a singular solution can occur if (23) and (27) hold. Compare this with Case II. The singular solution is given by (22), (24), and (25). The strong inequalities of Case II assure the existence of projects with positive value and with a steady state that can be achieved in finite time.

If $x_0 \neq x_s$, the singular solution will not be attained immediately. From (21), the only other possibilities are $u = \bar{u}$ or $u = 0$. If $x_0 < x_s$, then $u = \bar{u}$ for $0 \leq t \leq T$, where T is to be determined. Solve differential equations (16) and (19) with $u = \bar{u}$, using boundary conditions

$$x(0) = x_0, \qquad m(t) = m_s, \qquad x(T) = x_s$$

to determine the two constants of integration and the unknown time T. The first is given and the second and third are required by the continuity of the multiplier m and the state x. Alternatively, if $x_0 > x_s$, then put $u = 0$, $0 \leq t \leq T$ into (16) and (19) and solve with the same three boundary conditions.

Development of the other two cases is similar. Under some circumstance, $u = 0$ for all $t \geq 0$ may be the solution. Put $u = 0$ into (16) and (19), yielding

$$x(t) = x_0 e^{-bt}, \qquad m(t) = P/(r+b) + ce^{(r+bt)}.$$

If (20) is to be satisfied, $c = 0$. Now

$$ma(1-x) = Pa(1 - x_0 e^{-bt})/(r+b) < 1 \qquad \text{for all} \quad t \geq 0$$

is required for $u = 0$ to be optimal. This holds only if $Pa/(r+b) < 1$. This is Case I.

MOST RAPID APPROACH PATHS

There is a class of autonomous problems with one state variable and one control in which the optimal solution is to approach some stationary level of the state variable as fast as possible. This is called a most rapid approach path (MRAP) (recall Section I16). This class consists of problems that can be written as

$$\max \int_0^\infty e^{-rt}[P(x) + Q(x)f(x,u)]\, dt \tag{28}$$

$$\text{subject to} \qquad x' = F(x) + G(x)f(x,u), \quad x(0) = x_0, \tag{29}$$

$$a(x) \leq u \leq b(x). \tag{30}$$

This problem may, but need not, be linear in u. Constraint (29) can be used to eliminate $f(x, u)$ (and hence u) from (28):

$$\max \int_0^\infty e^{-rt}[M(x) + N(x)x']\, dt \tag{31}$$

$$\text{subject to} \qquad A(x) \leq x' \leq B(x), \tag{32}$$

where (32) is the constraint on x' corresponding to constraint (30) on u. The problem (31) and (32) has been discussed in Section I16.

Examples 1 and 2 are not autonomous and therefore the MRAP does not apply to them. However, Example 3 is amenable to solution by means of MRAP. Use (16) to eliminate u from (19):

$$\int_0^\infty e^{-rt}[Px - (x' + bx)/a(1-x)]\, dt.$$

In the notation of (31), $M(x) = Px - bx/a(1-x)$ and $N(x) = -1/a(1-x)$, so that

$$M'(x) + rN(x) = P - b/a(1-x)^2 - r/a(1-x) \equiv I(x).$$

Since $I'(x) = -2b/a(1-x)^3 - r/(1-x)^2a < 0$ and $I(0) = P - (b + r)/a$, the equation $I(x) = 0$ has a unique solution, x_s, in the interval $0 < x \leq 1$,

$$1 - x_s = \left[r + \left(r^2 + 4abP\right)^{1/2}\right]/2aP,$$

if and only if (23) holds. Hence, if (23) holds, set $u = \bar{u}$ until x_s is attained and then put $u = u_s = bx_s/(1-x_s)$ to maintain x_s thereafter. One can show that $x_s < \bar{x}$ implies $u_s < \bar{u}$, so the policy just given is feasible provided $x_s < \bar{x}$.

Similarly, if $x_0 > x_s > 0$ and $\bar{x} > x_s$, then put $u = 0$ until $x = x_s$ is attained (MRAP) and then set $u = u_s$ thereafter. Again $\bar{x} > x_s$ implies $\bar{u} > u_s$ and therefore the policy is feasible.

If (23) does not hold, then $I(x) < 0$ for all $0 \leq x \leq 1$ and profit is too low for any advertising to be worthwhile. $M(x) + rN(x)$ is maximized by $x = 0$, so one should set $u = 0$ for all t.

EXERCISES

1. Output $f(k)$ produced with capital k may be consumed or reinvested. Capital depreciates at proportionate rate b. The production function $f(k)$ satisfies

$$f(0) = 0, \qquad f'(k) > 0, \qquad f''(k) < 0, \qquad \lim_{k\to 0} f'(k) = \infty.$$

The fraction of output saved and reinvested is s, so the fraction consumed is $1 - s$. Maximize the discounted stream of consumption over a fixed time period

$$\max \int_0^T e^{-rt}(1-s)f(k)\,dt \tag{i}$$

$$\text{subject to} \quad k' = sf(k) - bk, \quad k(0) = k_0, \quad k(T) \geq 0, \tag{ii}$$

$$0 \leq s \leq 1. \tag{iii}$$

   a. Find the optimal savings plan in the special case $f(k) = k^a$, where $0 < a < 1$.
   b. Discuss the optimal savings plan for the general case where $f(k)$ is as specified in the problem above.
   c. Reconsider the above questions in case $T = \infty$.

2. A firm's output $Q = f(K, L)$ is a twice continuously differentiable concave linearly homogeneous function of capital K and labor L. Revenue is an increasing concave function $R(Q)$ of output. The wage rate of labor is w. The cost of investing in a unit of capital is c. Investment I must be nonnegative and cannot exceed an upper bound $\bar{I}$. Capital depreciates at a constant proportionate rate b. Characterize the plan $K(t)$, $I(t)$, $L(t)$ that maximizes the discounted stream of profits,

$$\max \int_0^\infty e^{-rt}\left[R(f(K,L)) - cI - wL\right]dt$$

$$\text{subject to} \quad K' = I - bK, \quad K(0) = K_0, \quad 0 \leq I \leq \bar{I},$$

where b, c, and w are given positive constants.

3. Suppose that $\bar{I} = aR(f(K, L))$ in Exercise 2, so that investment cannot exceed a given proportion of sales revenue.

FURTHER READING

Example 1 is discussed by Pontryagin, Example 2 by Arrow (1968) and Nickell (1974), and Example 3 by Sethi (1973). See also Section I16 and the references cited there. Exercise 1 is worked by Varaiya. For further examples with bounded controls and "averaging conditions" like Example 2 see Kamien and Schwartz (1977b) and Clark, Clarke, and Munro.

Section 14

The Pontryagin Maximum Principle, Existence

The Pontryagin maximum principle is stated somewhat differently from our usage. Our version is correct only under more stringent conditions than have been fully stated. We shall set forth the Pontryagin maximum principle and then note the differences between it and the version given in earlier solutions.

THE PROBLEM

Find a piecewise continuous control vector $\boldsymbol{u}(t) = [u_1(t), \ldots, u_m(t)]$ and an associated continuous and piecewise differentiable state vector $\boldsymbol{x}(t) = [x_1(t), \ldots, x_n(t)]$, defined on the fixed time interval $[t_0, t_1]$, that will

$$\max \int_{t_0}^{t_1} f(t, \boldsymbol{x}(t), \boldsymbol{u}(t))\, dt \tag{1}$$

subject to the differential equations

$$x_i'(t) = g_i(t, \boldsymbol{x}(t), \boldsymbol{u}(t)), \qquad i = 1, \ldots, n, \tag{2}$$

initial conditions

$$x_i(t_0) = x_{i0}, \qquad i = 1, \ldots, n \qquad (x_{i0} \text{ fixed}), \tag{3}$$

terminal conditions

$$\begin{aligned} &x_i(t_1) = x_{i1}, && i = 1, \ldots, p, \\ &x_i(t_1) \geq x_{i1}, && i = p+1, \ldots, q \qquad (x_{i1}, \quad i = 1, \ldots, q, \quad \text{fixed}), \\ &x_i(t_1) \text{ free}, && i = q+1, \ldots, n, \end{aligned} \tag{4}$$

and control variable restriction

$$u(t) \in U, \qquad U \text{ a given set in } R^m. \tag{5}$$

We assume that f, g_i, $\partial f/\partial x_j$, and $\partial g_i/\partial x_j$ are continuous functions of all their arguments, for all $i = 1, \ldots, n$ and $j = 1, \ldots, n$.

Theorem. *In order that $x^*(t)$, $u^*(t)$ be optimal for the above problem, it is necessary that there exist a constant λ_0 and continuous functions $\lambda(t) = (\lambda_1(t), \ldots, \lambda_n(t))$, where for all $t_0 \le t \le t_1$ we have $(\lambda_0, \lambda(t)) \neq (0, 0)$ such that for every $t_0 \le t \le t_1$*

$$H(t, x^*(t), u, \lambda(t)) \le H(t, x^*(t), u^*(t), \lambda(t)), \tag{6}$$

where the Hamiltonian function H is defined by

$$H(t, x, u, \lambda) = \lambda_0 f(t, x, u) + \sum_{i=1}^{n} \lambda_i g_i(t, x, u). \tag{7}$$

Except at points of discontinuity of $u^*(t)$,

$$\lambda_i'(t) = -\partial H(t, x^*(t), u^*(t), \lambda(t))/\partial x_i, \qquad i = 1, \ldots, n. \tag{8}$$

Furthermore

$$\lambda_0 = 1 \qquad \text{or} \qquad \lambda_0 = 0 \tag{9}$$

and, finally, the following transversality conditions are satisfied:

$$\begin{array}{lll} \lambda_i(t_1) \text{ no conditions}, & i = 1, \ldots, p, & \\ \lambda_i(t_1) \ge 0 & (= 0 \text{ if } x_i^*(t_1) > x_{i1}) & i = p+1, \ldots, q, \\ \lambda_i(t_1) = 0, & i = q+1, \ldots, n. & \end{array} \tag{10}$$

There are two types of differences between the above treatment and the approximation in preceding sections. One concerns technical accuracy and the other is stylistic.

We first take up the matter of technical accuracy. The variable λ_0 in the Hamiltonian in (7) is either 0 or 1 (see (9)). We have always taken $\lambda_0 = 1$ under the implicit supposition that the problem would have a solution in which the objective matters. Yet, this supposition need not always be satisfied. There are problems in which the optimal solution requires $\lambda_0 = 0$. We have disregarded that possibility and will continue to do so (despite the risk of technical inaccuracy). (But see the related discussion and example in Section 7.)

Second, we note the stylistic differences. The control region here is specified implicitly by (5), wherein it is required that the control vector $u(t) = [u_1(t), \ldots, u_m(t)]$ lie in some set U. That set may be the entire m-dimensional Euclidean space or may be a proper subset of it. As a result of the implicit specification of the control region, the choice of control that maximizes the Hamiltonian at each point of time t can be stated only implicitly; see (6).

If one specifies the control region U and also perhaps has some further information about the structure of f, g_i, $i = 1, \ldots, n$, then one can use the Kuhn-Tucker theorem, for example, to characterize further the value of $\mathbf{u}$, namely $\mathbf{u}^*$, that provides the maximum in (6). This was done in previous sections. The fundamental requirement is, however, (6).

We have largely avoided the question of *existence* of a solution of an optimal control problem. Conditions are known under which existence is assured. For instance, if the functions f, g are continuous and bounded with bounded derivatives and if f is strictly concave in the controls and the functions g are linear in the controls, then existence can be shown.

FURTHER READING

See Pontryagin, Lee and Markus, Hestenes, Fleming and Rishel, or Seierstad and Sydsaeter (1987), for careful treatments of the conditions under which the multiplier λ_0 can be set equal to 1.

On existence, see Cesari, Steinberg and Stalford, and Fleming and Rishel. Extensions to problems with unbounded time domain have been provided by Baum. See Long and Vousden and Seierstad and Sydsaeter (1977, 1987) for a summary of theorems. See also Gaines (1976, 1977).

Section 15

Further Sufficiency Theorems

For the problem

$$\max \int_{t_0}^{t_1} f(t, x, u)\, dt \tag{1}$$

$$\text{subject to} \quad x' = g(t, x, u), \tag{2}$$

$$x(t_0) = x_0, \quad t_0, t_1 \text{ fixed}, \tag{3}$$

it was shown that if the functions f and g are both concave in x and u, then the necessary conditions for optimality are also sufficient (provided also that $\lambda \geq 0$ if g is nonlinear in x, u). That sufficiency theorem (generalized in dimensionality, if necessary) is easy to apply. However, some interesting problems are not concave in x, u. A generalized version of Mangasarian's theorem, discovered by Arrow, applies to a broader range of problems. However, it may be more difficult to check whether it is applicable.

To state the Arrow sufficiency result, a couple of definitions are needed. Let $u = U(t, x, \lambda)$ denote the value of the control that maximizes

$$H(t, x, u, \lambda) = f(t, x, u) + \lambda g(t, x, u) \tag{4}$$

for given values of (t, x, λ). The notation $U(t, x, \lambda)$ reflects the dependence of the maximizing value of u upon the parameters of the maximization problem (4). Let

$$\begin{aligned} H^0(t, x, \lambda) &= \max_u H(t, x, u, \lambda) \\ &= f(t, x, U(t, x, \lambda)) + \lambda g(t, x, U(t, x, \lambda)). \end{aligned} \tag{5}$$

Thus H^0 is the value of the Hamiltonian when evaluated at the maximizing u; H^0 is called the *maximized Hamiltonian*.

Arrow's theorem, applied to (1)–(3), is as follows.

Theorem (Arrow). *If $H^0(t, x, \lambda)$ is a concave function of x for given λ, $t_0 \le t \le t_1$, and there exists $x^*(t), u^*(t), \lambda(t)$ with x^*, λ continuous satisfying* (2), (3), *and also*

$$u(t) = U(t, x,(t), \lambda(t)), \tag{6}$$

$$\lambda' = -(f_x + \lambda g_x), \tag{7}$$

$$f_u + \lambda g_u = 0, \tag{8}$$

$$\lambda(t_1) = 0, \tag{9}$$

then x^*, u^* will maximize (1) *subject to* (2) *and* (3).

Concavity of f and g in x, u has been replaced by the weaker condition that the maximized Hamiltonian H^0 be concave in x. Checking the properties of a derived function, such as H^0, can be more difficult than checking properties of f and g. However, if f and g are each concave in x and u, then it will be true that H^0 is concave in x. This is a result of the following lemma.

Lemma. *If a function $G(x, u)$ is concave in (x, u), then* $\max_u G(x, u)$ *is a concave function of x.*

PROOF. Let x_1, x_2 be two values of x and let u_i maximize $G(x_i, u)$ with respect to u, $i = 1, 2$. Then for any $0 \le a \le 1$,

$$\begin{aligned} &a \max_u G(x_1, u) + (1 - a) \max_u G(x_2, u) \\ &= aG(x_1, u_1) + (1 - a)G(x_2, u_2) \\ &\le G(ax_1 + (1 - a)x_2, au_1 + (1 - a)u_2) \\ &\le \max_u G(ax_1 + (1 - a)x_2, u), \end{aligned}$$

where the first equality holds by definition of u_i, the next relation holds by concavity of G, and the last by the property of a maximum. The inequality establishes the concavity of $\max_u G(x, u)$ in x.

If follows immediately from this lemma that if f and g are concave in x, u for all t and if $\lambda \ge 0$, then H^0 is concave and the present theorem may be applied. Since that special case constitutes Mangasarian's theorem, Arrow's theorem represents a direct generalization. The maximized Hamiltonian can be concave in x even if f and g are not concave in x and u, so Arrow's theorem is also useful in cases in which Mangasarian's theorem does not apply.

Finally, note that the current value Hamiltonian differs from the regular Hamiltonian only by a discount factor. Hence one can readily write a corresponding theorem in terms of concavity of the maximized current value Hamiltonian.

Example 1. Modify the problem in Exercise 13.1 so the objective is to maximize the discounted stream of utility $U((1 - s)f(k))$ of consumption where $U' > 0$, $U'' < 0$, and $\lim_{c \to 0} U'(c) = \infty$. We wish to show that the maximized current value Hamiltonian is concave in the state variable k, so that the necessary conditions for solution are sufficient. The current value Hamiltonian is

$$H = U((1-s)f(k)) + m[sf(k) - bk]. \tag{10}$$

The value of s that maximizes (10) satisfies

$$H_s = f(k)(m - U') = 0, \tag{11}$$

$$H_{ss} = f^2 U'' < 0. \tag{12}$$

Condition (12) is satisfied by our assumption on U, and condition (11) implies

$$U'((1-s)f(k)) = m. \tag{13}$$

Let g be the inverse function U'^{-1}, so that

$$1 - s = g(m)/f(k). \tag{14}$$

Substituting the maximizing control from (14) into (10) gives the maximized current value Hamiltonian

$$H^0 = U(g(m)) + m[f(k) - g(m) - bk].$$

The maximized Hamiltonian H^0 is clearly concave in the state variable k provided that $m > 0$, but $m > 0$ is assured by (11) and the assumption that $U' > 0$. Therefore, necessary conditions for solution are also sufficient for optimality.

Example 2. Limit Pricing

A firm wants to price its product to maximize the stream of discounted profits. If it maximizes *current* profits, the high price and profits may attract the entry of rivals, which in turn will reduce future profit possibilities. Let current profit $R_1(p)$ be a strictly concave function of price p with $R_1''(p) < 0$. The maximum profit the firm believes will be available to it after rival entry is $R_2 < \max_p R_1(p)$ (independent of current price and lower than current monopoly profits). Whether, or when, a rival will enter is not known, but let $F(t)$ denote the probability that entry will occur by time t, with $F(0) = 0$. The conditional probability density of entry at time t, given its nonoccurrence prior to t, is $F'(t)/[1 - F(t)]$. We assume that this conditional entry probability density is an increasing, convex function of product price. This specification reflects the supposition that as price rises, the profitability of potential entrants of a given size increases and so does their likelihood of entry. Thus, we assume

$$F'(t)/[1 - F(t)] = h(p(t))$$

where

$$h(0) = 0, \qquad h'(p) \geq 0, \qquad h''(p) \geq 0.$$

Discounting future profits at rate r, the firm seeks a price policy $p(t)$ to

$$\max \int_0^\infty e^{-rt}\{R_1(p(t))[1 - F(t)] + R_2 F(t)\}\, dt \tag{15}$$

$$\text{subject to} \quad F'(t) = h(p(t))[1 - F(t)], \tag{16}$$

$$F(0) = 0. \tag{17}$$

The integrand is the expected profits at t, composed of R_1 if no rival has entered by t, and otherwise R_2. The state variable is a probability F while the control function is price p.

The current value Hamiltonian is

$$H = R_1(p)(1 - F) + R_2 F + mh(p)(1 - F). \tag{18}$$

If F^*, p^* is optimal, then F^*, p^*, m satisfy the constraints (16) and (17) and also

$$\partial H/\partial p = R_1'(p)(1 - F) + mh'(1 - F) = 0, \tag{19}$$

$$\partial^2 H/\partial p^2 = [R_1''(p) + mh''(p)](1 - F) \le 0,$$

$$m' = rm - \partial H/\partial F = R_1 - R_2 + m[h(p) + r]. \tag{20}$$

The integrand and the constraint are *not* concave in F and p. However, the value of the control p that maximizes H satisfies (19) and therefore also

$$R_1'(p) + mh'(p) = 0 \tag{21}$$

and so is independent of F. Since the left side of (21) is monotone decreasing in p if $m < 0$, it associates a unique value of p with each negative value of m. Let $p = P(m)$ denote the function implicitly defined by (21). Then, the maximized current value Hamiltonian is

$$H^0 = [R_1(P(m)) + mh(P(m))](1 - F) + R_2 F, \tag{22}$$

which is linear and hence concave in the state variable F. Therefore, a solution to the foregoing necessary conditions (16)–(20) is also a solution to the optimization problem of (15) and (16).

We will show that the necessary conditions are satisfied by constant values of p and m. The constant m satisfying (20) is

$$m = -(R_1 - R_2)/[r + h(p)]. \tag{23}$$

Negativity of m reflects the fact that an increment in the state variable F raises the probability of rival entry and therefore *reduces* the optimal expected value (at a rate proportional to the difference in earnings without and with rivals). Substituting from (23) into (21) gives

$$R_1'(p)/[R_1(p) - R_2] = h'(p)/[r + h(p)], \tag{24}$$

an implicit equation for p.

Now assume that $R_1(p)$ is a strictly concave function of p with $\max_p R_1(p) = R_1(p^m) > R_2$, where p^m denotes the monopoly price, and that $R_1(p) = R_2$ has two distinct roots. Let

$$g(p) = R_1'(p)[r + h(p)] - h'(p)[R_1(p) - R_2]. \tag{25}$$

The p that solves (24) also solves $g(p) = 0$ and conversely. Call the smaller root of $R_1(p) = R_2$, $\underline{p}$ and the larger $\bar{p}$. Since $R_1(p)$ is strictly concave $R_1'(\underline{p}) > 0$, and $R_1'(\bar{p}) < 0$ so that $g(\underline{p}) > 0$ and $g(\bar{p}) < 0$. Also, $R_1(p) > R_2$ for $\underline{p} < p < \bar{p}$. Compute

$$\begin{aligned} g'(p) &= R_1''(p)[r + h(p)] + R_1'(p)h'(p) - h''(p)[R_1(p) - R_2] \\ &\quad - h'(p)R_1'(p) \\ &= R_1''(p)[r + h(p)] - h''(p)[R_1(p) - R_2] < 0, \\ &\qquad \text{for} \quad \underline{p} \le p \le \bar{p} \end{aligned} \tag{26}$$

where the negativity of (26) follows from the assumption that $R_1'' < 0$, $h'' > 0$ and $R_1(p) \ge R_2$. Thus, as $g(p)$ is positive at $\underline{p}$, negative at $\bar{p}$ and is continuous it follows by the intermediate value theorem that there must exist a p, say p^*, such that $g(p^*) = 0$ for $\underline{p} \le p^* \le \bar{p}$. Moreover, since $g'(p) < 0$ for $\underline{p} \le p \le \bar{p}$, it follows that p^* is unique. Finally, from (25) it follows that at $g(p^*) = 0$, $R_1'(p^*) > 0$ if $h'(p^*) > 0$ and that $R_1(p^m) > R_1(p^*) > R_2$ in this case. That is, if the conditional probability of entry is positive at p^* then $R_1'(p^*) > 0$. But then p^* occurs at the upward sloping part of $R_1(p)$, which means that $p^* < p^m$. Recall that $R'(p^m) = 0$ by the definition of p^m. Also, $p^* > \underline{p}$ because $g(\underline{p}) > 0$, while $g(p^*) = 0$ and $g'(p) < 0$. Equations (16) and (17) can be integrated after setting $p(t) = p^*$ to get

$$F(t) = 1 - e^{-h(p^*)t}. \tag{27}$$

So with p^* defined by (24), m, F are given in terms of p^* in (23) and (27). These functions satisfy (16)–(20) and are thus optimal. Sufficiency of the necessary conditions was established by means of Arrow's theorem even though Mangasarian's theorem could not be used directly.

EXERCISES

1. State the analog of Arrow's theorem for a minimization problem.

2. Show that in Example 2 neither the integrand of (15) nor the right side of (16) is concave or convex in F and p.

3. Show that the solution to the maintenance problem for given T in Section II.10 is optimal, that is, that the necessary conditions are also sufficient for optimality for a given T.

4. Show that the necessary conditions are sufficient for optimality in Exercise 13.1 (check the three cases of $s = 0$, $0 < s < 1$, and $s = 1$).

FURTHER READING

See Mangasarian for his full theorem and proof. Arrow and Kurz (1970) give a rough proof of Arrow's theorem while Kamien and Schwartz (1971d) provide another proof under restrictive conditions. See Seierstad and Sydsaeter (1977, 1987) for a full and careful treatment of sufficiency theorems. Also see Robson (1981) and Hartl.

Example 2 is discussed at length by Kamien and Schwartz (1971b). See also Leung (1991) for a discussion of the transversality conditions associated with this model.

Section 16

Alternative Formulations

Problems that look different may be equivalent.

ALTERNATE OBJECTIVE VALUE FORMULATION

It is possible to state

$$\max \int_{t_0}^{t_1} f(t, x, u)\, dt \tag{1}$$

$$\text{subject to} \quad x' = g(t, x, u), \tag{2}$$

$$x(t_0) = x_0 \tag{3}$$

as a terminal value problem, in which the objective is to maximize a function of the terminal state alone. To see this, define the function $y(t)$, $t_0 \le t \le t_1$, by

$$y'(t) = f(t, x, u), \tag{4}$$

$$y(t_0) = 0. \tag{5}$$

Then

$$y(t_1) = \int_{t_0}^{t_1} f(t, x, u)\, dt. \tag{6}$$

Thus, problem (1)–(3) can be stated as

$$\begin{aligned} &\max \quad y(t_1) \\ &\text{subject to} \quad x' = g(t, x, u), \quad y' = f(t, x, u), \\ &\qquad x(t_0) = x_0, \quad y(t_0) = 0. \end{aligned} \tag{7}$$

In the calculus of variations, (1) is known as the problem of *Lagrange* while the terminal value problem (7) is in the *Mayer* form.

A combined form with integral and terminal value term

$$\max \int_{t_0}^{t_1} f(t, x, u)\, dt + \phi(x(t_1)) \qquad \text{subject to (2) and (3)} \tag{8}$$

is known in the calculus of variations as the *Bolza* form of the problem. The three forms are equivalent. For instance, writing (8) as a terminal value problem is immediate, using the construction (4) and (5). To write (8) as an equivalent problem of Lagrange, let

$$Z' = 0, \quad Z(t_0) \text{ free}, \quad Z(t_1) = \phi(x(t_1)). \tag{9}$$

Then (8) is equivalent to

$$\max \int_{t_0}^{t_1} [f(t, x, u) + Z(t)/(t_1 - t_0)]\, dt \qquad \text{subject to (2), (3), and (9).} \tag{10}$$

Thus, the three forms of the objective are all of equal generality and equivalent. However, some authors appear to favor one form over the others. A theorem stated for any one form can be translated to apply to either of the other forms.

AUTONOMOUS FORMULATION

A nonautonomous problem (1)–(3) can be formally posed as an autonomous problem by letting $z'(t) = 1$, $z(t_0) = t_0$:

$$\max \int_{t_0}^{t_1} f(z, x, u)\, dt$$

$$\text{subject to} \quad x' = g(z, x, u), \quad x(t_0) = x_0, \quad z' = 1, \quad z(t_0) = t_0. \tag{11}$$

The problem does not formally depend on time. Theorems relating to properties of autonomous problems can be translated to nonautonomous problems.

ISOPERIMETRIC CONSTRAINTS

An isoperimetric constraint

$$\int_{t_0}^{t_1} G(t, x, u)\, dt = B \tag{12}$$

can be written in differential form by defining a new variable $y(t)$ by

$$y'(t) = G(t, x, u), \tag{13}$$

$$y(t_0) = 0, \tag{14}$$

$$y(t_1) = B. \tag{15}$$

Replace (12) by the equivalent specifications (13)–(15).

HIGHER DERIVATIVES

If higher derivatives enter, as in $f(t, x, x', x'', x''')$, one can let $x' = y$, $y' = z$, $z' = u$. Then successive substitution gives $f(t, x, y, z, u)$. This (part of the) problem has three state variables x, y, z and one control u.

EXERCISES

1. Write the necessary conditions for problem (1)–(3). Write the necessary conditions for problem (7). Show that the two sets of conditions are equivalent.

2. Show the equivalence between the necessary conditions for
   a. (8) and (10);
   b. (1)–(3) and (11).

3. Find necessary conditions for maximization of (1), subject to (2)–(3), and (13)–(15). Show that the multiplier associated with the state variable y is constant. Explain why this is so.

FURTHER READING

Ijiri and Thompson analyzed a commodity trading model in which the objective is to maximize the value of assets held at terminal time.

Section 17

State Variable Inequality Constraints

Variables may be required to be nonnegative for some expression to have meaning. Sometimes the nonnegativity of a state variable can be assured very simply. For instance, suppose the state variable moves according to a differential equation of the form

$$x' = g_1(x, u) - g_2(u), \qquad x(t_0) = x_0 > 0,$$

where $x \geq 0$ is required.

$$g_1(x, u) \geq 0, \qquad g_2(u) \geq 0 \qquad \text{for} \quad x \geq 0 \text{ and all admissible } u,$$

and also

$$g_1(x, u) \leq 0, \qquad \text{for} \quad x \leq 0 \text{ and all admissible } u.$$

These conditions imply that $g_1(0, u) = 0$. If we then merely require that $x(t_1) \geq 0$, we ensure that $x(t) \geq 0$, $t_0 \leq t \leq t_1$. To see this, note that once x falls to zero, it cannot increase thereafter. If x were to become negative, it could not increase later to satisfy the terminal nonnegativity requirement. Thus, the terminal nonnegativity restriction assures nonnegativity throughout. (See Exercise 1 below for another class of problems in which nonnegativity is assured.)

If such arguments cannot be employed, it may be necessary to incorporate the nonnegativity restriction directly. We develop the necessary conditions. It may be advisable to solve the problem without explicit cognizance of the nonnegativity restriction to see if perhaps it will be satisfied. If so, the problem is solved simply. If not, some hints may emerge regarding the structure of the solution to the constrained problem.

For the problem

$$\max \int_{t_0}^{t_1} f(t, x, u)\, dt + \phi(x(t_1)) \tag{1}$$

$$\text{subject to} \quad x' = g(t, x, u), \quad x(t_0) = x_0, \tag{2}$$

$$k(t, x) \geq 0, \tag{3}$$

we associate a multiplier function $\lambda(t)$ with (2) and a multiplier function $\eta(t)$ with (3). The Hamiltonian is

$$H = f(t, x, u) + \lambda g(t, x, u) + \eta k(t, x). \tag{4}$$

Necessary conditions for optimality include satisfaction of (2), (3), and

$$H_u = f_u + \lambda g_u = 0, \tag{5}$$

$$\lambda' = -H_x = -(f_x + \lambda g_x + \eta k_x), \tag{6}$$

$$\lambda(t_1) = \phi_x(x(t_1)), \tag{7}$$

$$\eta \geq 0, \qquad \eta k = 0. \tag{8}$$

Example 1.

$$\min_u \tfrac{1}{2} \int_0^T (x^2 + c^2u^2)\, dt$$

$$\text{subject to} \quad x' = u, \quad x(0) = x_0 > 0, \quad x(T) = 0,$$

$$h_1(t, x) = a_1 - b_1 t - x \leq 0,$$

$$h_2(t, x) = x - a_2 + b_2 t \leq 0,$$

where $a_i, b_i > 0$, $a_2 > x_0 > a_1$, and $a_2 / b_2 > a_1 / b_1$. The problem is illustrated graphically in Figure 17.1. The path begins at x_0 on the x axis, must stay in the shaded area, and must end on the t axis. The necessary conditions for the minimization problem are developed in Exercise 2.

The Hamiltonian is

$$H = (1/2)(x^2 + c^2u^2) + \lambda u + \eta_1(a_1 - b_1 t - x) + \eta_2(x - a_2 + b_2 t).$$

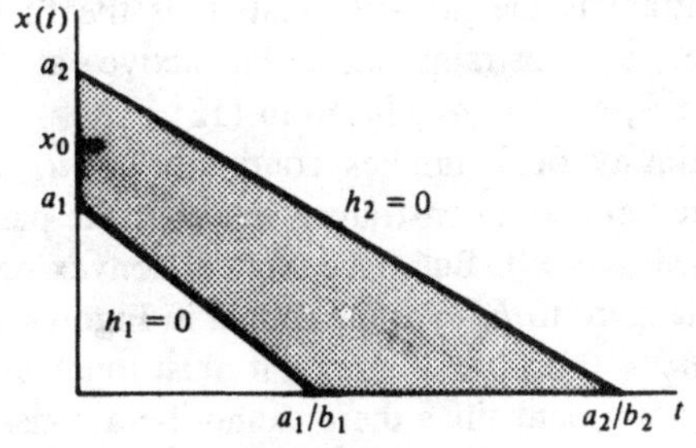

Figure 17.1

Necessary conditions for x, u, λ, η_1, η_2 in addition to the given constraints are

$$H_u = c^2 u + \lambda = 0, \qquad \text{so} \quad u = -\lambda / c^2; \tag{9}$$

$$\lambda' = -H_x = -x + \eta_1 - \eta_2; \tag{10}$$

$$\eta_i \geq 0, \qquad \eta_i h_i = 0, \qquad i = 1, 2. \tag{11}$$

Since initial and terminal values of x are specified, no boundary conditions on λ emerge. Since $x(T) = 0$ is required, the bounds at T require

$$a_1 / b_1 \leq T \leq a_2 / b_2.$$

From (7.30),

$$\begin{array}{llll} \text{either} & H(t) = 0 & \text{and} & a_1 / b_1 < T < a_2 / b_2, \\ \text{or} & H(T) \geq 0 & \text{and} & T = a_1 / b_1, \\ \text{or} & H(T) \leq 0 & \text{and} & T = a_2 / b_2. \end{array} \tag{12}$$

Initially $h_1 h_2 \neq 0$, so $\eta_1 = \eta_2 = 0$. On this (or any) free interval, $\lambda' = -x$ from (10) and $x' = u = -\lambda / c^2$. Hence,

$$x'' = x / c^2 > 0. \tag{13}$$

Therefore, the path of x on any free interval must be convex and of the form

$$x(t) = k_1 e^{t/c} + k_2 e^{-t/c} \tag{14}$$

for some constants k_1, k_2.

Suppose neither constraint were binding at T. Then, since $x(T) = 0$,

$$H(T) = (1/2) c^2 u^2 + \lambda u = -(1/2) c^2 u^2 = 0 \qquad \text{only if} \quad u(T) = 0.$$

But

$$u(T) = x'(T) = (k_1 e^{T/c} - k_2 e^{-T/c}) / c,$$

while

$$x(T) = k_1 e^{T/c} + k_2 e^{-T/c} = 0.$$

Then both $u(T) = x(T) = 0$ imply $k_1 = k_2 = 0$, so $x(t) = 0$ on a final free interval. This contradicts the property that T is the first moment for which $x = 0$. Therefore, a constraint must be active at T. Then $H(T) = -(1/2)c^2 u^2(T) < 0$, so $T = a_2 / b_2$, from (12).

From (9), continuity of λ implies continuity of u. Therefore at a point dividing a free interval and a constrained interval, the path must be tangent to the constraint (since $u = x'$). But since $x(t)$ is convex on every free interval, it cannot become tangent to $h_2 = 0$, as shown in Figures 17.1 and 17.2. Thus, the path touches $h_2 = 0$ only at T (since it must touch at T, as discovered in the previous paragraph, and since there cannot be a constrained interval along $h_2 = 0$; otherwise the tangency condition could not be met.)

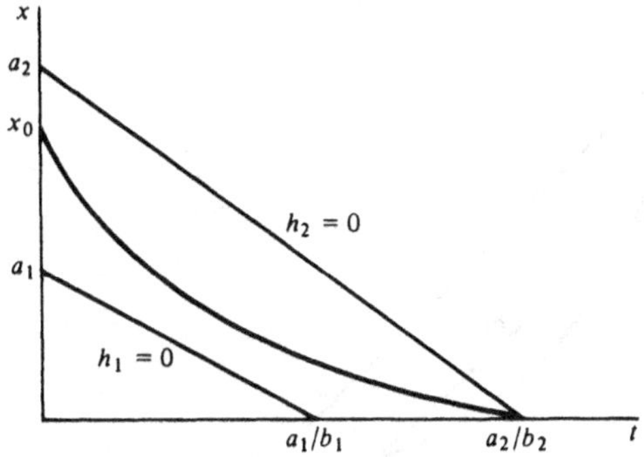

Figure 17.2

Therefore, the solution is (14), where k_1, k_2 are chosen so that

$$x(0) = k_1 + k_2 = x_0,$$
$$x(a_2/b_2) = k_1\exp(a_2/b_2c) + k_2\exp(-a_2/b_2c) = 0,$$

provided that this satisfies $h_1 \le 0$ (see Figure 17.2). If this is not feasible, then the solution follows a path of the form (14) from $(0, x_0)$ to a point of tangency with $h_1 = 0$, then follows $h_1 = 0$ for some interval, and finally leaves $h_1 = 0$ from a point of tangency and follows another path of the form (14) to $(a_2/b_2, 0)$. Thus,

$$x(t) = \begin{cases} k_1e^{t/c} + k_2e^{-t/c}, & 0 \le t \le t_1, \\ a_1 - b_1t, & t_1 \le t \le t_2, \\ k_3e^{t/c} + k_4e^{-t/c}, & t_2 \le t \le a_2/b_2, \end{cases}$$

where the values of k_1, k_2, and t_1 are determined by (1) the initial condition, (2) continuity at t_1, and (3) tangency at t_1, while the values of k_3, k_4 and t_2 are determined by (4) continuity at t_2, (5) tangency at t_2, and (6) the terminal condition $x(a_2/b_2) = 0$ (see Figure 17.4).

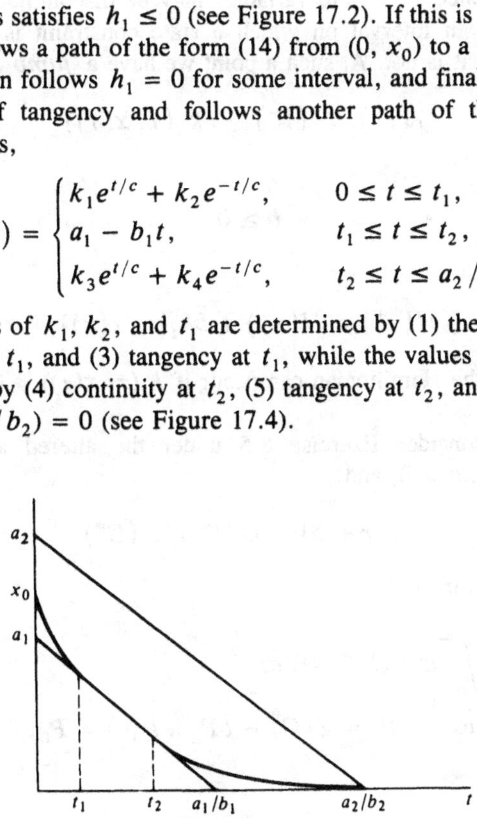

Figure 17.3

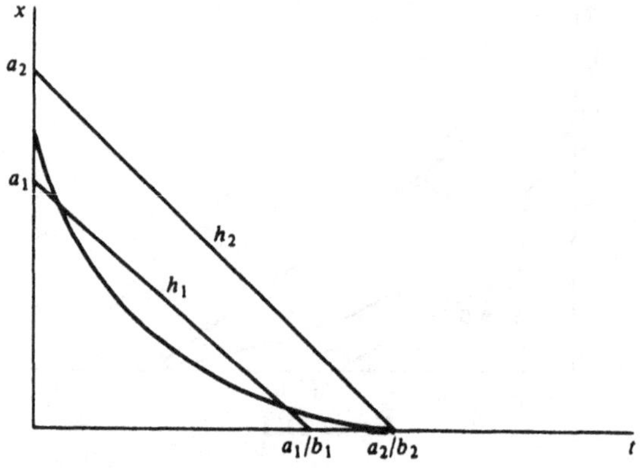

Figure 17.4

Seierstad and Sydsaeter (1977) have provided a sufficiency theorem for problems with constraints on the state variables. There is a new feature. The multipliers associated with state variables may be discontinuous at a junction point τ between an interval on which a state constraint is binding and an interval on which it is not. At such a point we have a *jump condition*:

$$\lambda(\tau^+) = \lambda(\tau^-) - bk_x(\tau, x(\tau)) \tag{15}$$

where

$$b \geq 0. \tag{16}$$

In addition,

$$H(\tau^+) = H(\tau^-) + bk_t(\tau, x(\tau)). \tag{17}$$

Thus, a jump in the Hamiltonian can occur if $k_t(\tau, x(\tau)) \neq 0$.

Example 2. Reconsider Exercise 8.5 under the altered assumptions that $\lim_{p \to 0} U_p = -a$, $a > 0$, and

$$a/(r+b) > U_c(C^*)/Z'(C^*). \tag{18}$$

The Hamiltonian for

$$\max \int_0^\infty e^{-rt} U(C, P)\, dt \tag{19}$$

$$\text{subject to} \quad P' = Z(C) - bP, \quad P(0) = P_0, \quad P \geq 0 \tag{20}$$

is

$$H = U(C, P) + m[Z(C) - bP] + \eta P,$$

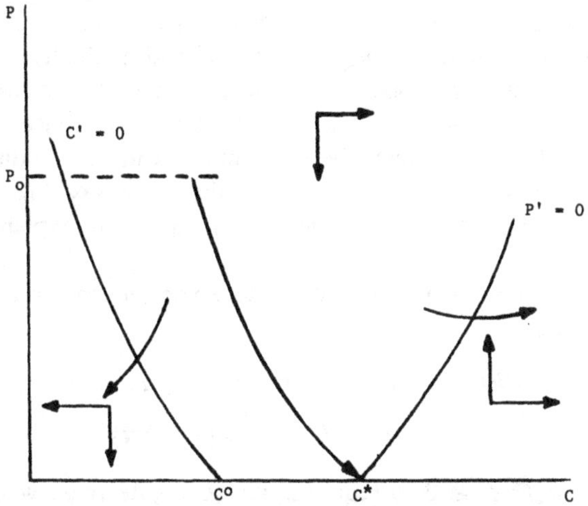

Figure 17.5

from which

$$U_c + mZ' = 0, \tag{21}$$

$$m' = (r+b)m - U_p - \eta, \qquad \eta \geq 0, \quad \eta P = 0. \tag{22}$$

The necessary conditions will be sufficient for optimality. We sketch the phase diagram in the C-P plane. Differentiating (21) totally and using (21) and (22) to eliminate m and m' from the result leads to the pair of differential equations, (20) and

$$(U_{cc} - U_c Z''/Z')C' = Z'\left[(r+b)U_c/Z' + U_p + \eta\right]. \tag{23}$$

The $P' = 0$ locus (Figure 17.5), $P = Z(C)/b$, is an increasing concave function with intercept $(C^*, 0)$, where C^* satisfies $Z(C^*) = 0$. P is decreasing above the locus and increasing below it.

The $C' = 0$ locus, where $P > 0$, obeys $(r+b)U_c/Z'(C) + U_p = 0$, a decreasing curve. Its intercept $(C^0, 0)$ obeys

$$U(C^0, 0)/Z'(C^0) = a/(r+b) > U_c(C^*)/Z'(C^*),$$

where the inequality follows by hypothesis (18). Since $U_c/Z'(C)$ is a decreasing function of C, it follows that $C^0 < C^*$. Above the $C' = 0$ locus, $C' > 0$ and below the $C' = 0$ locus, $C' < 0$.

Since the $C' = 0$ and $P' = 0$ loci have no intersection, there is no steady state with $P > 0$. The steady state is $(C^*, 0)$. C^* lies on $P' = 0$, and when $P = 0$, $\eta > 0$ so that $C' = 0$ is also possible. Indeed, from (23)

$$\eta = a - (r+b)U_c(C^*)/Z'(C^*) > 0 \quad \text{when} \quad P = 0, \tag{24}$$

where the inequality follows by hypothesis (18). Note that the required inequality of (22) is satisfied. Thus, the optimal path is the heavy path shown, with pollution decreasing monotonically to zero and then remaining at zero once attained. Consumption grows to C^* and stays at C^* once attained. Since $C < C^*$ before the steady state, the community is net pollution abating until the pollution is gone. The hypotheses in this case say that the marginal disutility of pollution, even at low levels, is high; this explains the pattern described.

There are other approaches to handling state variable constraints of the form $k(t, x) \geq 0$. Since

$$\begin{aligned} dk(t, x)/dt &= k_t(t, x) + k_x(t, x)x' \\ &= k_t + k_x g(t, x, u), \end{aligned}$$

one can assure $k(t, x) \geq 0$ by requiring that k not decrease whenever $k = 0$:

$$\phi(t, x, u) \equiv k_t(t, x) + k_x(t, x) g(t, x, u) \geq 0 \qquad \text{whenever} \quad k(t, x) = 0. \tag{25}$$

Now the results of Section 11 can be applied to the problem

$$\max \quad \int_{t_0}^{t_1} f(t, x, u)\, dt \tag{26}$$

$$\text{subject to} \qquad x' = g(t, x, u), \quad x(t_0) = x_0 \tag{27}$$

$$\phi(t, x, u) \geq 0, \qquad \text{whenever} \quad k(t, x) = 0, \tag{28}$$

where ϕ is as defined in (25). This is equivalent to problem (1)–(3).

To distinguish this notation, let $M(t)$ and $N(t)$ be the multipliers associated with (27) and (28), respectively. Form the Hamiltonian

$$\tilde{H}(t, x, u, M) = f(t, x, u) + M(t) g(t, x, u) \tag{29}$$

and the Lagrangian

$$\tilde{L}(t, x, u, M, N) = \tilde{H}(t, x, u, M) + N(t)\phi(t, x, u). \tag{30}$$

Necessary conditions for (x, u) to be optimal are that there exist functions M and N such that (x, u, M, N) satisfy not only (27) and (28) but also at points of continuity

$$\partial \tilde{L}/\partial u = 0, \tag{31}$$

$$M' = -\partial \tilde{L}/\partial x, \tag{32}$$

$$N(t) \geq 0, \qquad N'(t) \leq 0, \qquad Nk = 0. \tag{33}$$

The requirement $N'(t) \leq 0$ will be explained below.

To relate the necessary conditions arising from the two approaches, expand (31):

$$\tilde{L}_u = f_u + Mg_u + Nk_x g_u = 0. \tag{34}$$

Write

$$M = \lambda - Nk_x. \tag{35}$$

Substituting (35) into (34) gives

$$f_u + \lambda g_u = 0. \tag{36}$$

Expand (32), using (25)

$$M' = -f_x - Mg_x - N(k_{tx} + gk_{xx} + g_x k_x). \tag{37}$$

Differentiating (35) totally gives

$$M' = \lambda' - N'k_x - Nk_{xx}g - Nk_{xt}. \tag{38}$$

Equate (37) and (38), using (35) to eliminate M, and simplify:

$$\lambda' = -(f_x + \lambda g_x + \eta k_X), \tag{39}$$

where

$$\eta = -N'. \tag{40}$$

Thus, the two approaches give the same necessary conditions with the multipliers, λ, η related to M, N through (35) and (40). Note that since $\eta \geq 0$ by (8), (40) implies that $N' \leq 0$ is required on intervals of continuity of N, as is recorded in (33). Note also that continuity of λ implies continuity of $M + Nk_x$. However, from (15) λ is not continuous in general. Moreover, the jump conditions for M can be stated by combining (35) with (15). Neither M, N, nor $\tilde{H}$ is necessarily continuous in this formulation.

A modification of the second version results in continuous multipliers but at the cost of an unusual specification of the Hamiltonian. Associate multiplier $p(t)$ with (27) and $n(t)$ with (28). Form the Hamiltonian

$$\overline{H}(t) = f + pg + n\phi. \tag{41}$$

The multiplier p is continuous, n is piecewise continuous, and p, n, x, u satisfy (27), (28), and

$$\begin{aligned} p' &= -\overline{H}_x = -f_x - pg_x - n\phi_x, \\ \overline{H}_u &= 0. \end{aligned} \tag{42}$$

on each interval of continuity of u. The Hamiltonian is maximized with respect to u and the maximized Hamiltonian is continuous. The multiplier n is nonincreasing and is *constant* (not necessarily zero) on every interval on which $k(t, x) > 0$. It is continuous if and only if λ is continuous. Note that p, n, and $\overline{H}$ are continuous, but the product $n\phi$ is zero only when the constraint (3) is active. Further, as in (40), $\eta = -N'$.

EXERCISES

1. Consider a problem in which there is one state x and control u, and the state equation is

$$x' = g(u) - bx,$$

where $g(u) \geq 0$ for all admissible values of u, and where $x(0) \geq 0$. Show that if the above conditions are met, then

$$x(t) \geq 0 \quad \text{for} \quad 0 \leq t \leq T.$$

2. Using the results stated for problem (1)–(3), show that the necessary conditions for

$$\min \int_{t_0}^{t_1} f(t, x, u)\, dt$$

$$\text{subject to (2) and (3)}$$

are that x^*, u^*, λ, η satisfy (2), (3), (5)–(7), and

$$\eta(t) \leq 0, \quad \eta(t)k(t, x) = 0, \quad t_0 \leq t \leq t_1.$$

[Hint: Multiply the objective by -1 to get a maximization problem and then see what conditions are required.]

3. A firm wants to maximize the present value of its dividend stream

$$\int_0^\infty e^{-rt}u(t)\, dt$$

subject to

$$K' = R(K) - bK - u + y - rB, \qquad K(0) = K_0 > 0,$$

$$B' = y, \qquad B(0) = 0, \qquad aK - B \geq 0, \quad u \geq 0,$$

where $K(t)$ is its capital stock, $R(K)$ the revenue earned with K, b the decay rate of capital, $u(t)$ the dividend rate, $y(t)$ the rate of borrowing (repayment, if negative), and $B(t)$ the debt. The total debt cannot exceed a fraction a of the capital stock. Dividends must be nonnegative.

a. Verify the problem formulation

b. Show that if $a < 1$, then the optimal solution involves borrowing the maximum possible amount until K_s is attained. What is K_s?

c. Show that if $a \geq 1$, the pattern suggested in part b is impossible. Show that the only possible solution is an immediate jump at the initial moment to K_s. (See also Section I.8.)

4. Solve the example of Section I14 by optimal control.

5. Solve Example 1 of this section using an alternate approach.

6.

$$\min \int_0^5 (4x + u^2)\, dt$$

$$\text{subject to} \quad x' = u, \quad x(0) = 10, \quad x(5) = 0, \quad x(t) \geq 6 - 2t.$$

FURTHER READING

See, for instance, Jabobson, Lele, and Speyer for a full discussion of necessary conditions in the presence of state constraints and for a treatment of the jump conditions, as well as Hestenes, Bryson and Ho, Taylor (1972, 1974), Seierstad and Sydsaeter (1977, 1987), and Hartl (1984).

Applications to production and related problems have been made by Anderson (1970) and Pekelman (1974, 1979). The pollution problem is posed and discussed by Forster (1977).

Section 18

Jumps in the State Variable, Switches in State Equations

Thus far the state variable has been required to be continuous. Now imagine that a finite jump in the state may be permitted. For instance, if the state is the stock of a resource or capital, then a discrete jump in the stock could be achieved by a momentarily infinite investment. We consider the circumstances under which such a jump could be optimal and the necessary conditions accompanying that jump. The discussion follows that of Arrow and Kurz, who in turn describe the method of Vind for such problems.

Modify the standard form

$$\max \int_0^T f(t, x, u)\, dt$$

$$\text{subject to} \qquad x' = g(t, x, u)$$

by permitting a finite upward jump at some moment t_1. Let $x^+(t_1)$ be the state's value immediately after the jump, while $x^-(t_1)$ is the value before the jump. The time t_1 and the magnitude of the jump $x^+(t_1) - x^-(t_1)$ are to be determined optimally. This formulation permits, but does not require, a jump. Let $c(t_1)$ be the cost per unit increase in the state variable. The objective may be written

$$\max \int_0^T f(t, x, u)\, dt - c(t_1)\left[x^+(t_1) - x^-(t_1)\right]. \tag{1}$$

The state variable's evolution follows the typical pattern before t_1. Write it in integrated form:

$$x(t) = x(0) + \int_0^t g(s, x, u)\, ds, \qquad 0 \le t \le t_1. \tag{2a}$$

Following t_1, the discrete jump in x must be recorded, so

$$x(t) = x(0) + \int_0^t g(s, x, u)\, ds + x^+(t_1) - x^-(t_1), \qquad t_1 < t \leq T. \tag{2b}$$

The idea is to transform the problem (1) and (2) into an equivalent problem to which the usual methods can be applied. The necessary conditions for the equivalent problem can then be determined. Finally, these conditions are translated back to find their equivalent statement in terms of the original problem (1) and (2).

An equivalent problem is developed by viewing time as a clock that can run normally or can be stopped at will. Let w index artificial time. Artificial time runs at the same pace as natural time except at a jump. Natural time stops while the jump occurs, but artificial time continues running. The jump occurs smoothly in artificial time. After the jump, the clock resumes and artificial time runs apace with natural time.

Natural time t is now to be considered as a state variable whose movement in artificial time is governed by a new control u_0. Thus,

$$dt/dw = u_0(w) = \begin{cases} 0 & \text{during a jump in } x, \\ 1 & \text{otherwise}. \end{cases}$$

The development of x in artificial time follows the familiar form $dx/dw = g(w, x, u)$, except during a jump. During a jump, the rate of change in x is governed by another new control u_1. If the jump occurs during (w_1, w_2) in artificial time, then

$$x^+(t_1) - x^-(t_1) = x(w_2) - x(w_1) = \int_{w_1}^{w_2} u_1(w)\, dw = u_1(w_2 - w_1).$$

We can take u_1 to be a constant function over the interval that it is positive. Thus over a jump interval

$$dx/dw = u_1.$$

These two forms for dx/dw (the latter pertaining during the jump and the former at all other time) can be expressed in a single equation, using the control u_0 that is 0 during a jump and 1 elsewhere. Hence,

$$dx/dw = u_0(w) g(t, x(t), u(w)) + [1 - u_0(w)] u_1(w). \tag{3}$$

The reader should check that this single equation has the properties claimed for it. Also

$$dt/dw = u_0(w), \qquad 0 \leq u_0 \leq 1. \tag{4}$$

The objective increases at rate f when natural time is running. At other times, the cost of the jump is accumulated at a rate proportional to the speed of the jump. Both features can be combined into a single expression by the device just employed:

$$\max \int_0^W \{u_0(w)f(t, x(w), u(w)) - [1 - u_0(w)]c(t)u_1(w)\}\, dw. \quad (5)$$

The problem of maximizing (5) subject to (3) and (4) is an ordinary control problem, with state variables x and t and control variables u, u_0, u_1.

Let λ_1 and λ_0 be the multipliers associated with (3) and (4), respectively. The Hamiltonian is

$$\begin{aligned} H &= u_0 f - (1 - u_0)cu_1 + \lambda_1[u_0 g + (1 - u_0)u_1] + \lambda_0 u_0 \\ &= u_0 H_a + (1 - u_0)H_b, \end{aligned}$$

where

$$H_a \equiv f + \lambda_1 g + \lambda_0, \qquad H_0 \equiv (\lambda_1 - c)u_1.$$

We maximize H with respect to u, u_0, u_1 stepwise. Note that u appears only in H_a, u_1 appears only in H_b, and u_0 appears in neither. Thus, we maximize H_a and H_b separately, denoting

$$H_a^0 = \max_u H_a, \qquad H_b^0 = \max_{u_1} H_b.$$

Then

$$\max_{u, u_0, u_1} H = \max_{u_0}(u_0 H_a^0 + (1 - u_0)H_b^0) = \max(H_a^0, H_b^0). \quad (6)$$

The last step follows from the fact that u_0 will be either 0 or 1.

The multipliers obey

$$\begin{aligned} \lambda_1'(w) &= -H_x = -u_0(f_x + \lambda_1 g_x) \\ &= \begin{cases} -(f_x + \lambda_1 g_x) & \text{when } u_0 = 1. \\ 0 & \text{when } u_0 = 0; \end{cases} \end{aligned} \quad (7)$$

$$\begin{aligned} \lambda_0'(w) &= -H_t = -u_0(f_t + \lambda_1 g_t) + (1 - u_0)u_1 c'(t) \\ &= \begin{cases} -(f_t + \lambda_1 g_t) & \text{when } u_0 = 1, \\ u_1 c'(t) & \text{when } u_0 = 0. \end{cases} \end{aligned} \quad (8)$$

The next chore is analysis of (3)–(8) and translation of the results and their implications back to natural time. To that end, we identify λ_1 as the multiplier associated with (2) and let

$$H_n(t, x, u, \lambda) = f(t, x, u) + \lambda_1 g(t, x, u) \quad (9)$$

be the Hamiltonian in natural time for (1) and (2). First, since H_a is to be maximized with respect to u,

$$u \text{ maximizes } H_a \text{ at each } t. \quad (10)$$

Second, from (7), since λ_1 does not change during a jump, we have

$$\lambda_1' = -(f_x + \lambda_1 g_x) = -\partial H_n/\partial x. \tag{11}$$

Third, since H_b is linear in u_1 and u_1 is not bounded above,

$$\lambda_1 - c \le 0 \tag{12}$$

since otherwise an infinite value in (5) could be achieved. Furthermore, a jump occurs only when $u_1 > 0$ and this cannot happen when (12) holds with strict inequality (since u_1 is chosen to maximize H_b). Thus,

$$\lambda_1 = c \qquad \text{at a jump (i.e., at } t_1). \tag{13}$$

Fourth, the argument just given assures that $H_b \le 0$. But, in fact, $H_b = 0$ can always be attained by setting $u_1 = 0$ so that $H_b^0 = 0$. This implies

$$\max H = \max(H_a^0, 0) = \begin{cases} H_a^0 & \text{when } u_0 = 1, \\ 0 & \text{when } u_0 = 0. \end{cases} \tag{14}$$

Since max H must be a continuous function of w, it must be zero at the end of a jump. Furthermore, if the jump occurs after the first moment then max H is also zero at the beginning of a jump. Thus, if $t_1 > 0$ then

$$\begin{aligned} \max H &= \max_u [f(t_1, x^-(t_1), u) + \lambda_1(t_1) g(t_1, x^-(t_1), u) + \lambda_0^-(t_1)] \\ &= 0 \\ &= \max_u [f(t_1, x^+(t_1), u) + \lambda_1(t_1) g(t_1, x^+(t_1), u) + \lambda_0^+(t_1)]. \end{aligned} \tag{15}$$

To determine the implication of (15), the behavior of λ_0 over a jump is needed. Reviewing (8), we see that λ_0' is constant on a jump since u_1 is then constant and since natural time is stopped. Further, since λ_0 changes at a constant rate $u_1 c'$ over the interval and since the duration of the jump must be $[x^+(t_1) - x^-(t_1)]/u_1$ (recall the total gain in x to be made during the jump and the rate at which it is accumulated), the total change in λ_0 over the jump interval $[w_1, w_2]$ is

$$\begin{aligned} \lambda_0^+(t_1) - \lambda_0^-(t_1) = \lambda_0(w_2) - \lambda_0(w_1) &= c'(t_1) u_1 (w_2 - w_1) \\ &= c'(t_1)[x^+(t_1) - x^-(t_1)] \end{aligned} \tag{16}$$

since $x(w_2) - x(w_1) = u_1(w_2 - w_1)$. Combining (15) and (16) gives

$$\begin{aligned} &\max_u H_n(t_1, x^+(t_1), u, \lambda(t_1)) - \max_u H_n(t_1, x^-(t_1), u, \lambda(t_1)) \\ &\quad = c'(t_1)[x^-(t_1) - x^+(t_1)] \qquad \text{if } t_1 > 0. \end{aligned} \tag{17}$$

In sum, if the solution to (1) and (2) involves a jump at t_1, then, with the Hamiltonian given by (9), necessary conditions are (10)–(13) and (17).

Arrow and Kurz give some indications of when jumps may and may not occur. If $\max_u H_n$ is *strictly* concave in x for given λ_1 and t, then a jump can never be optimal except possibly at the initial moment. They argue as follows: Suppose that $\max_u H_n$ is strictly concave in x and that there is a jump at $t_1 > 0$; then H_a^0 is strictly concave in x as H_a^0 and $\max_u H_n$ only differ by a constant. Since λ_0 is a linear function of w ($\lambda_0(w) = u_1 c'(t)w + k$, where k is a constant) and since x changes during a jump, H_a^0 is strictly concave in w. This follows from the observation that if H_a^0 is regarded as a function of x only, $H_a^0 = H_a^0(x(w))$, then $dH_a^0/dw = H_{ax}^0\, dx/dw$. But $dx/dw = u_1$, a constant during the jump. So $d^2H_a^0/dw^2 = H_{axx}^0 u_1^2 < 0$ by virtue of $H_{axx}^0 < 0$. Since H_a^0 is zero at both the beginning and the end of the jump, it must be positive in the middle. It is here that the strict concavity of H_a^0 plays a crucial role. If H_a^0 were not strictly concave in w then the function could be zero at the beginning of the jump, and coincide with the horizontal axis until the end of the jump. But according to (14), $H_a^0 \le 0$ during a jump. We have a contradiction, indicating there cannot be a jump at some $t_1 > 0$. However, a jump at 0 is not ruled out by this argument since there is no requirement that H_a^0 be zero at the beginning of the jump.

Example. A monopolist produces a quality-differentiated spectrum of a good for a market of consumers with differing tastes. Let q be the index of quality of a unit of product and let t be the index of a consumer's taste, his marginal utility for quality. Each individual buys at most one unit. A consumer with taste t, faced with a range of qualities and corresponding unit prices $P(q)$ will equate marginal cost with marginal utility and choose the unit with that quality q satisfying

$$P'(q) = t \tag{18}$$

if there is such a quality available. The monopolist knows the cost $C(q)$ of producing a unit with quality q (with $C' > 0$, $C'' > 0$) and knows the proportion $F(t)$ of the population whose taste index does not exceed t. F is assumed continuously differentiable, with $F' > 0$, $F(0) = 0$, $F(T) = 1$. The monopolist chooses the quality of product $q(t)$ to induce a person with taste t to buy and associated price $p(t)$ to maximize profit:

$$\int_0^T F'(t)[p(t) - C(q(t))]\, dt, \tag{19}$$

which is price less cost for each person with taste t, weighted by the density of consumers with taste t, and summed over all tastes. (The choice of $q(t)$ and $p(t)$ imply the price and quality must be chosen with individual's utility maximizing behavior (18) in mind.) Since $dp/dq = p'/q' = t$, write $u = p'$. Then, (19) is to be maximized subject to

$$p' = u, \qquad p(0) = C(0), \tag{20}$$

$$q' = u/t, \qquad q(0) = 0, \tag{21}$$

$$u \ge 0. \tag{22}$$

The control variable u must be nonnegative so price and quality do not decrease with taste. The quality index begins at zero and the price begins at the fixed cost of producing a unit of lowest quality (so profit will be nonnegative).

Associate multipliers λ_1, λ_2, and w with constraints (20)–(22), respectively. The Lagrangian is

$$L = F'[p - C(q)] + \lambda_1 u + \lambda_2 u/t + wu.$$

An optimal solution obeys (20)–(22) and

$$L_u = \lambda_1 + \lambda_2/t + w = 0, \qquad w \geq 0, \qquad wu = 0, \tag{23}$$

$$\lambda_1' = -L_p = -F', \qquad \lambda_1(T) = 0, \tag{24}$$

$$\lambda_2' = -L_q = F'C'(q), \qquad \lambda_2(T) = 0. \tag{25}$$

The maximizing Hamiltonian is strictly concave in q (since $-C'' < 0$), so there can be no jumps in q. Thus u is bounded. Either $\lambda_1 + \lambda_2/t < 0$ and $u = 0$, or else $\lambda_1 + \lambda_2/t = 0$ and $u > 0$.

Integrating (24) and (25) gives

$$\lambda_1(t) = 1 - F(t), \qquad \lambda_2(t) = -\int_t^T F'(s)C'(q(s))\,ds. \tag{26}$$

A unit increase in price for a person with taste t has value $1 - F$ since all whose taste index is at least t will pay a similarly higher price. A unit increase in quality for a person with taste t raises the quality for all with higher taste indices and therefore raises production and cost accordingly.

From (23), $t\lambda_1 + \lambda_2 \leq 0$. Substituting from (26) gives

$$t[1 - F(t)] \leq \int_t^T F'(s)C'(q(s))\,ds. \tag{27}$$

By the Fundamental Theorem of Calculus

$$t[1 - F(t)] = -\int_t^T (d\{s[1 - F(s)]\}/ds)\,ds$$

$$= \int_t^T \{sF'(s) - [1 - F(s)]\}\,ds.$$

Substitution in (27) gives

$$0 \leq \int_t^T F(s)\{C'(q(s)) - s + [1 - F(s)]/F'(s)\}\,ds. \tag{28}$$

On an interval for which $u > 0$, (28) holds with equality and so

$$C'(q(t)) = t - [1 - F(t)]/F'(t) \qquad \text{when} \quad u > 0 \tag{29}$$

implicitly specifies $q(t)$ (so long as $q(t)$ satisfies $q' > 0$) for $t \geq t_0$ where t_0 is implicitly defined by

$$C'(0) = t_0 - [1 - F(t_0)]/F'(t_0). \tag{30}$$

Individuals with tastes in the range $0 \leq t \leq t_0$ are assigned $q = 0$ and hence do not buy the good. Determination of u and then p follows from (21) and then (20) once q is known.

There could be an interior interval $t_1 \leq t \leq t_2$, where $0 < t_1 < t_2 < T$, on which $u = 0$ so that p and q are constant there. All individuals with tastes in this range are assigned the same quality item (and same price). This could happen only if the right side of (29) were decreasing in t over some interval. Then (28) holds as an equality when evaluated at $t = t_1$ and at $t = t_2$, so that

$$\int_{t_1}^{t_2} F'(t)\{C'(q(t)) - t + [1 - F(t)]/F'(t)\}\, dt = 0. \tag{31}$$

The condition (29) that holds whenever $u > 0$ also holds on average over an interval with $u = 0$. (Compare with the average condition in Example 13.2 during periods of no investment in the capital model.)

Specifically, suppose $F(t) = t/T$ (uniform distribution) and $C(q) = aq + bq^2/2$, where $0 \leq a < T$ and $0 < b$. From (30), $t_0 = (a + T)/2$, and from (29) $q(t) = (2t - T - a)/b$ for $t_0 \leq t \leq T$. Since $t = (bq + T + a)/2$, and $P'(q) = t$, we get $P'(q) = (bq + T + a)/2$, and therefore consumers will be presented with the price schedule $P(q) = bq^2/4 + (a + T)q/2$. Consumers optimize according to (18) with this price schedule and select the quality

$$q(t)\begin{cases} 0, & 0 \leq t \leq (a + T)/2 \\ (2t - T - a)/b, & (a + T)/2 \leq t \leq T, \end{cases}$$

designed by the monopolist.

A problem somewhat analogous to a jump in the state variable involves an optimal change in the state equation during the period of optimization. Such an opportunity can arise naturally in, say, switching from one mode of extracting oil from a well to another. Formally, the problem now becomes

$$\max \quad \int_{t_0}^{t_1} f^1(t, x(t), u(t))\, dt + \int_{t_1}^{t_2} f^2(t, x(t), u(t))\, dt$$
$$-\phi(t, x(t_1), u(t_1)) \tag{32}$$

$$\text{s.t.} \quad x'(t) = \begin{cases} g^1(t, x(t), u(t)), & t_0 \leq t \leq t_1 \\ g^2(t, x(t), u(t)), & t_1 \leq t \leq t_2 \end{cases} \tag{33}$$

$$x(t_0) = x_0, \quad t_1, \quad x(t_1), \quad t_2, \quad x(t_2) \text{ free.}$$

In (32), f^1 and f^2 are two possibly different objective functions, and ϕ is the cost of changing the state equation from g^1 to g^2 at t_1. Solution of (32)–(33) involves forming Hamiltonians $H^1 = f^1 + \lambda_1 g^1$ for $t_0 \le t \le t_1$, and $H^2 = f^2 + \lambda_2 g^2$ for $t_1 \le t \le t_2$. The necessary conditions within each time interval are the usual ones, namely,

$$H_u^1 = 0, \qquad \lambda_1' = -H_x^1 \qquad \text{for} \quad t_0 \le t \le t_1$$
$$H_u^2 = 0, \qquad \lambda_2' = -H_x^2 \qquad \text{for} \quad t_1 \le t \le t_2. \tag{34}$$

The new conditions are

$$H^1(t_1) - \phi_t(t_1) = H^2(t_1) \qquad \text{if} \quad t_0 < t_1 < t_2 \tag{35}$$
$$H^1(t_1) - \phi_t(t_1) \le H^2(t_1) \qquad \text{if} \quad t_0 = t_1 < t_2 \tag{36}$$
$$H^1(t_1) - \phi_t(t_1) \ge H^2(t_1) \qquad \text{if} \quad t_0 < t_1 = t_2 \tag{37}$$
$$\lambda_1^-(t_1) + \phi_x(t_1) = \lambda_2^+(t_1) \tag{38}$$
$$H^2(t_2) = 0, \quad \lambda_1(t_2) = 0 \qquad \text{or} \quad \lambda_2(t_2) = 0. \tag{39}$$

Condition (39) is just the ordinary condition for a problem with free terminal time and free terminal value. Conditions (35)–(38) are new. According to (35), if there is a time t_1 at which H^1 less the marginal cost of switching from g^1 to g^2 equals H^2 then it is optimal to switch at t_1. If such a switch occurs then according to (38) the marginal valuation of the state variable evaluated according to f^1 and g^1 plus the marginal cost, with respect to the state variable, must equal the valuation of the state variable evaluated according to f^2 and g^2. Finally, if there is no t_1 that satisfies (35) then we should either skip directly to the second optimization problem involving f^2 and g^2, if (36) holds, or stick entirely with the first optimization problem involving f^1 and g^1 if (37) holds.

FURTHER READING

See Vind and Arrow and Kurz for the theoretical developments. Mussa and Rosen provide an extensive discussion of the product quality example.

Another example in which the optimizer takes the maximizing behavior of another agent as a constraint is the selection of the tax schedule, subject to individual behavior. Mirrlees wrote the seminal paper; see Brito and Oakland and also Cooter.

The analysis of switching from one state equation to another is based on Amit, who derives the necessary conditions. See also Tomiyama and Tomiyama and Rossana.

Section 19

Delayed Response

Thus far we have considered situations in which the value of the state variable responds immediately to the control variable. This may not accurately describe many economic and management problems. For example, suppose sales is regarded as the state variable and price as the firm's control variable; sales may not respond immediately to a reduction in price; for example, it might take time for customers to become aware of the new price. Likewise, a firm's decision to invest in new capital equipment does not result in new productive capacity until the equipment has been ordered, delivered, installed, and tested.

Such situations can be handled by optimal control methods, but not as easily as the immediate response situations. For this reason, delayed response problems are often modeled as immediate response problems.

Formally, however, necessary conditions for optimal control of a problem involving a delayed response can be derived. We consider the analog to the simplest problem considered in Section 2 and seek the necessary conditions satisfied by the control that can

$$\max \quad \int_{t_0}^{t_1} f(t, x(t), u(t))\, dt \tag{1}$$

$$\text{subject to} \quad x' = g(t, x(t), x(t-\tau), u(t), u(t-\tau)), \tag{2}$$

$$x(t) = x_0 \quad \text{for} \quad t_0 - \tau \le t \le t_0, \tag{3}$$

$$u(t) = u_0 \quad \text{for} \quad t_0 - \tau \le t \le t_0, \tag{4}$$

$$x(t_1) \text{ free}. \tag{5}$$

Several features of this formulation merit attention. First, only contemporaneous values of the state variable and the control variable enter the maximand. Second, the rate of change in the state variable depends on the current and past values of the state and control variables. The delay in the response is, moreover, a fixed interval of time, τ. Third, the values of the control variable and the state variable are given constants for time τ prior to the initial time t_0 from which the optimal solution is sought.

To derive the necessary conditions for this problem we proceed as before. Append (2) to (1) with continuously differentiable multiplier function $\lambda(t)$, perform the usual integration by parts, and compute the first variation

$$\begin{aligned}\delta J = \int_{t_0}^{t_1} &[(\partial f/\partial x_t + \lambda \partial g/\partial x_t + \lambda')\,\delta x_t + (\lambda \partial g/\partial x_{t-\tau})\,\delta x_{t-\tau}\\ &+ (\partial f/\partial u_t + \lambda \partial g/\partial u_t)\,\delta u_t + (\lambda \partial g/\partial u_{t-\tau})\,\delta u_{t-\tau}]\,dt\\ &+ \lambda(t_1)\,\delta x(t_1),\end{aligned} \tag{6}$$

where the subscripts indicate whether the contemporaneous or lagged variable is meant.

By setting $s = t - \tau$, so that $t = s + \tau$, one has

$$\begin{aligned}&\int_{t_0}^{t_1} \lambda(t)[\partial g(t, x(t), x(t-\tau), u(t), u(t-\tau))/\partial x_{t-\tau}]\,\delta x_{t-\tau}(t)\,dt\\ &\quad = \int_{t_0-\tau}^{t_1-\tau} \lambda(s+\tau)[\partial g(s+\tau, x(s+\tau), x(s), u(s+\tau), u(s)/\partial x_s]\\ &\quad \times \delta x_{t-\tau}(s+\tau)\,ds,\end{aligned} \tag{7}$$

and similarly for the term involving $\delta u_{t-\tau}$. Since x_t and u_t are fixed prior to t_0, $x_{t-\tau}$ and $u_{t-\tau}$ are fixed prior to $t_0 + \tau$, so that $\delta x_{t-\tau} = \delta u_{t-\tau} = 0$ for $t < t_0 + \tau$. Thus, we may increase the lower limit of integration in (7) by τ since the integrand is zero on the deleted interval. Substituting from (7) into (6) with the change in the lower limit of integration, using the analogous expression for the term involving $\delta u_{t-\tau}$, and collecting terms gives

$$\begin{aligned}&\int_{t_0}^{t_1-\tau} \{[\partial f/\partial x_t + \lambda \partial g/\partial x_t + \lambda' + (\lambda \partial g/\partial x_{t-\tau})|_{t+\tau}]\,\delta x_t\\ &\qquad + [\partial f/\partial u_t + \lambda \partial g/\partial u_t + (\lambda \partial g/\partial u_{t-\tau})|_{t+\tau}]\,\delta u_t\}\,dt\\ &\qquad + \int_{t_1-\tau}^{t_1} [(\partial f/\partial x_t + \lambda \partial g/\partial x_t + \lambda')\,\delta x_t\\ &\qquad + (\partial f/\partial u_t + \lambda \partial g/\partial u_t)\,\delta u_t]\,dt + \lambda(t_1)\,\delta x(t_1).\end{aligned} \tag{8}$$

Note that in the first integral $\lambda \partial g/\partial x_{t-\tau}$ and $\lambda \partial g/\partial u_{t-\tau}$ are to be evaluated at $t + \tau$. Also note that $\delta u_{t-\tau}$ and $\delta x_{t-\tau}$ do not appear in the second integral because they would be generated beyond the horizon t_1.

To render the first variation (8) equal to zero, we choose λ so the coefficient of δx_t is zero. We then get the optimality conditions in the usual way, resulting in the necessary conditions

$$\lambda' = -\partial f/\partial x_t - \lambda \partial g/\partial x_t - (\lambda \partial g/\partial x_{t-\tau})|_{t+\tau}, \qquad t_0 \le t < t_1 - \tau, \tag{9}$$

$$\partial f/\partial u_t + \lambda \partial g/\partial u_t + (\lambda \partial g/\partial u_{t-\tau})|_{t+\tau} = 0, \qquad t_0 \le t < t_1 - \tau, \tag{10}$$

$$\lambda' = -\partial f/\partial x_t - \lambda \partial g/\partial x_t, \qquad t_1 - \tau \le t \le t_1, \tag{11}$$

$$\partial f/\partial u_t + \lambda \partial g/\partial u_t = 0, \qquad t_1 - \tau \le t \le t_1, \tag{12}$$

$$\lambda(t_1) = 0. \tag{13}$$

If the Hamiltonian is written as

$$H(t, x(t), x(t-\tau), u(t), u(t-\tau), \lambda(t))$$
$$= f(t, x_t, u_t) + \lambda_t g(t, x_t, x_{t-\tau}, u_t, u_{t-\tau}),$$

then it can easily be verified that

$$\lambda' = -\partial H/\partial x_t - \partial H/\partial x_{t-\tau}|_{t+\tau}, \qquad t_0 \le t < t_1 - \tau, \tag{14}$$

$$\partial H/\partial u_t + \partial H/\partial u_{t-\tau}|_{t+\tau} = 0, \qquad t_0 \le t < t_1 - \tau, \tag{15}$$

$$\lambda' = -\partial H/\partial x_t, \qquad t_1 - \tau \le t \le t_1, \tag{16}$$

$$\partial H/\partial u_t = 0, \qquad t_1 - \tau \le t \le t_1. \tag{17}$$

The presence of a delayed response adds terms to the necessary conditions that would obtain if the response were immediate, as (9) and (10) or (14) and (15) indicate. These conditions reduce to the ones that obtain in the case of immediate response if the lagged variables are not arguments, so the partial derivatives with respect to them vanish.

The interpretation of these conditions is analogous to earlier interpretations. The total impact of a slight change in the control at any time t should be zero in an optimal program. That impact is partially realized contemporaneously at $(\partial H/\partial u_t)$ and partially, τ periods later, through its appearance as a lagged variable $(\partial H/\partial u_{t-\tau}|_{t+\tau})$. The reasoning is similar for the marginal impact of the state variable.

Example. In the limit pricing example of Section 15, the probability of rival entry at any time, conditional on no prior entry, depends on the incumbent firm's current price. It may be more realistic to suppose that entry cannot be achieved immediately but takes a certain amount τ of time, as acquisition of the productive and distributive facilities takes some time. Thus, the incumbent firm's current price may determine the conditional probability of entry τ periods in the future.

Retaining all the other assumptions of Section 15 regarding the relevant functions,

$$\max \int_0^\infty e^{-rt}\{R_1(p(t))[1 - F(t)] + R_2 F(t)\}\,dt \tag{18}$$

$$\text{subject to} \quad F'(t) = h(p(t-\tau))[1 - F(t)], \quad t \geq \tau, \tag{19}$$

$$p(t) = 0, \quad -\tau \leq t < 0, \tag{20}$$

$$F(t) = 0, \quad -\tau \leq t < 0. \tag{21}$$

According to (19) the price at time t influences the probability density of entry at time $t + \tau$.

The current value Hamiltonian is

$$H = R_1(p(t))[1 - F(t)] + R_2 F(t) + m(t)h(p(t-\tau))[1 - F(t)]. \tag{22}$$

Employing (9) and (10), we obtain

$$R_1'(p(t))[1 - F(t)] + m(t+\tau)h'(p(t))[1 - F(t+\tau)] = 0 \tag{23}$$

and

$$m' = R_1(p(t)) - R_2 + m(t)[h(p(t-\tau)) + r]. \tag{24}$$

Trying constant values for $p(t)$ and $m(t)$ yields

$$m = -(R_1 - R_2)/[r + h(p)]. \tag{25}$$

as before, and

$$R_1'(p)/[R_1(p) - R_2] = h'(p)e^{-h(p)\tau}/[r + h(p)]. \tag{26}$$

The last expression follows from (23) and integration of (19) with $p(t)$ constant, which yields

$$F(t) = 1 - e^{-h(p)(t-\tau)}, \qquad t \geq \tau, \tag{27}$$

so that

$$[1 - F(t+\tau)]/[1 - F(t)] = e^{-h(p)\tau}. \tag{28}$$

This solution satisfies the necessary conditions and reduces to the solution found earlier in case $\tau = 0$. The entry lag can be shown to raise the incumbent firm's current price, compared to that in the absence of a lag. The reason is the further away a future loss, the lower its present cost. This in turn means that the incumbent firm should sacrifice less of its current profits and therefore price higher than in the absence of an entry lag.

EXERCISES

1. Show that if the time lag τ is a differentiable function of time τ(t), then in going from (6) to (8) we get

$$\int_{t_0}^{t_1-\tau(t_1)} [(\partial f/\partial x_t + \lambda \partial g/\partial x_t + \lambda'$$
$$+ (\partial g/\partial x_{t-\tau(t)}/(1 - \dot{\tau}(t)) \,|_{t+\tau(t)}) \,\delta x_t$$
$$+ (\partial f/\partial u_t + \lambda \partial g/\partial u_t + (\lambda \partial g/\partial u_{t-\tau}/(1 - \dot{\tau}(t)) \,|_{t+\tau(t)}) \,\delta u_t] \,dt$$
$$+ \int_{t_1-\tau(t_1)}^{t_1} [(\partial f/\partial x_t + \lambda \partial g/\partial x_t + \lambda') \,\delta x_t$$
$$+ (\partial f/\partial u_t + \lambda \partial g/\partial u_t) \,\delta u_t] \,dt + \lambda(t_1) x(t_1),$$

where $\dot{\tau}(t) = d\tau/dt$. Derive the necessary conditions corresponding to (14)–(17). [Hint: $\delta u_{t-\tau(t)} = 0$ and $\delta x_{t-\tau(t)} = 0$ for $t < t_0 + \tau(t_0)$, while $\delta u_{t-\tau(t)} = \delta u_t (1 - \dot{\tau})$, $\delta x_{t-\tau(t)} = \delta x_t(1 - \dot{\tau})$ for $t \geq t_0 + \tau(t_0)$.]

2. Characterize the optimal investment plan if investment cost is proportional to the investment rate but investment affects the capital stock with a fixed lag of τ.

$$\max \quad \int_0^{\infty} e^{-rt} [P(x(t)) - cu(t)] \,dt$$

$$\text{subject to} \quad x'(t) = u(t - \tau) - bx(t), \quad x(0) = x_0,$$
$$u(t) = 0, \quad -\tau \leq t < 0.$$

FURTHER READING

The derivation of (14)–(17) and Exercise 1 is based on Budelis and Bryson (1970). Further analysis of the example appears in DeBondt (1976). Other applications of this technique appear in El-Hodiri, Loehman, and Whinston (1972) and Sethi and McGuire (1977). The latter article involves variable (rather than fixed) lags. See also Mann (1975), for a discussion of Exercise 2 and further extensions of lags applied to advertising.

Section 20

Optimal Control with Integral State Equations

Just as some situations in economics and management science are most appropriately described by state equations involving a delayed response, so others are best described by having the state equation be an integral equation. Although any state equation of the form

$$x' = g(t, x(t), u(t)) \tag{1}$$

can be written as an integral equation

$$x(t) = \int_{t_0}^{t} g(s, x(s), u(s))\, ds + x(t_0). \tag{2}$$

there are state equations such as

$$x(t) = \int_{t_0}^{t} g(t, x(s), u(s), s)\, ds + x(t_0) \tag{3}$$

that cannot be put in the form of (1) by merely differentiating both sides with respect to t, as (2) can.

An example involving a firm's optimal selection of a product's durability through time will serve to illustrate when the use of an integral state equation, such as (3), is useful. But first we will derive the necessary conditions for the problem

$$\max \quad \int_{t_0}^{t_1} f(t, x(t), u(t))\, dt \tag{4}$$

subject to (3), and $x(t_0) = x_0$, $x(t_1)$ free. To do this we form the Lagrangian

$$L(x, u, \lambda) = \int_{t_0}^{t_1} f(t, x(t), u(t)) + \int_{t_0}^{t_1} \lambda(t)\left[\int_{t_0}^{t} g(t, x(s), u(s), s)\, ds + x(t_0) - x(t)\right] dt. \tag{5}$$

Upon changing the order of integration in the terms involving the double integral we get

$$L = \int_{t_0}^{t_1} f(t, x(t), u(t))\, dt + \int_{t_0}^{t_1}\left[\int_{t}^{t_1} \lambda(s) g(s, x(t), u(t), t)\, ds + x(t_0)\lambda(t) - x(t)\lambda(t)\right] dt. \tag{6}$$

Note that in going from (5) to (6) the first and fourth arguments of g have been interchanged. Now define the Hamiltonian as

$$H(t, x, u, \lambda) = f(t, x(t), u(t)) + \int_{t}^{t_1} \lambda(s) g(s, x(t), u(t), t)\, ds. \tag{7}$$

The Lagrangian (6) can now, upon substitution from (7) into (6), be written as

$$L(x, u, \lambda) = \int_{t_0}^{t_1} \{H(t, x, u, \lambda) - \lambda(t)[x(t) - x(t_0)]\}\, dt. \tag{8}$$

We can now regard (8) as a maximization problem in the calculus of variations involving two functions, $x(t)$ and $u(t)$, but in which $x'(t)$, and $u'(t)$ do not appear (see Section I.18). The Euler equations are then

$$\partial L/\partial u = f_u(t, x(t), u(t) + \int_{t}^{t_1} \lambda(s) g_u(s, x(t), u(t), t)\, ds = 0 \tag{9}$$

and

$$\partial L/\partial x = f_x(t, x(t), u(t)) + \int_{t}^{t_1} \lambda(s) g_x(s, x(t), u(t), t)\, ds - \lambda(t) = 0. \tag{10}$$

But (9) and (10) together with the definition of H in (7) give

$$\partial H/\partial u = 0 \tag{11}$$

and

$$\partial H/\partial x = \lambda(t) \tag{12}$$

as the necessary conditions.

Now suppose that $\partial g(t, x(s), u(s), s)/\partial t \equiv 0$, then $\partial(\partial g/\partial t)/\partial u = \partial(\partial g/\partial u)/\partial t \equiv 0$ by virtue of the equality of cross-partial derivatives. Next recall that in going from (5) to (6) the first and fourth arguments of g were interchanged so that $\partial g/\partial t \equiv 0$ implies that $\partial g_u/\partial s \equiv 0$, that is, that g_u is constant with respect to the variable of integration s in (9). Thus (9) becomes

$$f_u(t, x(t), u(t)) + g_u(x(t), u(t), t)\int_t^{t_1}\lambda(s)\, ds = 0. \tag{13}$$

Expression (13) is almost the same as the necessary condition for an optimal control problem in which the state equation is a differential equation such as (1) except that in (13) g_u is multiplied by an integral of the multiplier $\lambda(s)$ instead of single multiplier, say, $\lambda_0(t)$, that appears in the standard optimal control problem. However, this suggests that the multiplier $\lambda(s)$ appearing in (13) is just the negative time derivative of $\lambda_0(t)$, as we can write $\lambda_0(t) = \lambda_0(t_1) - \int_t^{t_1}\lambda_0'(\tau)\, d\tau$ and identify $\lambda(s)$ in (13) as $-\lambda_0'(\tau)$. Thus, $\int_t^{t_1}\lambda(s)\, ds = -\int_t^{t_1}\lambda_0'(\tau)\, d\tau$ and (13) can be rewritten as,

$$f_u + g_u(\lambda_0(t) - \lambda_0(t_1)) = 0. \tag{14}$$

But for the simplest optimal control problem in which $\lambda_0(t_1)$ can be chosen freely $\lambda_0(t_1) = 0$, and therefore (9) finally reduces to

$$f_u + \lambda_0(t) g_u = 0 = \partial H/\partial u. \tag{15}$$

Similarly, (10) reduces to

$$-\lambda_0'(t) = f_x + \lambda_0(t) g_x = \partial H/\partial x \tag{16}$$

upon recalling that $\lambda(t) = -\lambda_0'(t)$.

Thus, what has been shown is that the necessary conditions for the optimal control problem (4) subject to the integral state equation (3) reduce to the necessary conditions for the standard optimal control problem with state equation (1). Similarly, it can be shown that just as in the standard optimal control problem that if the maximized Hamiltonian is concave in the state variable then the necessary conditions (9) and (10) are also sufficient for a maximum.

Example. A firm produces a durable good, the durability of which, as measured by its decay rate, can be chosen at each instant of time. That is, each vintage of the good can have a different durability.

Let $x(t)$ denote the quantity of the durable good produced at time t, and $b(t)$ denote its decay rate. The total quantity of the durable at time t, denoted by $Q(t)$, is

$$Q(t) = \int_0^t e^{-b(t)(t-s)} x(s)\, ds, \tag{17}$$

and $Q(0) = 0$. If $b(t) = b$, that is, the decay rate is constant through time, then

$$Q'(t) = x(t) - b\int_0^t e^{-b(t-s)} x(s)\, ds = x(t) - bQ(t) \tag{18}$$

and (18) could be rewritten as an equivalent differential equation.

$P(b(t), t)$ denotes the product's selling price at time t given that its decay rate is $b(t)$, and $c(x(t), b(t))$ denotes the total cost of producing the quantity $x(t)$ of the good with decay rate $b(t)$. It is supposed that total production cost has the particular form

$$C(x, b) = xg(x)h(b) + f(x). \tag{19}$$

That is, the quantity of the good produced and its decay rate interact multiplicatively in the total cost function. However, $h'(b) > 0$, $h''(b) \geq 0$, $f'(x) > 0$, and all other second derivatives are positive. The sign of $g'(x)$ is not specified at this point. The good's selling price, $P(b(t), t)$, can be related to its durability as reflected by $b(t)$, a higher $b(t)$ meaning a lower durability, through the observation that it must equal the discounted stream of rental prices of the service provided by the good. If this were not so there would be an opportunity either for the buyer or the seller to profit from the difference in the selling price and the rental price of the durable good. The product's selling price is the discounted stream of rental prices of the service provided, where the discounting takes into account the discount rate r, assumed to be a constant, and the deterioration of the service provided at the decay rate $b(t)$. Thus,

$$P(b(t), t) = \int_t^\infty p((Q(s))e^{-(r+b(t))(s-t)}\, ds, \tag{20}$$

where $p(Q(s))$ refers to the rental price at time s when the stock of the durable good is Q.

The firm's objective is to choose how much to produce $x(t)$ at each point in time and to set the decay rate $b(t)$, so as to maximize its discounted profit stream, namely,

$$\pi = \int_0^\infty e^{-rt}\left[P(b(t), t)x(t) - C(x(t), b(t))\right] dt, \tag{21}$$

subject to (17). Now (21) can be rewritten as

$$\pi = \int_0^\infty e^{-rt}\left\{x(t)\left[\int_t^\infty p(Q(s))e^{-(r+b(t))(s-t)}\,ds\right] - C(x, b)\right\} dt \quad (22)$$

upon substitution for $P(b(t), t))$ from (20) into (21). The terms in (22) involving the double integral can, upon changing the order of integration, be rewritten as

$$\int_0^\infty e^{-rs}p(Q(s))\left[\int_0^s e^{-b(t)(s-t)}x(t)\,dt\right] ds. \quad (23)$$

But from (17) it follows that the inner integral is just $Q(s)$, the stock of the durable good at time s. Substituting from (23) back into (22) and switching t for s as the variable of integration we get

$$\pi = \int_0^\infty e^{-rt}(p(Q(t))Q(t) - C(x(t), b(t)]\,dt. \quad (24)$$

The Hamiltonian formed from the objective function (24) and the state equation (17) is

$$H = e^{-rt}[pQ - C(x(t), b(t))] + \int_0^t e^{-b(t)(s-t)}x(t)\lambda(s)\,ds. \quad (25)$$

Note that as, in going from (3) to (7), the roles of t and s are interchanged, so also are they here in going from (17) to (25). The necessary conditions are

$$\partial H/\partial x = -e^{-rt}C_x + \int_0^t e^{-b(t)(s-t)}\lambda(s)\,ds = 0 \quad (26)$$

$$\partial H/\partial b = -e^{-rt}C_b - \int_0^t (s-t)e^{-b(t)(s-t)}x(t)\lambda(s)\,ds = 0 \quad (27)$$

$$\lambda(t) = \partial H/\partial Q(t) = e^{-rt}[p(Q(t)) + p'(Q(t))Q(t)] \quad (28)$$

where C_x and C_b refer to the partial derivatives of C with respect to x and b, respectively, while p' refers to the derivative of p with respect to Q. Substituting from (19) and (28) into (26) gives

$$[g(x) + xg'(x)]h(b) + f'(x)$$
$$= -\int_t^\infty e^{-(r+b(t))(s-t)}[p(Q(s)) + p'(Q(s))Q(s)]\,ds \quad (29)$$

and substituting from (19) and (28) into (27) gives

$$g(x)h'(b) = -\int_t^{\infty} (s-t)e^{-(r+b(t))(s-t)}\left[p(Q(s)) + p'(Q(s))Q(s)\right] ds. \tag{30}$$

In both (29) and (30) the transversality condition

$$\lim_{t\to\infty} \int_0^t e^{-b(t)(s-t)}\lambda(s)\, ds = 0$$

has been employed to set $\int_0^t e^{-b(t)(s-t)}\lambda(s)\, ds = -\int_t^{\infty} e^{-b(t)(s-t)} L(s)\, ds$.

FURTHER READING

The derivation of the necessary conditions (11) and (12) is based on Kamien and Muller. A more rigorous derivation appears in Bakke, Nerlove and Arrow, and Mann provide examples of problems involving integral state equations. The example is based on Muller and Peles. They also indicate when the quality of the good optimally declines through time.

Section 21

Dynamic Programming

A third approach to dynamic optimization problems, called *dynamic programming*, was developed by Richard Bellman. It has been fruitfully applied to both discrete and continuous time problems. We discuss only the latter.

The basic principle of dynamic programming, called the *principle of optimality*, can be put roughly as follows. An optimal path has the property that whatever the initial conditions and control values over some initial period, the control (or decision variables) over the remaining period must be optimal for the remaining problem, with the state resulting from the early decisions considered as the initial condition.

Consider

$$\begin{aligned} &\max \quad \int_0^T f(t, x, u)\, dt + \phi(x(T), T) \\ &\text{subject to} \quad x' = g(t, x, u), \quad x(0) = a. \end{aligned} \tag{1}$$

Define the optimal value function $J(t_0, x_0)$ as the best value that can be obtained starting at time t_0 in state x_0. This function is defined for all $0 \le t_0 \le T$ and for any feasible state x_0 that may arise. Thus,

$$\begin{aligned} &J(t_0, x_0) = \max_u \int_{t_0}^T f(t, x, u)\, dt + \phi(x(T), T) \\ &\text{subject to} \quad x' = g(t, x, u), \quad x(t_0) = x_0. \end{aligned} \tag{2}$$

Then, in particular,

$$J(T, x(T)) = \phi(x(T), T). \tag{3}$$

We break up the integral in (2) as

$$J(t_0, x_0) = \max_u \left(\int_{t_0}^{t_0+\Delta t} f\,dt + \int_{t_0+\Delta t}^{T} f\,dt + \phi \right), \tag{4}$$

where Δt is taken to be very small and positive. Next, by the dynamic programming principle, we argue that the control function $u(t)$, $t_0 + \Delta t \leq t \leq T$ should be optimal for the problem beginning at $t_0 + \Delta t$ in state $x(t_0 + \Delta t) = x_0 + \Delta x$. The state $x_0 + \Delta x$, of course, depends on the state x_0 and on the control function $u(t)$ chosen over the period $t_0 \leq t \leq t_0 + \Delta t$. Hence,

$$J(t_0, x_0) = \max_{\substack{u \\ t_0 \leq t \leq t_0+\Delta t}} \left[\int_{t_0}^{t_0+\Delta t} f\,dt + \max_{\substack{u \\ t_0+\Delta t \leq t \leq T}} \left(\int_{t_0+\Delta t}^{T} f\,dt + \phi \right) \right]$$

$$\text{subject to} \quad x' = g, \quad x(t_0 + \Delta t) = x_0 + \Delta x; \tag{5}$$

that is,

$$J(t_0, x_0) = \max_{\substack{u \\ t_0 \leq t \leq t_0+\Delta t}} \left[\int_{t_0}^{t_0+\Delta t} f\,dt + J(t_0 + \Delta t, x_0 + \Delta x) \right]. \tag{6}$$

The return over $t_0 \leq t \leq T$ can be thought of as a sum of the return over $t_0 \leq t \leq t_0 + \Delta t$ plus the return by continuing optimally from the resulting position $(t_0 + \Delta t, x_0 + \Delta x)$. Both the immediate return and the future return are affected by the control $u(t)$, $t_0 \leq t \leq t_0 + \Delta t$, which is to be chosen optimally.

To put (5) in a more useful form, we approximate the first integral on the right by $f(t_0, x_0, u)\,\Delta t$, the height of the curve at the lower limit of integration times the width of the interval. Since Δt is very small we can consider the control to be constant over $t_0 \leq t \leq t_0 + \Delta t$. Further, we *assume* J is a twice continuously differentiable function and expand the second term on the right by Taylor's theorem. Hence,

$$J(t_0, x_0) = \max_u [f(t_0, x_0, u)\,\Delta t + J(t_0, x_0) + J_t(t_0, x_0)\,\Delta t + J_x(t_0, x_0)\,\Delta x + h.o.t.].$$

Subtracting $J(t_0, x_0)$ from each side, dividing through by Δt, and letting $\Delta t \to 0$ gives

$$0 = \max_u [f(t, x, u) + J_t(t, x) + J_x(t, x) x'].$$

Zero subscripts have been dropped since there is no ambiguity. Since $x' = g(t, x, u)$, we have finally

$$-J_t(t, x) = \max_u [f(t, x, u) + J_x(t, x) g(t, x, u)]. \tag{7}$$

This is the fundamental partial differential equation obeyed by the optimal value function $J(t, x)$. It is often referred to as the Hamilton-Jacobi-Bellman equation. One finds the maximizing u in terms of t, x, and the unknown function J_x and then substitutes the result into (7) to get the partial differential equation to be solved with boundary condition (3).

Note that the expression on the right of (7), to be maximized with respect to u, looks like the Hamiltonian, except that $J_x(t, x)$ is playing the role of the multiplier $\lambda(t)$. However, recall our interpretation of $\lambda(t)$ as the marginal valuation of the state variable x! Indeed $\lambda(t) = J_x(t, x(t))$, and the right side of (7) is exactly our old directive that the Hamiltonian be maximized with respect to u. Note also that (7) is exactly the same as (I.12.29).

The differential equation of optimal control for $\lambda(t)$ can be derived from (7) as well. Let u provide the maximum in (7), so $f_u + J_x g_u = 0$. Also, since (7) must hold for any x, it must hold if x is modified slightly. Thus, the partial derivative of (7) with respect to x must be zero (with u chosen optimally in terms of t, x, and J_x):

$$-J_{tx} = f_x + J_{xx} g + J_x g_x. \tag{8}$$

The total derivative of J_x is

$$dJ_x(t, x)/dt = J_{xt} + J_{xx} g. \tag{9}$$

Writing $\lambda = J_x$ and substituting for J_{tx} from (9) into (8) gives the familiar differential equation

$$-\lambda' = f_x + \lambda g_x.$$

Hence, with sufficient differentiability, the dynamic programming approach can be used to develop the necessary conditions of optimal control. Alternatively, one can also use the fundamental partial differential equation (7) directly (although we do not recommend it for most problems).

Example. To illustrate, consider

$$\begin{aligned} &\min \int_0^\infty e^{-rt}(ax^2 + bu^2)\,dt \\ &\text{subject to} \qquad x' = u, \quad x(0) = x_0 > 0, \end{aligned} \tag{10}$$

where $a > 0$, $b > 0$. With $f = e^{-rt}(ax^2 + bu^2)$ and $g = u$, (7) becomes

$$-J_t = \min_u\left[e^{-rt}(ax^2 + bu^2) + J_x u\right]. \tag{11}$$

Differentiate to find the optimal u:

$$2e^{-rt}bu + J_x = 0, \qquad \text{so} \quad u = -J_x e^{rt}/2b. \tag{12}$$

Substituting for u from (12) into (11) yields

$$-J_t = e^{-rt}\left(ax^2 + J_x^2 e^{2rt}/4b\right) - J_x^2 e^{rt}/2b.$$

Collecting terms and multiplying through by e^{rt} gives

$$ax^2 - J_x^2 e^{2rt}/4b + e^{rt} J_t = 0. \tag{13}$$

To solve the partial differential equation, we propose a general form of the solution to see if there is some set of parameter values for which the proposed solution satisfies the partial differential equation. Let us "try"

$$J(t, x) = e^{-rt} A x^2, \tag{14}$$

where A is a constant to be determined. From the problem statement (10), the optimal value must be positive, so $A > 0$. For (14) compute

$$J_t = -re^{-2rt} A^2 x^2 \text{ and } J_x = 2e^{-rt} Ax$$

Substituting into (13) gives

$$A^2/b + rA - a = 0. \tag{15}$$

Thus, (14) solves (13) if A is the positive root of the quadratic equation (15); that is,

$$A = \left[-r + \left(r^2 + 4a/b\right)^{1/2}\right] b. \tag{16}$$

The optimal control is now determined from (14) and (12) to be

$$u = -Ax/b. \tag{17}$$

Note that (17) gives the optimal control in terms of the state variable; this is the so-called *feedback* form. If solution in terms of t is desired, one can recall that $x' = u = -Ax/b$ and solve this differential equation for $x(t) = x_0 e^{-At/b}$, from which u is readily determined.

A simpler form of the optimality conditions is available for infinite horizon autonomous problems, such as in the preceding example. An infinite horizon autonomous problem can be written in the form

$$\max \int_0^\infty e^{-rt} f(x, u)\, dt \qquad \text{subject to} \quad x' = g(x, u). \tag{18}$$

Hence,

$$J(t_0, x_0) = \max \int_{t_0}^\infty e^{-rt} f(x, u)\, dt$$

$$= e^{-rt_0} \max \int_{t_0}^\infty e^{-r(t-t_0)} f(x, u)\, dt.$$

The value of the integral on the right depends on the initial state, but is independent of the initial time, i.e. it only depends on elapsed time. Now let

$$V(x_0) = \max \int_{t_0}^\infty e^{-r(t-t_0)} f(x, u)\, dt.$$

Then

$$J(t, x) = e^{-rt}V(x),$$
$$J_t = -re^{-rt}V(x),$$
$$J_x = e^{-rt}V'(x).$$

Substituting into (7) and multiplying through by e^{rt} yields the basic ordinary differential equation

$$rV(x) = \max_u[f(x, u) + V'(x)g(x, u)] \tag{19}$$

obeyed by the optimal current value function $V(x)$ associated with problem (18).

EXERCISES

1. Solve the example of this section by using (19).

2. Solve by dynamic programming:

$$\min \int_0^T (c_1u^2 + c_2x)\, dt$$

$$\text{subject to} \quad x' = u, \quad x(0) = 0, \quad x(T) = B.$$

[Hints: $c_2x - J_x^2/4c_1 + J_t = 0$. Try a solution of the form

$$J(t, x) = a + bxt + hx^2/t + kt^3,$$

where a, b, h, and k are constants to be determined. Compare the solution with that found earlier by other methods.]

FURTHER READING

For an introduction to dynamic programming, see Bellman, Howard, or Nemhauser. Bellman and Dreyfus shows the relationships among dynamic programming, the calculus of variations, and optimal control. Beckmann discusses some applications of dynamic programming to economics.

Our discussion has focused on continuous time dynamic programming. In applications, dynamic programming's greater strength may be in discrete problems, particularly where the underlying functions are not smooth and "nice." The dynamic programming approach permits very efficient computer algorithms to be developed for such problems. This topic is of great interest but aside from the primary thrust of this book. See Stokey and Lucas and any of the references just mentioned for an introduction to this area.

Section 22

Stochastic Optimal Control

Stochastic features have appeared in many of our examples, including uncertain consumer or machine lifetime and uncertain rival behavior. The literature contains further examples, with uncertainty regarding, for example, the timing of illness, catastrophe, expropriation, or technical breakthrough. These applications involve a known probability distribution function that is typically a state variable and occasionally a function of a state variable. The widespread use of this approach attests to its serviceability, but it will not do for all stochastic problems of interest. Another approach to stochastic modeling, which has become especially prevalent in modern finance, is the topic of this section.

The movement of the state variable may not be fully deterministic but it may be subject to stochastic disturbance. To consider such problems, we make some assertions about the stochastic calculus of Itô, which forms the basis for the analysis. The dynamic programming tools of the last section will also be used; since a random element enters into the movement of the system, the optimal control must be stated in feedback form, in terms of the state of the system, rather than in terms of time alone (because the state that will be obtained cannot be known in advance, due to the stochastic disturbance).

Instead of the usual differential equation $x' = g(t, x, u)$, we have the formal stochastic differential equation

$$dx = g(t, x, u)\, dt + \sigma(t, x, y)\, dz, \tag{1}$$

where dz is the increment of a stochastic process z that obeys what is called *Brownian motion* or *white noise* or is a *Wiener process*. The expected rate of change is g, but there is a disturbance term. Briefly, for a Wiener process z, and for any partition $t_0, t_1, t_2, \ldots$ of the time interval, the random variables $z(t_1) - z(t_0), z(t_2) - z(t_1), z(t_3) - z(t_2), \ldots$ are independently and normally distributed with mean zero and variances $t_1 - t_0, t_2 - t_1, t_3 - t_2, \ldots$, respectively. It turns out that the differential elements dt and dz have the following multiplication table:

$$\begin{array}{c|cc} & dz & dt \\ \hline dz & dt & 0 \\ dt & 0 & 0 \end{array} \tag{2}$$

Since $(dz)^2 = dt$, the differential of a function

$$y = F(t, z),$$

where z is a Wiener process, will include a second partial derivative. In particular, in expanding by Taylor series, we get

$$dy = F_t\, dt + F_z\, dz + (1/2)F_{tt}(dt)^2 + F_{tz}\, dt\, dz + (1/2)F_{zz}(dz)^2 + h.o.t.$$

so

$$dy = \left(F_t + \tfrac{1}{2}F_{zz}\right) dt + F_z\, dz \tag{3}$$

on using the multiplication table (2). Subscripts indicate partial derivatives.

Similarly, if

$$y = F(t, x), \tag{4}$$

where x obeys (1), then y is stochastic since x is. The stochastic differential of y is found using Taylor's theorem and (2). We get

$$dy = F_t\, dt + F_x\, dx + (1/2)F_{xx}(dx)^2, \tag{5}$$

where dx is given by (1). This rule (5) is known as *Itô*'s theorem. Of course, (3) is a special case of (5). One can substitute from (1) into (5) and use (2) to simplify the results, obtaining the equivalent statement

$$dy = \left(F_t + F_x g + (1/2)F_{xx}\sigma^2\right) dt + F_x \sigma\, dz. \tag{6}$$

The Itô stochastic calculus extends to many variables, with many stochastic processes. For instance, let $\boldsymbol{x} = [x_1, \ldots, x_n]$ and

$$dx_i = g_i(t, x)\, dt + \sum_{j=1}^{n} \sigma_{ij}(t, x)\, dz_j, \qquad i = 1, \ldots n. \tag{7}$$

Let the Wiener processes dz_i and dz_j have correlation coefficient ρ_{ij}. Finally, let $y = F(t, x)$. Then Itô's theorem gives the rule for the stochastic differential:

$$dy = \sum_{i=1}^{n} (\partial F/\partial x_i)\, dx_i + (\partial F/\partial t)\, dt$$
$$+ (1/2)\sum_{i=1}^{n}\sum_{j=1}^{n} (\partial^2 F/\partial x_i\, \partial x_j)\, dx_i\, dx_j, \tag{8}$$

where the dx_i are given in (7) and the products $dx_i\, dx_j$ are computed using (7) and the multiplication table

$$dz_i\, dz_j = \rho_{ij}\, dt, \qquad i, j = 1, \ldots, n,$$

$$dz_i\, dt = 0, \qquad i = 1, \ldots n, \tag{9}$$

where the correlation coefficient $\rho_{ii} = 1$ for all $i = 1, \ldots, n$.

The rules for integration of a stochastic differential equation are different from the ordinary calculus rules. For example, in the usual case, if $dy = y\, dx$, then $y = e^x$. But in the stochastic calculus, the differential equation $dy = y\, dz$ has solution $y = e^{z - t/2}$. The method of verification is the same in each case; one differentiates the proposed solution by the appropriate rules and checks whether the differential equation is satisfied. To verify the stochastic example, write

$$y = e^{z - t/2} = F(z, t)$$

and differentiate using (3). Since

$$F_z = y, \qquad F_{zz} = y, \qquad \text{and} \qquad F_t = -y/2,$$

we get

$$dy = (-y/2 + y/2)\, dt + y\, dz = y\, dz$$

as claimed.

As a second example, the stochastic differential equation

$$dx = ax\, dt + bx\, dz$$

has solution

$$x(t) = x_0 e^{(a - b^2/2)t + bz}.$$

To verify, we denote the right side $F(t, z)$, compute

$$F_t = (a - b^2/2)x,$$

$$F_z = bx, \qquad F_{zz} = b^2 x,$$

and plug into (3) to get

$$dx = \left[(a - b^2/2)x + b^2 x/2\right] dt + bx\, dz$$

$$= ax\, dt + bx\, dz$$

as claimed.

As a third example, we seek the stochastic differential of $x \equiv Q/P$ where Q and P obey

$$dP/P = a\, dt + b\, dz, \tag{10}$$

$$dQ/Q = c\, dt, \tag{11}$$

with a, b, c given constants. In the ordinary deterministic case, we should have

$$dx/x = dQ/Q - dP/P.$$

However, using Itô's theorem (8), we compute

$$\begin{aligned} dx &= x_P\, dP + x_Q\, dQ + (1/2)x_{PP}(dP)^2 + x_{PQ}\, dP\, dQ + (1/2)x_{QQ}(dQ)^2 \\ &= -(Q/P^2)\, dP + dQ/P + (Q/P^3)(dP)^2 - dPdQ/P^2. \end{aligned}$$

Multiplying through by $1/x = P/Q$ gives

$$dx/x = -dP/P + dQ/Q + (dP/P)^2 - (dP/P)(dQ/Q).$$

Substituting from (10) and (11) and simplifying gives

$$dx/x = (c + b^2 - a)\, dt - b\, dz.$$

Now consider the stochastic optimal control problem.

$$\begin{aligned} &\max \quad E\left\{\int_0^T f(t, x, u)\, dt + \phi(x(T), T)\right\} \\ &\text{subject to} \qquad dx = g(t, x, u)\, dt + \sigma(t, x, u)\, dz, \quad x(0) = x_0, \end{aligned} \tag{12}$$

where the function E refers to the expected value. To find necessary conditions for solution, we follow the method of the preceding section. Define $J(t_0, x_0)$ to be the maximum expected value obtainable in a problem of the form of (12), starting at time t_0 in state $x(t_0) = x_0$:

$$\begin{aligned} J(t_0, x_0) &= \max{}_u E\left(\int_{t_0}^T f(t, x, u)\, dt + \phi(x(T), T)\right) \\ &\text{subject to} \qquad dx = g\, dt + \sigma\, dz, \quad x(t_0) = x_0. \end{aligned} \tag{13}$$

Then, as in (20.2)–(20.6), we obtain

$$J(t, x) \cong \max{}_u E(f(t, x, u)\, \Delta t + J(t + \Delta t, x + \Delta x)). \tag{14}$$

Assuming that J is twice continuously differentiable, we expand the function on the right around (t, x):

$$\begin{aligned} J(t + \Delta t, x + \Delta x) &= J(t, x) + J_t(t, x)\, \Delta t + J_x(t, x)\, \Delta x \\ &\quad + (1/2)J_{xx}(t, x)(\Delta x)^2 + h.o.t. \end{aligned} \tag{15}$$

But recalling the differential constraint in (12), we have approximately

$$\begin{aligned} &\Delta x = g\, \Delta t + \sigma\, \Delta z, \\ &(\Delta x)^2 = g^2(\Delta t)^2 + \sigma^2(\Delta z)^2 + 2g\sigma\, \Delta t\, \Delta z = \sigma^2\, \Delta t + h.o.t., \end{aligned} \tag{16}$$

where use has been made of (2) and the foresight that we will soon divide by Δt and then let $\Delta t \to 0$. Substitute from (16) into (15) and then put the result into (14) to get

$$J(t, x) = \max_u E(f\Delta t + J + J_t \Delta t + J_x g \Delta t + J_x \sigma \Delta z + (1/2) J_{xx}\sigma^2 \Delta t + h.o.t.). \tag{17}$$

Note that the stochastic differential of J is being computed in the process. Now take expectation in (17); the only stochastic term in (17) is Δz and its expectation is zero by assumption. Also, subtract $J(t, x)$ from each side, divide through by Δt, and finally let $\Delta t \to 0$ to get

$$-J_t(t, x) = \max_u(f(t, x, u) + J_x(t, x) g(t, x, u) + (1/2)\sigma^2 J_{xx}(t, x)). \tag{18}$$

This is the basic condition for the stochastic optimal control problem (12). It has boundary condition

$$J(T, x(T)) = \phi(x(T), T). \tag{19}$$

Conditions (18) and (19) should be compared with (21.7) and (21.3).

To illustrate the use of these necessary conditions, consider a stochastic modification of the example of Section 21:

$$\begin{aligned} &\min \quad E\int_0^\infty e^{-rt}(ax^2 + bu^2)\,dt \\ &\text{subject to} \quad dx = u\,dt + \sigma\,dz, \quad a > 0, \quad b > 0, \quad \sigma > 0. \end{aligned} \tag{20}$$

Note that σ is a constant parameter here. The function $\sigma(t, x, u) = \sigma x$. Substituting the special form of (20) into (18) yields

$$-J_t = \min_u(e^{-rt}(ax^2 + bu^2) + J_x u + (1/2)\sigma^2 x^2 J_{xx}). \tag{21}$$

The minimizing u is

$$u = -J_x e^{rt}/2b. \tag{22}$$

Substituting (22) into (21) and simplifying gives

$$-e^{rt}J_t = ax^2 - J_x^2 e^{2rt}/4b + (1/2)\sigma^2 x^2 J_{xx} e^{rt}. \tag{23}$$

Now try a solution to this partial differential equation of the same form as that which worked in the deterministic case:

$$J(t, x) = e^{-rt}Ax^2. \tag{24}$$

Compute the required partial derivatives of (24), substitute into (23), and simplify to find that A must satisfy

$$A^2/b + (r - \sigma^2)A - a = 0. \tag{25}$$

Since only the positive root made sense before, we take the positive root here:

$$A = \left\{\sigma^2 - r + \left[(r-\sigma^2)^2 + 4a/b\right]^{1/2}\right\} b/2. \tag{26}$$

Again using (24) in (22) gives the optimal control,

$$u = -Ax/b, \tag{27}$$

where A is given by (26). Compare this with the findings of Section 21.

A simpler form of (18) is available for problems that are autonomous and have an infinite time horizon. The optimal expected return can then be expressed in current value terms independently of t. The procedure was followed in the preceding section and the results are analogous. Let

$$V(x_0) = \max E \int_{t_0}^{\infty} e^{-r(t-t_0)} f(x,u)\, dt$$

$$\text{subject to} \qquad dx = g(x,u)\, dt + \sigma(x,u)\, dz, \quad x(t_0) = x_0, \tag{28}$$

so

$$J(t,x) = e^{-rt} V(x). \tag{29}$$

Substituting from (29) into (18) gives

$$rV(x) = \max_u \left(f(x,u) + V'(x) g(x,u) + (1/2)\sigma^2(x,u) V''(x)\right), \tag{30}$$

which should be compared with (21.19).

The next example, based on work by Merton, concerns allocating personal wealth among current consumption, investment in a sure or riskless asset, and investment in a risky asset in the absence of transaction costs. Let

W = total wealth,
w = fraction of wealth in the risky asset,
s = return on the sure asset,
a = expected return on the risky asset, $a > s$,
σ^2 = variance per unit time of return on risky asset,
c = consumption,
$U(c) = c^b/b$ = utility function, $b < 1$.

The change in wealth is given by

$$dW = \left[s(1-w)W + awW - c\right] dt + wW\sigma\, dz. \tag{31}$$

The deterministic portion is composed of the return on the funds in the sure asset, plus the expected return on the funds in the risky asset, less consumption. The objective is maximization of the expected discounted utility stream.

For convenience, we assume an infinite horizon:

$$\max \quad E\int_0^\infty (e^{-rt}c^b/b)\,dt \qquad\qquad (32)$$
$$\text{subject to } (31) \quad \text{and} \quad W(0) = W_0.$$

This is an infinite horizon autonomous problem with one state variable W and two controls c and w. Although (30) was developed for a problem with just one state variable and one control, it is readily extended to the present case. Using the specifications of (31) and (32), (30) becomes

$$rV(W) = \max_{c,w}\big(c^b/b + V'(W)\left[s(1-w)W + awW - c\right]$$
$$+ (1/2)w^2W^2\sigma^2V''(W)\big). \qquad (33)$$

Calculus gives the maximizing values of c and w in terms of parameters of the problem, the state W, and the unknown function V:

$$c = \left[V'(W)\right]^{1/(b-1)}, \qquad w = V'(W)(s-a)/\sigma^2WV''(W). \quad (34)$$

We assume the optimal solution involves investment in both assets at all times. Substituting from (34) and (33) and simplifying gives

$$rV(W) = (V')^{b/\,b-1)}(1-b)/b + sWV' - (s-a)^2(V')^2/2\sigma^2V''. \quad (35)$$

Let us "try" a solution to this nonlinear second order differential equation of the form

$$V(W) = AW^b, \qquad (36)$$

where A is a positive parameter to be determined. Compute the required derivatives of (36) and substitute the results into (35). After simplification, one gets

$$Ab = \left\{\left[r - sb - (s-a)^2b/2\sigma^2(1-b)\right]/(1-b)\right\}^{b-1}. \qquad (37)$$

Hence the optimal current value function is (36), with A as specified in (37). To find the optimal control functions, use (36) and (37) in (34):

$$c = W(Ab)^{1/(b-1)}, \qquad w = (a-s)/(1-b)\sigma^2. \qquad (38)$$

The individual consumes a constant fraction of wealth at each moment. The optimal fraction depends on all the parameters; it varies directly with the discount rate and with the riskiness of the risky asset. The optimal division of wealth between the two kinds of assets is a constant, independent of total wealth. The portion devoted to the risky asset varies directly with the expected return of the risky asset and inversely with the variance of that return.

EXERCISES

1. Solve problem (20) using the current optimal value function $V(x)$.

2. Find a control function $c(t)$ to

$$\max \quad E\int_0^\infty e^{-rt}c^a(t)\,dt$$

$$\text{subject to} \qquad dx = (bx - c)\,dt + hx\,dz, \quad x(0) = x_0 > 0,$$

where $z(t)$ is Wiener.
SOLUTION: $c(t) = [(r - ab)/(1 - a) + ah^2/2]x(t)$.

FURTHER READING

For stochastic problems in this text, review examples and exercises in Section I8, I9, I11, and II10 and II15, for instance. See also Cropper (1976, 1977), Dasgupta and Heal, Kamien and Schwartz (1971a, 1974a, 1977a), Long (1975), Raviv, and Robson.

Dreyfus (1965, pp. 215–224) gives a derivation of the necessary conditions of this section and some examples. Arnold provides a good readable treatment of the stochastic calculus. See Merton (1969) for a more thorough discussion of the example of this section and the methodology of solution and for analysis of the more realistic case of an individual with a finite planning horizon. Brock (1976) is an excellent "user's manual." Exercise 2 is discussed fully by Brock. Malliaris and Brock is the expanded sequel, that provides a proof of Itô's theorem.

For further applications of the stochastic optimal control, see Merton (1971) (consumption and asset management), Fischer (index bonds), Gonedes and Lieber (production planning). Constantinides and Richard (cash management), and Tapiero (advertising).

Section 23

Differential Games

Up to this point the dynamic optimization methods studied related to problems involving a sole decision maker carrying out the optimization. In particular, in the case of an optimal control problem it is a single decision maker's choice of the control variable that advances the state of the system. However, there are many situations in which the state of the system is determined by more than a sole individual. For example, if the stock of fish in a body of water represents the state of the system, then the amount of fish harvested by a number of individuals determines the state. Similarly, the probability that someone will successfully develop a patentable new product or production process depends on how much each one of many firms invests in research and development. Finally, unless its seller is a monopolist, a product's price at a point in time depends on the output level of each of its producers.

Situations in which the joint actions of several individuals, each acting independently, either effect a common state variable or each other's payoffs through time, are modeled as differential games. In particular, we confine our attention to noncooperative games, those in which the individuals, referred to as players, do not explicitly cooperate in selecting the values of their control variables, and for which the state of the system changes according to one or more differential equations. Thus, in a differential game, the players interact repeatedly through time. However, their interaction is not a simple repetition of the original game, as the initial conditions for each game differ through the continuous change in the state.

In the case of two players, 1 and 2, a typical differential game is posed as player 1 choosing his control, u_1, to maximize

$$J^1(u_1, u_2) = \int_{t_0}^{t_1} f^1(t, x_1(t), x_2(t), u_1(t), u_2(t))\, dt \tag{1}$$

while player 2 chooses u_2 to maximize

$$J^2(u_1, u_2) = \int_{t_0}^{t_1} f^2(t, x_1(t), x_2(t), u_1(t), u_2(t))\, dt \tag{2}$$

with both maximization problems subject to the state equations

$$x_i'(t) = g^i(t, x_1(t), x_2(t), u_1(t), u_2(t)) \tag{3}$$
$$x_i(t_0) = x_{i0}, \quad x_i(t_i) \text{ free}, \quad i = 1, 2.$$

As in a standard optimal control problem, the controls are assumed to be piecewise continuous, while f^i and g^i are assumed to be known and continuously differentiable functions of their four independent arguments. Each player is aware of the other's presence and how the other's choice of his control variable affects the state equations. As a result, the individual player must take into account the other player's choice of his control variable in choosing his own. That is, each player chooses his control variable so as to maximize his payoff for every possible choice of the other player's control variable. The two players are assumed to choose their control variables simultaneously. But this appears to require that each player guess what the other is going to do in order for him to make his optimal choice. This guessing in turn leads to the possibility that one or both of the players would like to revise his decision given what was actually done by the other. When there is no incentive for either player to revise the choice of his control variable, then the choices are said to be in equilibrium. In particular, they are said to be in a *Nash equilibrium* if

$$J^1(u_1^*, u_2^*) \geq J^1(u_1, u_2^*) \tag{4a}$$

and

$$J^2(u_1^*, u_2^*) \geq J^2(u_1^*, u_2) \tag{4b}$$

where u_1^*, u_2^* are referred to as player 1's and player 2's respective equilibrium strategies. According to 4(a, b), each player is doing the best he can given the other's strategy, the choice of the control variable, and neither has an incentive to deviate from his choice. This definition of equilibrium together with each player's knowledge of both objective functions in fact enables each player to avoid having to guess what the other will do by figuring out what the equilibrium has to be on his own and choosing his equilibrium strategy accordingly. In effect, each player can compute not only what his best response is for each of his rival's strategies but also what his rival's best response is to each of his own strategies. Moreover, each player knows that the other has also conducted this thought experiment and so they can both select their equilibrium strategies immediately. The obvious strong assumption in this line of reasoning is that there is only one Nash equilibrium. If there are several Nash equilibria then neither player can be certain about which one the other will focus on. There is intense research and an extensive literature on the question of how players might both choose one Nash equilibrium among several.

The strategies, in the form of selecting the control variables u_1 and u_2, most commonly employed in the application of the theory of differential games are either *open-loop* or *feedback*. Open-loop strategies are ones for which each player chooses all the values of his control variable for each point in time at the outset of the game. That is, $u_1 = u_1(t)$; the value of the control at each point in time is only a function of time. This type of strategy implies that each player has committed to his entire course of action at the beginning of the game and will not revise it at any subsequent point in time. The Nash equilibrium open-loop strategies are relatively easy to determine as they involve a straightforward application of the standard optimal control methods. That is, for each player form a Hamiltonian:

$$\begin{aligned} H^i(t, x_1(t), x_2(t), u_1(t), u_2(t), \lambda_1^i(t), \lambda_2^i(t)) \\ = f^i + \lambda_1^i g^1 + \lambda_2^i g^2, \qquad i = 1,2 \end{aligned} \tag{5}$$

corresponding to (1), (2) and (3). The Nash equilibrium conditions (4a, b) in terms of the Hamiltonians are:

$$\begin{aligned} &H^1(t, x_1(t), x_2(t), u_1^*(t), u_2^*(t), \lambda_1^1(t), \lambda_2^1(t)) \\ &\quad \geq H^1(t, x_1(t), x_2(t), u_1(t), u_2^*(t), \lambda_1^1(t), \lambda_2^1(t)) \\ &H^2(t, x_1(t), x_2(t), u_1^*(t), u_2^*(t), \lambda_1^2(t), \lambda_2^2(t)) \\ &\quad \geq H^2(t, x_1(t), x_2(t), u_1^*(t), u_2(t), \lambda_1^2(t), \lambda_2^2(t)) \end{aligned}$$

Applying the standard necessary conditions of optimal control theory to (5) gives rise to the equations

$$H^i_{u_i} = 0, \quad \lambda_j^{i\prime}(t) = -\partial H^i/\partial x_1, \qquad i = 1,2, \quad j = 1,2 \tag{6}$$

which together with (3) yield eight equations for determining the eight functions $u_1^*(t)$, $u_2^*(t)$, $\lambda_1^{i*}(t)$, $\lambda_2^{i*}(t)$, $x_1^*(t)$, $x_2^*(t)$, $i = 1,2$.

The relative simplicity of finding or characterizing open-loop strategies has caused them to be frequently employed. However, the implicit requirement that each player commit to his entire sequence of actions through time at the outset is thought to be rarely satisfied in real world situations. It is commonly held that a player will have the option and incentive to revise his actions through time as the game evolves. Thus, it is thought that a more appropriate way of modeling a player's behavior is to suppose that he can condition his action at each point in time on the basis of the state of the system at that point in time. This class of strategies is referred to as *feedback* strategies and is characterized by the requirement that $u_i = u_i(t, x_1(t), x_2(t))$, $i = 1, 2$. That is, the control at each point in time be a function of both time and the state of the system $x_1(t)$, $x_2(t)$ at that time. Conceptually, what distinguishes feedback strategies from open-loop strategies is that a feedback strategy consists of a contingency plan that indicates what the best thing to do is for each value of the state variable at each point in time rather than just what the best thing to do is at each point in time at the outset of the game. While there are circumstances for which the two types of strategies coincide in that the actions taken are identical at each point in time, in general they do not. Moreover, feedback strategies have the property of being *subgame perfect*. This means that after each player's actions have caused the state of the system to evolve from its initial state to a new state, the continuation of the game with this new state thought of as the initial state may be regarded as a subgame of the original game. A feedback strategy allows the players to do their best in this subgame even if the initial state of the subgame evolved through prior suboptimal actions. Thus, a feedback strategy is optimal not only for the original game as specified by its initial conditions but also for every subgame evolving from it

While the concept of feedback strategy is more appealing and more general in the sense it could be one that depends only on time and not on the state of the system at that time—that is, it subsumes open-loop strategies, it is more difficult to compute. As with open-loop strategies, feedback strategies must satisfy the Nash equilibrium conditions (4a, b). The computational difficulty arises in connection with the co-state variables $\lambda_1^i(t)$, $\lambda_2^i(t)$, $i = 1, 2$. To compute the feedback strategies the Hamiltonians corresponding to problems (1), (2) and (3) are formed.

$$H^i(t, x_1(t), x_2(t), u_1(t, x_1(t), x_2(t)), u_2(t, x_1(t)x_2(t)), \lambda_1^i(t), \lambda_2^i(t))$$
$$= f^i + \lambda^i g^1 + \lambda_2^i g^2, \qquad i = 1, 2. \tag{7}$$

The first set of necessary conditions are

$$H_{u_i}^i = 0, \qquad i = 1, 2. \tag{8}$$

However, the second set of necessary conditions are

$$\lambda_j^{1\prime}(t) = -\partial H^1/\partial x_j - (\partial H^1/\partial u_2)(\partial u_2^*/\partial x_j)$$
$$\lambda_j^{2\prime}(t) = -\partial H^2/\partial x_j - (\partial H^2/\partial u_1)(\partial u_1^*/\partial x_j)$$
$$j = 1, 2 \tag{9}$$

The second term on the right side of (9) appears because it has been posited that the players employ feedback strategies and therefore that u_j is function of $x_1(t)$, $x_2(t)$. The term $(\partial H^i/\partial u_i)(\partial u_i^*/\partial x_i)$ does not appear on the right side of (9) because $\partial H^i/\partial u_i = 0$ according to the necessary condition (8). The feedback strategies are difficult to compute because of the presence of the last term, referred to as the interaction term, on the right side of (9). Intuitively, finding player 1's optimal feedback strategy $u_1^*(t, x_1(t), x_2(t))$ requires that player 2's optimal feedback strategy $u_2^*(t, x_1(t), x_2(t))$ be known which, in turn, requires that player 1's be known, and so on. Yet, despite this difficulty, feedback strategies have been computed as the examples below will illustrate. A further requirement for a feedback strategy is that it be optimal for any initial conditions t_0, x_0. This, of course, is consistent with the requirement that a feedback strategy be optimal from any subsequent time and value of the state variable forward.

The nature of a feedback strategy is in the spirit of the principle of optimality of dynamic programming. Recall that in derivation of the fundamental partial differential equation, the Hamilton-Jacobi-Bellman equation, obeyed by the optimal value function (21.7), the requirement that the objective function be optimal for each time and value of the state variable forward was employed. Thus, (21.7) is commonly employed to find feedback strategies. This, of course, involves the simultaneous solution of two partial differential equations, a tricky task indeed.

Example 1. This example will serve to illustrate both how open-loop and feedback solutions are computed and the difference between the two. Suppose two firms produce an identical product. The cost of producing it is governed by the total cost function

$$C(u_i) = cu_i + u_i^2/2, \qquad i = 1,2 \tag{10}$$

where $u_i(t)$ refers to the i-th firm's production level at time t. Each firm supplies all it produces at time t to the market. At each point in time the firms face a common price $p(t)$ at which they can sell their product. However, the amount they jointly supply at time t, $u_1(t) + u_2(t)$, determines the rate at which price changes at time t, $dp(t)/dt = p'(t)$. The relationship between the total amount supplied and the change in price at time t is described by the differential equation,

$$p'(t) = s[a - u_1(t) - u_2(t) - p(t)], \qquad p(0) = p_0. \tag{11}$$

Thus, $p(t)$ is the state variable. The parameter s refers to the speed at which the price adjusts to the price corresponding to the total quantity supplied on the demand function. The full meaning of this will become apparent shortly. At this point it should be observed that if $s = 0$, then $p'(t) = 0$.

Each firm chooses its level of output $u_i(t)$ so as to maximize

$$J^i = \int_0^\infty e^{-rt}(p(t)u_i(t) - C_i(u_i(t))]\, dt, \qquad i = 1,2 \tag{12}$$

subject to (10) and (11). The role of s can now be further understood by solving for $p(t)$ in (11) and substituting into (12) to get

$$J^i(u_1, u_2) = \int_0^\infty e^{-rt}[(a - u_1 - u_2)u_i - p'u_i/s - C(u_i)]\, dt. \tag{13}$$

From (13) it is evident that in the limit as $s \to \infty$ the second term in the integral vanishes and each firm in fact faces a downward sloping linear demand function $a - u_1 - u_2 = \tilde{p}(t)$. Thus, if the speed of adjustment is instantaneous, then the firms are engaged in a simple Cournot duopoly situation repeated at each instant of time. If, on the other hand, there is some lag in the adjustment of price to the contemporaneous quantity produced, then each firm faces a horizontal demand function at each instant of time. That is what the formulation in (12) implies, together with the adjustment equation (11).

To find the open-loop Nash equilibrium for this game we form the current value Hamiltonians

$$H^i = p(t)u_i(t) - C(u_i(t)) + m_i(t)s(a - u_1(t) - u_2(t) - p(t)),\quad i = 1,2 \tag{14}$$

and obtain the necessary conditions

$$H^i_{u_i} = p(t) - c - u_i(t) - m_i(t)s = 0, \qquad i = 1,2 \tag{15}$$

and

$$-m_i'(t) = u_i(t) - m_i(t)(s + r) \qquad \lim_{t\to\infty} e^{-rt}m_i(t) = 0 \tag{16}$$

where (10) was employed in obtaining (15). Now solving (15) for $u_i(t)$, plugging into (16), integrating (16) with the aide of the integrating factor $e^{-(2s+r)t}$, and using the transversality condition to determine the constant of integration, yields

$$m_i(t) = \int_t^\infty e^{-(2s+r)(\tau - t)}(p(\tau) - c)\,d\tau, \qquad i = 1,2. \tag{17}$$

From (17) it is evident that $m_1(t) = m_2(t)$ as both are equal to the same right hand side. It then follows from (15) that $u_1(t) = u_2(t) = u(t)$. Differentiating (15) with respect to time and substituting for $sm'(t)$ from this operation, and for $sm(t)$ from (15), into (16) multiplied by s, yields

$$u'(t) = s[a - u(t) - p(t)] - (s + r)[p(t) - c - u(t)]. \tag{18}$$

Now the stationary open-loop Nash equilibrium strategies are the ones for which $u'(t) = p'(t) = 0$. From setting the right sides of (11) and (18) equal to zero it follows that the stationary open-loop strategies and the product price are

$$u^* = (a - c)(s + r)/[4s + 3r] \tag{19}$$

and

$$p^* = [2(a + c)s + (a + 2c)r]/[4s + 3r], \tag{20}$$

respectively.

The price p^* in (20) that results from the open-loop strategies has an interesting interpretation. Suppose the two firms were engaged in a static

Cournot duopoly game in which the demand function was $\bar{p} = a - u_1 - u_2$, and with cost functions given by (10). Then it is not difficult to show that the equilibrium quantity each firm produces is $u_0 = (a - c)/4$ and the equilibrium price is $p_D = (a + c)/2$. On the other hand, if each firm were to behave naively in the sense of believing that it faces a horizontal demand function, i.e., that neither its own level of output nor its rival's influences the product's price, then each will produce $u_N = (a - c)/3$ and the equilibrium price will be $p_N = (a + 2c)/3$. Now with some algebra applied to (20) it is possible to show that the stationary open-loop equilibrium price is

$$p^* = (4sp_D + 3rp_N)/(4s + 3r). \tag{21}$$

Thus, p^* is a convex combination of the two prices p_D and p_N. Moreover, it follows that

$$\lim_{s\to\infty} p^* = p_D, \qquad \lim_{r\to\infty} p^* = p_N \tag{22}$$

where r and s are held fixed when the respective limits are taken. Thus, when price adjusts instantaneously, $s \to \infty$, the stationary open-loop Nash equilibrium price becomes the duopoly price of the static Cournot duopoly. On the other hand, when $r \to \infty$, each firm discounts future profits completely, and the stationary open-loop equilibrium price approaches the static equilibrium price p_N that would prevail if each firm acted naively. It is also easy to see that $p^* = p_D$ if $r = 0$, the firm's value future profits the same as present profits, and $p^* = p_N$ if $s = 0$, the price is insensitive to the total quantity supplied.

This concludes our analysis of the open-loop Nash equilibrium strategies. We now turn to a characterization of the feedback Nash equilibrium strategies.

To find the feedback strategies $u_i(t, p(t))$ we form for each player the value function

$$rV^i(p) = \max_{u_i}\{(p - c)u_i - u_i^2/2 + sV_p^i(p)[a - p - u_i - u_j]\},$$
$$i = 1,2, \quad i \neq j \tag{23}$$

by recalling (20.19), and where V_p^i refers to the derivative of V^i with respect to p. The maximization with respect to u_i yields

$$u_i(p) = p - c - sV_p^i, \qquad i = 1,2. \tag{24}$$

Note that the control u_i is just a function of p as time does not enter explicitly. Substitution from (24) into (23) yields

$$\begin{aligned} rV^i &= (p - c)(p - c - sV_p^i) - (p - c - sV_p^i)^2/2 \\ &\quad + sV_p^i[a - p - (2p - 2c - sV_p^i - sV_p^j)], \qquad i = 1,2, \quad i \neq j. \end{aligned} \tag{25}$$

Note that in (25) both V_p^i and V_p^j appear. We shall only be concerned here with interior solutions, i.e., those for which $u_i(p) > 0$. This will be assured if

$a > p_0$. Solving the system of differential equations represented by (25) means finding value functions $V^i(p)$ that satisfy them. The following value functions are proposed as solutions,

$$V^i(p) = g_i - E_i p + K_i p^2/2, \qquad i = 1,2 \tag{26}$$

which implies

$$V_p^i = K_i p - E_i. \tag{27}$$

In order for the value functions proposed in (26) to be solutions to (25), the coefficients E_i, K_i and the constant g must have the "right" values. To find them, substitute from (27) into (25) to get

$$\begin{aligned}(1/2)rK_i p^2 &- rE_i p + rg_i \\ &= \left(1/2 - 3sK_i + s^2K_iK_j + (1/2)s^2K_i^2\right)p^2 \\ &\quad + \left[3sE_i - s^2K_iE_i - 2s^2K_iE_j - c + sK_i(a+2c)\right]p \\ &\quad + (1/2)c^2 + \left((1/2)sE_i + sE_j - a - 2c\right)E_i, \qquad i = 1,2.\end{aligned} \tag{28}$$

Now the left and right sides of (28) must be equal for all possible values of p. This means that the coefficients of p^2 and p on both side of (28) must be equal, as well as the constant terms on both sides. Equating the coefficients of p^2 and collecting terms yields

$$s^2K_i^2 + \left(2s^2K_j - 6s - r\right)K_i + 1 = 0, \qquad i = 1,2, \quad i \neq j. \tag{29}$$

By equating the coefficients of p an expression for E_i as a function of K_1, K_2 and c can be found. Finally, equating the constant terms gives g_i as a function of E_1 and E_2.

The next step is to establish that $K_1 = K_2$, from which it will follow that $E_1 = E_2$, $g_1 = g_2$, and ultimately that the optimal feedback strategies are symmetric, i.e., $u_1^*(p) = u_2^*(p)$. To show that $K_1 = K_2$, subtract the two equations in (29) to get

$$(K_1 - K_2)\left[s^2(K_1 + K_2) - (r + 6s)\right] = 0. \tag{30}$$

From (30) it follows that either $K_1 = K_2$ or

$$s^2(K_1 + K_2) = r + 6s. \tag{31}$$

Substituting from (27) into (24) and then into the state equation (11) gives

$$p'(t) - sp\left[s(K_1 + K_2) - 3\right] = s\left[a + 2c - s(E_1 + E_2)\right]. \tag{32}$$

A particular solution to this first order differential equation is obtained by setting $p'(t) = 0$ and solving for $p(t) = \tilde{p}$, a constant. The solution to the homogeneous part is

$$p(t) = ce^{Dt} \tag{33}$$

where $D = s[s(K_1 + K_2) - 3]$, and c is the constant of integration. The complete solution to (31) is

$$p(t) = \tilde{p} + (p_0 - \tilde{p})e^{Dt}. \tag{34}$$

where $c = p_0 - \tilde{p}$ is obtained by setting $t = 0$ in (34). Now in order for $p(t)$ to converge to $\tilde{p}$ as $t \to \infty$, $D < 0$ or $s(K_1 + K_2) < 3$ is required. But by substitution from (31) for $s^2(K_1 + K_2)$ this means that $r + 3s < 0$ is required. However, this cannot hold as r and s are both nonnegative. Thus, the requirement that the product's price converge through time, that a stationary solution exists, rules out the possibility of an asymmetric Nash equilibrium with feedback strategies that are asymmetric, $u_1^*(p) \neq u_2^*(p)$.

Having established that $K_1 = K_2 = K$ we can now solve for the roots of (29). Specifically,

$$\overline{K}, \underline{K} = \left\{r + 6s \pm \left[(r + 6s)^2 - 12s^2\right]^{1/2}\right\}/6s^2. \tag{35}$$

To distinguish between the two roots $\overline{K}$, $\underline{K}$ given in (35) we return to (34), with $K_1 = K_2$ where now $D = s(2sK - 3)$. The requirement that $D < 0$ means that $K < 3/2s$ is required. Now the larger root $\overline{K}$ takes on its smallest value when $r = 0$ in (35). But for $r = 0$, $\overline{K} = (3 + \sqrt{6})/3s > 3/2s$. Thus, the larger root of (35) prevents convergence of $p(t)$. On the other hand, the smaller root $\underline{K}$ achieves its highest value at $r = 0$, as $\partial\underline{K}/\partial r < 0$. At $r = 0$, $\underline{K} = (3 - \sqrt{6})/3s < 3/2s$. Thus, only the smaller root allows for the convergence of $p(t)$.

With $K = \underline{K}$ it follows that

$$E = \left[c - sK(a + c)\right]/(r - 3s^2K + 3s) \tag{36}$$

and

$$u^*(p) = (1 - sK)p + (sE - c). \tag{37}$$

Expression (37) gives the Nash equilibrium feedback strategies. Now, from (32), the particular solution with $K_1 = K_2 = K$, and $E_1 = E_2 = E$ is

$$\tilde{p} = \left[a + 2(c - sE)\right]/\left[2(1 - sK) + 1\right]. \tag{38}$$

This is the stationary feedback Nash equilibrium price.

Our last step is to see what happens to (38) as $s \to \infty$, price adjustment is immediate. Let $\lim_{s\to\infty} sK = \beta$ and $\lim_{s\to\infty} sE = \gamma$. Now for $K = \underline{K}$, $\beta = 1 - \sqrt{2/3}$. From (36), $\gamma = (c - a\beta - 2c\beta)/(3 - 3\beta)$. Thus, from (38)

$$\lim_{s\to\infty} \tilde{p} = (a + 2(c - \gamma))/(3 - 2\beta). \tag{39}$$

Finally, substituting for β and γ in (39) and rearranging terms gives

$$\begin{aligned}\lim_{s\to\infty} \tilde{p} &= \left[(a + 2c)(1 - \beta) + 2(a + c)\right]/3(3 - 2\beta)(1 - \beta) \\ &= \left(p_N + 2p_D\sqrt{2/3}\right)/\left(1 + 2\sqrt{2/3}\right)\end{aligned} \tag{40}$$

where p_N and p_D were defined above.

Recalling (22), and comparing with (40), it becomes evident that the open-loop Nash equilibrium and the feedback Nash equilibrium of this game give significantly different answers when $s \to \infty$. In the open-loop case the equilibrium price converges to the static duopoly price while in the feedback case it converges to a price below it. The reason for this is because when the players play feedback strategies they produce more than when they play open-loop strategies, or the static Cournot duopoly game. The intuitive reason for this is that in (37) as $s \to \infty$ the coefficient of p is $\sqrt{2/3}$, a positive number. Thus, each player increases (decreases) his output with an increase (decrease) in price. What this means is that if one player increases his output, thereby causing price to decline, the other will decrease his output. This tends to offset some but not all of the price decline caused by the first player's output expansion. Therefore, the first player has greater incentive to expand his output than he would if the price decline were not partially offset by the other's output contraction. However, both players know this about the other's behaviour and therefore both expand their respective outputs above the equilibrium levels that would prevail in a static Cournot duopoly equilibrium.

Example 2. Two players are engaged in a race to develop a new product. The winner of the race will receive a patent that will enable her to realize positive profits until the patent expires and imitators drive profits to zero. The present value of the winner's profits that commence at the moment the race ends are denoted by P. Neither player knows the total amount of knowledge, z, it will take to successfully develop the new product. Knowledge is accumulated through time according to the differential equation

$$dz_i/dt = z_i'(t) = u_i(t), \qquad z_1(0) = 0, \quad i = 1,2, \tag{41}$$

where $u_i(t)$ is the level of effort devoted to accumulation of knowledge by firm i at time t, and neither firm has accumulated any knowledge at the outset. Thus, (41) represents the state equations for this game, with state variables $z_i(t)$ and controls $u_i(t)$. The cost of accumulating knowledge at rate $u_i(t)$ is $u_i^2/2$.

As knowledge is accumulated over time and the total amount for successful development of the new product is unknown, its successful development date is unknown. However, the probability of successful development by a given date is known to both firms and supposed to be given by

$$F_i(z_i(t)) = 1 - e^{-\lambda z_i(t)}, \qquad F_i(0) = 0, \quad i = 1, 2, \tag{42}$$

with λ a constant.

According to (42) the probability of sucessful development is zero if no knowledge is accumulated and λ is a constant. From (42) it follows that

$$\begin{aligned} dF_i(z_i(t))/dt &= \lambda z_i'(t) e^{-\lambda z_i(t)} \\ &= \lambda u_i(t) e^{-\lambda z_i(t)} = f_i(z_i) u_i(t) \end{aligned} \tag{43}$$

where (41) was employed to get the final expression and $f_i(z_i)$ refers to the derivative of F_i with respect to z_i. Now from (42) and (43) it follows that the conditional probability of successful development immediately beyond time t given no successful development until time t, the hazard rate, is

$$f_i(z_i)z_i'(t)/(1-F_i) = \lambda u_i(t), \qquad i = 1,2. \tag{44}$$

The problem faced by each player is to maximize with respect to her u_i

$$J^i(u_1,u_2) = \int_0^T [P(1-F_j(z_j))\,dF_i(z_i)/dt$$
$$-e^{-rt}(1-F_1(z_1))(1-F_2(z_2))u_i^2/2]\,dt,$$
$$i \neq j, \quad i = 1,2. \tag{45}$$

The interpretation of (45) is that with probability dF_i/dt player i realizes the payoff P at time t provided the rival has not successfully developed the product by then, probability $(1-F_j)$, less the development costs that are incurred as long as neither she nor her rival has succeeded in developing the product, probability $(1-F_1)(1-F_2)$.

Now we can integrate the first term under the integral in (45) by parts, letting $dv = (dF_i/dt)\,dt$ and $u = 1 - F_j$ to get

$$\int_0^T P(1-F_j)(dF_i/dt)\,dt = P[(1-F_j(z_j(T))]F_i(z_i(T))$$
$$+\int_0^T PF_i(dF_j/dt)\,dt. \tag{46}$$

Substituting from (43) and (46) into (45) gives

$$J^1(u_1,u_2) = \int_0^T [P(1-e^{-\lambda z_i})e^{-\lambda z_j}\lambda u_j - e^{-rt}e^{-\lambda(z_1+z_2)}u_i^2/2]\,dt$$
$$+P[e^{-\lambda z_j(T)} - e^{-\lambda(z_i(T)+z_2(T))}], \qquad i \neq j, \quad i = 1,2 \tag{47}$$

We can now begin to find the feedback Nash equilibrium controls by forming the value function for each player. As we are not dealing with an infinite horizon problem we have to use form (21.7) rather than (21.19). Thus, for the first player

$$-J_t^1(t,z_1,z_2) = \max_{u_1(t,z_1,z_2)}[P(1-e^{-\lambda z_1})e^{-\lambda z_2}\lambda\bar{u}_2$$
$$-e^{-rt}e^{-\lambda(z_1+z_2)}u_1^2/2 + J_{z_1}^1 u_1 + J_{z_2}^1 \bar{u}_2] \tag{48}$$

where J_t^1, $J_{z_1}^1$, $J_{z_2}^1$ refer to the respective partial derivatives of J^1 with respect to t, z_1 and z_2. Note that both $J_{z_1}^1$ and $J_{z_2}^1$ appear in (48) because there are two states equations, (41). Also, the maximization with respect to u_1 is taken with a given value of $u_2 = \bar{u}_2$. Obviously, there is a counterpart to (48) for the second player.

The first order condition for (48) is

$$-e^{-rt}e^{-\lambda(z_1+z_2)}u_1 + J^1_{z_1} = 0. \tag{49}$$

Solving for u_1 in (49) and for u_2 in the counterpart of (49), and substituting for u_1 and u_2 back into (48) gives

$$J^1_t + \left[\left(J^1_{z_1}\right)^2 e^{rt}e^{\lambda(z_1+z_2)}\right]/2 + J^1_{z_2}J^2_{z_2}e^{rt}e^{\lambda(z_1+z_2)}$$
$$+ PJ^2_{z_2}e^{rt}\lambda(e^{\lambda z_1} - 1) = 0. \tag{50}$$

We now propose

$$J^i(t, z_1, z_2) = b(t)e^{-\lambda(z_1+z_2)} + k(t)e^{-\lambda z_j}, \qquad i \neq j, \quad i = 1,2 \tag{51}$$

as a solution to the partial differential equation (50) and its counterpart for player 2. From (51) it follows that

$$J^1_{z_1} = J^2_{z_2} = -\lambda be^{-\lambda(z_1+z_2)} \tag{52.a}$$

$$J^i_{z_j} = -\lambda be^{-\lambda(z_1+z_2)} - \lambda ke^{-\lambda z_j}, \qquad i \neq j, \quad i = 1,2 \tag{52.b}$$

and

$$J^i_t = b'(t)e^{-\lambda(z_1+z_2)} + k'(t)e^{-\lambda z_2} \tag{52.c}$$

where $b'(t)$ and $k'(t)$ refer to the time derivatives of $b(t)$ and $k(t)$, respectively. Substituting from (52.a.b.c.) back into (50) and collecting terms gives

$$e^{-\lambda(z_1+z_2)}\left[b' + b\lambda^2Pe^{rt} + 3b^2\lambda^2e^{rt}/2\right]$$
$$+ e^{-\lambda z_2}\left[k' + (k - P)b\lambda^2e^{rt}\right] = 0. \tag{53}$$

As both $e^{-\lambda(z_1+z_2)}$ and $e^{-\lambda z_2}$ are always positive it follows that each of the bracketed expressions must be zero. This gives rise to two ordinary differential equations

$$b' + b\lambda^2Pe^{rt} + 3b^2\lambda^2e^{rt}/2 = 0 \tag{54}$$

and

$$k' + (k - P)b\lambda^2e^{rt} = 0. \tag{55}$$

Now, from the proposed value function (51) and (47) together with the requirement that

$$J^1(T, z_1(T), z_2(T)) = -Pe^{-\lambda(z_1(T)+z_2(T))} + Pe^{-\lambda z_2(T)}, \tag{56}$$

recall (21.3), it follows that

$$b(T) = -P, \qquad k(T) = P. \tag{57}$$

Thus, (57) provides the boundary conditions for (54) and (55). Expression (54) is known as a Bernoulli equation and can be solved by letting $b(t) = -1/q(t)$, so $b'(t) = q'(t)/q^2$, where $q'(t)$ is the time derivative of $q(t)$. Substituting into (54) and multiplying through by q^2 gives

$$q' - q\lambda^2 Pe^{rt} + 3\lambda^2 e^{rt}/2 = 0, \tag{58}$$

a first order linear differential equation. A particular solution to (58) is $q = 3/2P$. The homogeneous part of the equation has the form $q'/q = \lambda^2 Pe^{rt}$ or $dq/q = \lambda^2 Pe^{rt}\,dt$. Thus, $\ln q = \lambda^2 P\int_0^t e^{rs}\,ds + c_0$, where c_0 is the constant of integration. Carrying out the integration, taking antilogs and combining the solution to the homogeneous equation with the particular solution gives

$$q(t) = 3/2P + c_0 \exp \lambda^2 P(e^{rt} - 1)/r. \tag{59}$$

Using the boundary condition (57) to evaluate the constant of integration gives $q(T) = 1/P$, which upon substitution into (59) yields $c_0 = -(\frac{1}{2}P)\exp[-\lambda^2 P(e^{-rt} - 1)]$. Substituting back into (59) and recalling that $q = -1/b$ finally gives as a solution to (54)

$$b(t) = -2P/\left[3 - \exp\left(P\lambda^2(e^{rt} - e^{rT})/r\right]. \tag{60}$$

Turning to (55) the particular solution is $k = P$. The solution to the homogeneous equation is $k(t) = c_1\exp[-\int_0^t b(s)\lambda^2 e^{rs}\,ds]$, where c_1 is the constant of integration. Combining the solution to the homogeneous equation with this particular solution gives

$$k(t) = c_1\exp\left[-\int_0^t b(s)\lambda^2 e^{rs}\,ds\right] + P. \tag{61}$$

The boundary condition $k(T) = P$, (57), implies that $c_1 = 0$. Thus, $k(T) = P$, i.e., $k(t)$ is constant with respect to time.

Now to find the explicit expression for u_1^*, from (49) we need to know $J_{z_1}^1$. To determine $J_{z_1}^1$ we substitute for $b(t)$ from (60) and $k(t) = P$ into the value function (51), to get

$$J^1 = \frac{-2Pe^{-\lambda(z_1+z_2)}}{3 - \exp m(t)} + Pe^{-\lambda z_2} \tag{62}$$

where $m(t) = P\lambda^2(e^{rt} - e^{-rT})/r$. There is a counterpart to (62) for player 2. From (62) it follows that

$$J_{z_1}^1 = 2\lambda Pe^{-\lambda(z_1+z_2)}/(3 - \exp m(t)]. \tag{63}$$

Finally, combining (63) and (49) gives

$$u_i^*(t, z) = 2P\lambda e^{rt}/\left[3 - \exp\left(P\lambda^2(e^{rt} - e^{rT})/r\right)\right], \qquad i = 1,2 \tag{64}$$

the feedback Nash equilibrium strategies. It is evident from (64) that $u_1^* = u_2^*$, and that u_i^* is independent of the state variables z_1 and z_2, i.e., it only depends on time. Thus, in this case the feedback controls coincide with the open-loop controls.

Example 3. Suppose the evolution of the stock of fish, in a particular body of water is governed by the differential equation

$$dx/dt = x'(t) = \alpha x(t) - bx(t)\ln x(t), \tag{65}$$

where $x(t)$ and $x'(t)$ refer to the stock of fish and its rate of change at time t, respectively, and $\alpha > 0$, $b > 0$. It is also assumed that $x(t) \geq 2$. We need at least two fish, one of each gender, for the fish population to survive. According to (65), the stock $x(t)$ generates $\alpha x(t)$ births and $bx(t)\ln x(t)$ deaths at each point in time. The steady state fish stock, $x'(t) = 0$, consistent with (65) is $x(t) = e^{\alpha/b}$.

Now suppose that two fishermen harvest fish from this stock. Each fisherman's catch is directly related to the level of effort he devotes to this activity and the stock of fish. Thus,

$$c_i(t) = w_i(t)x(t), \qquad i = 1,2, \tag{66}$$

where c_i refers to the i-th fisherman's catch and $w_i(t)$ the effort level he expends, at time t. The fisherman's harvesting activity reduces the fish stock and is incorporated into the state equation (65) to yield

$$x'(t) = (\alpha - w_1 - w_2)x - bx\ln x. \tag{67}$$

Each fisherman derives satisfaction from his catch according to the utility function

$$u_i(c_i(t)) = a_i \ln c_i(t) = a_i \ln w_i(t)x(t), \qquad i = 1,2, \tag{68}$$

where $w_i(t)x(t) \geq 0$ and $a_i > 0$. It will prove convenient for computation to let

$$y(t) = \ln x(t) \tag{69}$$

so that $dy/dt = y'(t) = x'(t)/x(t)$. Dividing (67) through by $x(t)$ yields

$$y'(t) = \alpha - w_1 - w_2 - by \tag{70}$$

as the transformed state equation.

The interaction between the players is described by an infinite horizon differential game in which each seeks to

$$\max_{w_i} J^i(w_1, w_2) = a_i \int_0^\infty e^{-rt}[y(t) + \ln w_i(t)]\,dt, \qquad i = 1,2 \tag{71}$$

subject to (70), and where the utility function (68) was transformed according to (69). The feedback Nash equilibrium strategies, $w_i^*(t, y(t))$, are sought. It turns out in this case too that they coincide with the open-loop strategies. The open-loop strategies are found by forming the Hamiltonians

$$H^i = e^{-rt}a_i(y + \ln w_i) + \lambda_i(\alpha - w_1 - w_2 - by), \qquad i = 1,2. \quad (72)$$

The necessary conditions yield

$$w_i = a_i e^{-rt}/\lambda_i \quad (73a)$$

$$\lambda_i'(t) = -e^{-rt}a_i + b\lambda_i(t), \quad \lim_{t\to\infty}\lambda_i(t) = 0, \qquad i = 1,2, \quad (73b)$$

where $\lambda_i'(t)$ refers to the time derivative of $\lambda_i(t)$. Note that the state variable $y(t)$ does not appear explicitly in (73a). However, it may appear implicitly through $\lambda_i(t)$. Now from (73b) and the use of the integrating factor e^{-bt} and the transversality condition it follows that

$$\lambda_i(t) = a_i e^{-rt}/(r + b), \quad (74)$$

which is also independent of $y(t)$. Substituting from (74) back into (73a) gives

$$w_1^*(t) = w_2^*(t) = r + b \quad (75)$$

as the Nash equilibrium effort levels of the two players at each point in time. It is the absence of the state variable in both (73a) and (74) that makes the open-loop and feedback strategies coincide. Roughly speaking, open-loop and feedback Nash equilibrium strategies coincide whenever the necessary condition with respect to the control variable after substitution for the solution of the co-state variable does not contain the state variable.

EXERCISES

1. Show that for the finite horizon version of Example 1 that the Nash equilibrium feedback strategies $u_i^*(t,p(t))$ are symmetric and $u^*(t,p(t)) = (1 - sK(t))p(t) - c + sE(t)$, where $K(t)$ is the solution to the Riccati equation

$$K'(t) = -3s^2K^2(t) + (6s + r)K(t) - 1.$$

(Hint: assume the value function

$$V^i(t,p) = K_i(t)p^2/2 - E_i(t)p + g_i(t).)$$

2. Suppose two firms are complementary monopolists. That is, each firm is the sole producer of an intermediate product which, together with the other firm's product, makes up a final good. The price of the final good is the sum of the prices of the intermediate goods, under the assumption that one unit of the final good requires one unit of each of them. Suppose that the quantity demanded of the final good adjusts according to the differential equation

$$dx/dt = x'(t) = s(a - p_1(t) - p_2(t) - x(t))$$

where $x(t)$, $x'(t)$ denote the quantity of the good and its time derivative at time t, respectively, and s is a speed of adjustment parameter as in Example 1. Each firm faces the identical cost functions $C(x) = cx + x^2/2$ and seeks to

$$\max_{p_i} J^i(p_1, p_2) = \int_0^{\infty} e^{-rt}[p_i x - cx - x^2/2]\, dt$$

subject to the state equation above and $x(0) = x_0$. Show that

$$p = [(c + 2a)s + ra]/(5s + 2r)$$

is the open-loop Nash equilibrium strategy for each player and that it is also the feedback Nash equilibrium strategy.

3. Suppose two firms produce an identical product, the respective outputs of which at a point in time are $u_1(t)$, $u_2(t)$. The cost of producing this product depends on the quantity produced and the rate of change in the quantity produced, $du_i(t)/dt = u_i'(t)$, according to the function $c_i = cu_i + bu_i^2 + Au_i'^2/2$, i = 1, 2. The demand function for the product is linear, $p(t) = a - (u_1 + u_2)$. Each firm seeks to

$$\max_{x_i} J^i(x_1, x_2) = \int_0^{\infty} e^{-rt}(pu_i - C_i]\, dt$$

subject to

$$u_i'(t) = x_i(t), \qquad u_i(0) = u_{i0,} \qquad i \neq j$$
$$u_j'(t) = x_j(u_i, u_j), \qquad u_j(0) = u_{j0}, \qquad i = 1, 2,$$

where $x_i(t)$, the rate of change of output at t is the control variable and u is the interest rate. Show that the Nash equilibrium feedback strategies are of the form

$$x_1^* = K + k_1 u_1 + k_2 u_2$$
$$x_2^* = K + k_2 u_1 + k_1 u_2$$

(Hint: see Driskell and McCafferty.)

4. Suppose two firms compete in the sale of an identical product, the demand function for which is

$$p(t) = a - y_1(t) - y_2(t)$$

where $p(t)$, $y_1(t)$, $y_2(t)$ refer to the product's price, the output of the first firm, and the output of the second firm, respectively, at time t. Each firm's marginal cost of production is zero up to its production capacity level, $k_i(t)$ at time t, and infinite beyond it. The firms can expand their capacity levels through net investment $I_i(t)$ according to the differential equation

$$k_i'(t) = I_i(t) - \delta k_i(t), \qquad k_i(0) = k_{i0}, \qquad i = 1, 2,$$

where δ refers to the rate at which capacity decays and $k_i'(t)$ the time derivative of $k(t)$. The cost of net investment is governed by the functions

$$C(I_i) = qI_i + cI_i^2/2, \qquad i = 1, 2.$$

Each firm produces up to its capacity level at each point in time. Thus, each firm competes through its capacity investment over time. Namely, each firm seeks to

$$\max_{I_i} \int_0^\infty e^{-rt}\left[\left(a - y_1(t) - y_2(t)\right) y_i(t) - C\left(I_i(t)\right)\right] dt$$

subject to the state equation governing capital accumulation.

i. Show that the open-loop steady-state Nash equilibrium capacity levels are

$$k_i = \left[a - q(r+\delta)\right] / \left(3 + c\delta(r+\delta)\right].$$

ii. Find the feedback Nash equilibrium investment strategies. (Hint: let the value function for the i-th player be $V^i(k) = u + vk_i + wk_j + xk_i^2/2 + yk_ik_j + zk_j^2/2$, where u, v, x, y, z are parameters to be determined.)

FURTHER READING

The formal theory of differential games was introduced by Isaacs, although Roos analyzed a differential game back in 1925, without calling it that. The heavy-duty mathematical text on the subject is Friedman's. Basar and Olsder's, and Mehlmann's are more recent and more intuitive expositions. Feichtinger and Jorgensen, and Clemhout and Wan provide surveys of recent applications of differential games in economics and management science. Example 1 is based on Fershtman and Kamien (1987, 1990). Example 2 is based on Reinganum (1981, 1982). Example 3 is based on Plourde and Yeung, who present a continuous time version of the game introduced by Levhari and Mirman.

The necessary conditions for Nash equilibrium feedback strategies are based on Starr and Ho. Reinganum and Stokey explain the relationship between commitment in open-loop and feedback strategies. The latest word on when open-loop and feedback strategies coincide is by Fershtman. There is a third class of strategies called closed-loop in which the control $u = u(t, x(t_0), x(t))$, depends on the initial condition as well as the state and time. For a discussion see Basar and Oldser or Mehlmann. Problem 2 is based on Ohm; Problem 3 on Driskell and McCafferty; and Problem 4 on Reynolds. See also Dockner, and Tsutsui and Mino for further analysis of Example 1.

Obviously Nash is the source of the equilibrium concept bearing his name. Selten is the source of the concept of subgame perfection. Luce and Raiffa is the classic game theory text following the introduction of the subject by von Neumann and Morgenstern. Myerson provides the latest word on the subject, including the literature on refinements.

APPENDIX A

CALCULUS AND NONLINEAR PROGRAMMING

Section 1

Calculus Techniques

We assume familiarity with the calculus of several variables and the usual algebra prerequisite. Some previous exposure to mathematical programming is also helpful. Certain results from these areas are reviewed in this appendix.

Fundamental Theorem of Integral Calculus. If $f(x)$ is continuous in the interval $a \le x \le b$, and if $F(x) = \int f(x)\,dx$ is an indefinite integral of $f(x)$, then the definite integral

$$\int_a^b f(x)\,dx = \int_a^b F'(x)\,dx = F(b) - F(a), \tag{1}$$

where the derivative of

$$F(x) = \int_a^x f(u)\,du \tag{2}$$

is

$$dF/dx = F'(x) = f(x). \tag{3}$$

The *chain rule* for differentiation of a function of a function is

$$df(y(x))/dx = f'(y)y'(x). \tag{4}$$

It extends readily to functions of several variables. For example,

$$dg(y(x), z(x))/dx = g_y(y, z)y'(x) + g_z(y, z)z'(x), \tag{5}$$

where subscripts label the partial derivative with respect to the indicated argument.

The counterpart to the chain rule for differentiation is the change of the variable of integration. Suppose $x = h(t)$ on the interval $t_0 \le t \le t_1$, such that $h(t_0) = a$, $h(t_1) = b$, and that x increases continuously or decreases continuously from a to b as t goes from t_0 to t_1, then

$$\int_a^b f(x)\,dx = \int_{t_0}^{t_1} f[h(t)]h'(t)\,dt. \tag{6}$$

The *product rule* for differentiation is

$$d[u(x)v(x)]/dx = u(x)v'(x) + v(x)u'(x). \tag{7}$$

Integrate and rearrange to get the formula for *integration by parts*:

$$\int u(x)v'(x)\,dx = u(x)v(x) - \int v(x)u'(x)\,dx$$

or, briefly,

$$\int v\,dv = uv - \int v\,du, \tag{8}$$

where $dv = v'(x)\,dx$ and $du = u'(x)\,dx$. For example, to evaluate $\int xe^x\,dx$, let $u(x) = x$ and $v'(x) = e^x$. Then $u'(x) = 1$ and $v(x) = e^x$, and therefore,

$$\int xe^x\,dx = xe^x - \int e^x\,dx = (x-1)e^x.$$

For definite integrals, (7) becomes

$$\int_a^b u(x)v'(x)\,dx = u(b)v(b) - u(a)v(a) - \int_a^b v(x)u'(x)\,dx. \tag{9}$$

The rule for differentiating an integral with respect to a parameter is

Liebnitz's Rule. Let $f(x, r)$ be continuous with respect to x for every value of r, with a continuous derivative $df(x, r)/\partial r$ with respect to x and r in the rectangle $a \le x \le b$, $\underline{r} \le r \le \bar{r}$ of the x–r plane. Let the functions $A(r)$ and $B(r)$ have continuous derivatives. If $V(r) = \int_{A(r)}^{B(r)} f(x, r)\,dx$, then

$$\begin{aligned} V'(r) &= f(B(r), r)B'(r) - f(A(r), r)A'(r) \\ &\quad + \int_{A(r)}^{B(r)} (\partial f(x, r)/\partial r)\,dx. \end{aligned} \tag{10}$$

For example, if

$$V(r) = \int_{r^2}^{r} e^{-rs}P(s)\,ds,$$

then

$$dV(r)/dr = P(r)e^{-r^2} - 2P(r^2)re^{-r^3} - \int_{r^2}^{r} se^{-rs}P(s)\,ds.$$

A function $f(x, y)$ is said to be homogeneous of degree n if

$$f(kx, ky) \equiv k^n f(x, y). \tag{11}$$

For example, $x^2 y$ is homogeneous of degree 3, $x^a y^{1-a}$ and $2x - 5y$ are homogeneous of degree 1, and $4x/y$ is homogeneous of degree 0. Differentiating (11) with respect to k and evaluating at $k = 1$ gives

$$f_x x + f_y y = nf, \tag{12}$$

which is known as *Euler's theorem* on homogeneous functions.

If $f(x, y)$ is homogeneous of degree one and we take $k = 1/x$ in (11) and rearrange, we get

$$f(x, y) = xf(1, y/x) = xg(y/x). \tag{13}$$

The value of the function is the product of a scale factor x and the value of a function of the ratio y/x.

Section 2

Mean-Value Theorems

Mean-Value Theorem. *If $f(x)$ is continuous on an interval $a \le x \le b$, then there exists a number $\bar{x}$, such that*

$$\int_a^b f(x)\,dx = f(\bar{x})(b-a), \qquad \text{where} \quad a < \bar{x} < b. \tag{1}$$

Recalling (1.1) *and* (1.3), *one can rewrite* (1)

$$[F(b) - F(a)]/(b-a) = F'(\bar{x}) \qquad \text{for some} \quad a < \bar{x} < b. \tag{2}$$

Expression (2) is the mean-value theorem of differential calculus, while (1) is the mean-value theorem of integral calculus.

Mean-value theorem (2) has an interesting geometric interpretation. The left side is the slope of the hypotenuse of the right triangle ABC in Figure 2.1, with base of length $b - a$ and height $F(b) - F(a)$. Then (2) indicates that there is at least one point, say $\bar{x}$, at which the curve $F(x)$ has the same slope as the hypotenuse AB of the triangle joining the end points.

In (2), $\bar{x}$ is strictly between a and b. The mean-value theorem can be used to show that

$$\text{if } F(b) - F(a) = F'(a)(b-a), \text{ then either } a = b \text{ or else } F''(r) = 0 \text{ for some } r \text{ strictly between } a \text{ and } b. \tag{3}$$

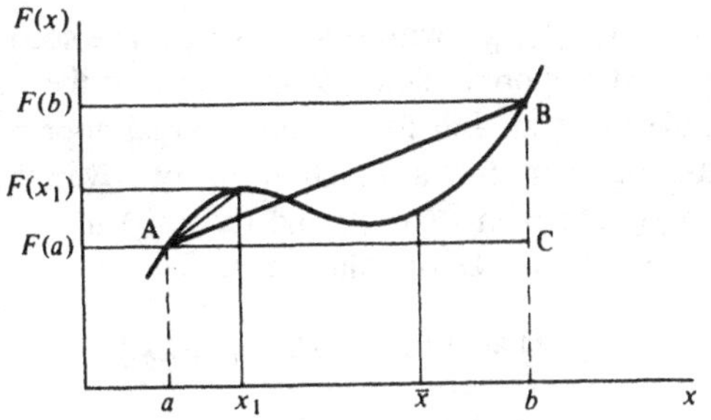

Figure 2.1

This is clear from the graph. Algebraically, if $a = b$, the result is immediate. So suppose $a < b$. By the mean-value theorem, there is a number q, $a < q < b$, such that

$$F(b) - F(a) = (b - a)F'(q).$$

Combining this with the hypothesis of (3) gives

$$[F'(a) - F'(q)](b - a) = 0.$$

Again, by the mean-value theorem, there is a number r, $a < r < q$, such that

$$F'(a) - F'(q) = (a - q)F''(r),$$

and therefore,

$$(a - q)(b - a)F''(r) = 0.$$

Since $a < q < b$, $F''(r) = 0$, as was to be shown. The case of $a > b$ is analogous.

Any curve can be approximated arbitrarily well in the neighborhood of a point by a polynomial of sufficiently high order, according to

Taylor's Theorem. *If the function $f(x)$ and its first $n - 1$ derivatives are continuous in the interval $a \leq x \leq b$ and its nth derivative for each x, $a < x < b$ exists, then for $a < x_0 < b$, the Taylor series expansion of $f(x)$ about the point x_0 is*

$$f(x) = f(x_0) + \sum_{i=1}^{n} (x - x_0)^i f^{(i)}(x_0)/i! + R_n, \tag{4}$$

where $f^i \equiv d^i f/dx^i$ and $R_n = (x - x_0)^n f^n(\bar{x})/n!$, for some $\bar{x}$, $a < \bar{x} < b$.

This Taylor series expansion of $f(x)$ extends (2). We call R_{n+1} the *remainder* and $(n - 1)$th the *order* of the approximation obtained by deleting R_{n+1}. The special case (2) gives the zero order approximation and $R_1 = (b - a)f'(\bar{x})$. The first order approximation is $f(x) = f(x_0) + (x - x_0)f'(x_0)$, a straight line through $(x_0, f(x_0))$ with slope $f'(x_0)$. The second order approximation gives a quadratic approximation to the curve at the point x_0. Deleting the remainder in (4) gives an nth degree polynomial approximation to $f(x)$.

The mean-value theorem for a function of two variables, $F(x, y)$, is developed as follows. For points (x, y) and (x_0, y_0) in the domain of F, let $h = x - x_0$ and $k = y - y_0$. Define the function

$$f(t) = F(x_0 + th, y_0 + tk) \tag{5}$$

of the single variable t on the interval $0 \leq t \leq 1$. Then

$$f(1) - f(0) = F(x, y) - F(x_0, y_0).$$

But from (2),

$$f(1) - f(0) = f'(\bar{t}) \qquad \text{for some} \quad \bar{t}, \quad 0 < \bar{t} < 1. \tag{6}$$

Differentiating (5) with respect to t gives

$$f'(t) = hF_x + kF_y, \tag{7}$$

where F_x and F_y are evaluated at $(x_0 + th, y_0 + tk)$. Combining (5)–(7) gives the mean-value theorem for a function of two variables:

$$F(x, y) - F(x_0, y_0) = (x - x_0)F_x(\bar{x}, \bar{y}) + (y - y_0)F_y(\bar{x}, \bar{y}) \tag{8}$$

for some $\bar{x}, \bar{y}$, where $x_0 < \bar{x} < x$ and $y_0 < \bar{y} < y$.

The Taylor series expansion or generalized mean-value theorem (3) can likewise be extended to a function of two variables. Let $h = x - x_0$ and $k = y - y_0$. Then,

$$F(x, y) = F + (hF_x + kF_y) + R_2 \tag{9}$$

where $R_2 = h^2F_{xx}/2 + hkF_{xy} + k^2F_{yy}/2$. On the right side of (9), F, F_x, and F_y are evaluated at x_0, y_0 and the second partial derivatives in R_2 are evaluated at some point between x, y and x_0, y_0. Expanding further, we get

$$F(x, y) = F + (hF_x + kF_y) + (h^2F_{xx}/2 + hkF_{xy} + k^2F_{yy}/2) + R_3, \tag{10}$$

where F and its partial derivatives on the right side are all evaluated at (x_0, y_0) and R_3 is the remainder.

More generally, the first order Taylor series expansion of a function $F(x_1, \ldots, x_n)$ of n variables about the point $x^0 = (x_1^0, \ldots, x_n^0)$ is

$$F(x_1, \ldots, x_n) = F + \sum_{i=1}^{n} h_i F_i + R_2 \tag{11}$$

where

$$R_2 = (1/2)\sum_{i=1}^{n}\sum_{j=1}^{n} h_i h_j F_{ij}, \qquad h_i = x_i - x_i^0, \quad i = 1, \ldots, n.$$

On the right side of (11), F and F_i are evaluated at x^0 and the F_{ij} are evaluated at some point between x and x^0. The second order expansion is

$$F(x_1, \ldots, x_n) = F + \sum_{i=1}^{n} h_i F_i + (1/2)\sum_{i=1}^{n}\sum_{j=1}^{n} h_i h_j F_{ij} + R_3, \quad (12)$$

where F and its first and second partial derivatives F_i, F_{ij} on the right are all evaluated at x^0, and R_3 is the remainder term. The series can be expanded to as many terms as desired.

Section 3

Concave and Convex Functions

A function $f(x)$ is said to be *concave* on the interval $a \leq x \leq b$ if for all $0 \leq t \leq 1$ and for any $a \leq x_1 \leq x_2 \leq b$

$$tf(x_1) + (1 - t)f(x_2) \leq f(tx_1 + (1 - t)x_2). \qquad (1)$$

A weighted average of the values of the function at any two points is no greater than the value of the function at the same weighted average of the arguments. If the inequality in (1) is strong, then the function $f(x)$ is said to be *strictly concave*. The chord joining any two points on the graph of a concave function is not above the graph. If the function is strictly concave, the chord is strictly below the graph. Linear functions are concave, but not strictly concave (see Figure 3.1).

A function $f(x)$ is said to be *convex* if the inequality in (1) is reversed. It is *strictly convex* if the reversed inequality is strong. For example, $f(x) = -x^2$ is a strictly concave function, whereas $f(x) = x^2$ is strictly convex. These functions illustrate the general principle that if $f(x)$ is a concave function, then $-f(x)$ is a convex function.

Concave functions have several important properties. First, *a concave function $f(x)$ is continuous on any open interval $a < x < b$ on which it is defined.*

Second, *if f is concave and differentiable,*

$$(x_2 - x_1)f'(x_1) \geq f(x_2) - f(x_1) \geq (x_2 - x_1)f'(x_2). \qquad (2)$$

The slope of the line joining two points on its graph is less than the slope of the tangent line at its left endpoint and greater than the slope of its tangent line on its right endpoint (see Figure 3.1). For a convex function, the inequalities in (2) are reversed.

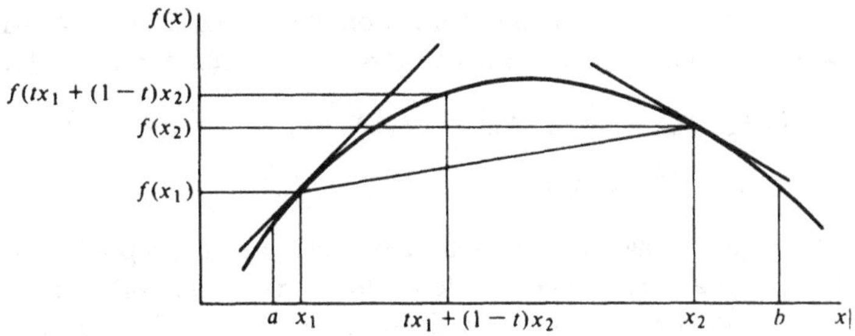

Figure 3.1

Third, *the second derivative of a twice differentiable concave function is nonpositive*. To see this, use (2.4) to write

$$f(x_2) - f(x_1) - (x_2 - x_1)f'(x_1) = (x_2 - x_1)^2 f''(\bar{x})/2 \qquad (3)$$

for some $\bar{x}$ where $x_1 < \bar{x} < x_2$. But from (2) or Figure 3.1, it is apparent that the left side of (2) is nonpositive. Therefore, the right side of (3) must also be nonpositive. Since $(x_2 - x_1)^2$ is positive, $f''(\bar{x})$ must be nonpositive. Further, since (3) holds for any $a \le x_1 < x_2 \le b$, $f''(x) \le 0$ for all $a \le x \le b$. Similarly, the second derivative of a convex function is nonnegative.

The condition $f''(x) \le 0$ for all x in the domain is both necessary and sufficient for a twice differentiable function f to be concave. If $f''(x) < 0$, then f is strictly concave. The converse of this, however, is not true. The function $f(x) = -x^4$ appears to be strictly concave when plotted but, $f''(0) = 0$. Despite this possibility, we assume that $f''(x) < 0$ for strictly concave functions. The reader should keep in mind the possibility of exceptional points.

The definition of a *concave function of two variables* $f(x, y)$ is a direct extension of the definition (1):

$$tf(x_1, y_1) + (1 - t)f(x_2, y_2) \le f(tx_1 + (1 - t)x_2, ty_1 + (1 - t)y_2) \qquad (4)$$

for all $0 \le t \le 1$ and any pair of points $(x_1, y_1), (x_2, y_2)$ in the domain of f. The extension of (2) is

$$f(x_2, y_2) - f(x_1, y_1) \le (x_2 - x_1)f_x(x_1, y_1) + (y_2 - y_1)f_y(x_1, y_1), \qquad (5)$$

which holds for a differentiable concave function of two variables at any pair of points in its domain. Inequalities (4) and (5) are reversed for a convex function.

To find the analog of the sign condition on the second derivative, write the counterpart to (3), employing (2.9). Let $h = x_2 - x_1$ and $k = y_2 - y_1$. Then,

$$f(x_2, y_2) - f(x_1, y_1) - hf_x(x_1, y_1) - kf_y(x_1, y_1)$$
$$= h^2 f_{xx}/2 + hkf_{xy} + k^2 f_{yy}/2 \tag{6}$$

where the second partial derivatives are evaluated at an appropriate point $\bar{x}$, $\bar{y}$ between (x_1, y_1) and (x_2, y_2). The left side of (6) is nonpositive by (5), and therefore, the *quadratic form* on the right side must be nonpositive as well. (A *quadratic form* is a function of the form $f(x_1, \ldots, x_n) = \sum_{j=1}^{n} \sum_{n=1}^{n} a_{ij} x_i x_j$ where $a_{ij} = a_{ji}$, $i, j = 1, \ldots, n$.) If $f_{xx} \neq 0$, add and subtract $k^2 f_{xy}^2 / f_{xx}$ and collect terms to write the right side of (6) equivalently

$$f_{xx}\left[\left(h + kf_{xy}/f_{xx}\right)^2 + \left(f_{xx}f_{yy} - f_{xy}^2\right)k^2/f_{xx}^2\right] \leq 0. \tag{7}$$

Since (7) must hold for any choice of h and k, including $k = 0$, it follows that

$$f_{xx} \leq 0. \tag{8}$$

In addition, (7) must hold in case $h = -kf_{xy}/f_{xx}$, so in view of (8), we must have

$$f_{xx}f_{yy} - f_{xy}^2 \geq 0. \tag{9}$$

Thus, if $f(x, y)$ is concave and if $f_{xx} \neq 0$, then (8) and (9) hold. Note that (8) and (9) together imply that

$$f_{yy} \leq 0. \tag{10}$$

In case $f_{xx} = 0$, one can conduct the argument under the supposition that $f_{yy} \neq 0$, to conclude that (9) and (10) hold. These conditions are sufficient for concavity. If the inequalities are strict, f is strictly concave. If $f(x, y)$ is a convex function, then $f_{xx} \geq 0$, $f_{yy} \geq 0$, and (9) holds as well.

A function $f(x_1, \ldots, x_n)$ is *concave* if

$$tf(x^*) + (1 - t)f(x^0) \leq f(tx^* + (1 - t)x^0) \tag{11}$$

for all $0 \leq t \leq 1$ and any pair of points $x^* = [x_1^*, \ldots, x_n^*]$, $x^0 = [x_1^0, \ldots, x_n^0]$ in the domain of f. The extension of (2) may be stated that if $f(x_1, \ldots, x_n)$ is concave and differentiable, then

$$f(x^*) - f(x^0) \leq \sum_{i=1}^{n} (x_i^* - x_i^0) f_i(x^0), \tag{12}$$

where x^* and x^0 are any two points in the domain of f. And, letting $h_i = x_i^* - x_i^0$, $i = 1, \ldots, m$,

$$f(x^*) - f(x^0) - \sum_{i=1}^{n} h_i f_i(x^0) = \tfrac{1}{2} \sum_{i=1}^{n} \sum_{j=1}^{n} h_i h_j f_{ij} \tag{13}$$

by the Taylor series expansion (2.11), where the second partial derivatives are evaluated at an appropriate point $\bar{x}$ between x^* and x^0. Since the left side of (13) is nonpositive by (12), the quadratic form on the right of (13) must be nonpositive.

To state an equivalent condition for the *quadratic form* in h_i on the right of (13) to be nonpositive, we need some definitions. The coefficient matrix of second partial derivatives of a function f,

$$H = \begin{bmatrix} f_{11} & f_{12} & \cdots & f_{1n} \\ f_{21} & f_{22} & \cdots & f_{2n} \\ \vdots & & & \vdots \\ f_{n1} & f_{n2} & \cdots & f_{nn} \end{bmatrix},$$

is called a *Hessian matrix* of f. The quadratic form on the right on (13) can be written

$$hHh^T \le 0 \tag{14}$$

where $h = [h_1, \ldots, h_n]$, and h^T is the transpose of h. The quadratic form hHh^T is said to be *negative semidefinite* if (14) holds for all h. (It is *negative definite* if (14) holds with strict inequality for all $h \neq 0$.) Equivalently, we say that the Hessian matrix H is negative semidefinite if hHh^T is negative semidefinite. The matrix is negative semidefinite if its principal minors alternate in sign, beginning with negative:

$$f_{11} < 0, \qquad \begin{vmatrix} f_{11} & f_{12} \\ f_{21} & f_{22} \end{vmatrix} > 0, \qquad \begin{vmatrix} f_{11} & f_{12} & f_{13} \\ f_{21} & f_{22} & f_{23} \\ f_{31} & f_{32} & f_{33} \end{vmatrix} < 0, \ldots$$

$$(-1)^n |H| \ge 0. \tag{15}$$

The last principal minor, namely the determinant of H itself, may be zero. (If H is negative definite, then the principal minors alternate in sign and none may be zero.) It is clear from (12) and (13) that H is negative semidefinite for all x if f is concave.

If $f(x_1, \ldots, x_n)$ is twice continuously differentiable and convex, then the sign in (12) is reversed, and thus the Hessian must be positive semidefinite. The matrix H is positive semidefinite if all its principal minors are positive, except possibly $|H|$ which may be zero. It is positive semidefinite if and only if f is convex.

The notion of concavity has been generalized in several ways. A function $f(x_1, \ldots, x_n)$ is said to be quasiconcave if

$$f(tx^* + (1-t)x^0) \ge \min[f(x^*), f(x^0)] \tag{16}$$

for any x^*, x^0 in the domain of f and for all $0 \leq t \leq 1$. Equivalently, $f(x_1, \ldots, x_n)$ is quasiconcave if and only if the set

$$A_a = \{x^*: f(x) \geq a\} \tag{17}$$

is convex for every number a. A function g is *quasiconvex* if $-g$ is quasiconcave. Every concave function is quasiconcave, but a quasiconcave function need not be concave, nor even continuous.

FURTHER READING

See Arrow and Enthoven for a discussion of quasiconcavity and the uses of the concept in optimization.

Section 4

Maxima and Minima

The Weierstrass theorem assures us that a continuous function assumes a maximum and a minimum on a closed bounded domain. If the hypotheses are not satisfied, then there may be no maximum and/or minimum. For instance, $f(x) = 4x - x^2$ has no maximum on $0 \leq x < 2$ since the interval is not closed; the function attains values arbitrarily close to 4, but the value 4 is not achieved on the interval. The function $f(x) = 4 + 1/x$ is not continuous on $-1 \leq x \leq 1$ and has no maximum on that interval; it becomes arbitrarily large as x approaches zero from the right. The maximum may occur on the interior of the domain or at a boundary point. It may be attained at just one or at several points in the domain.

If $f(x^*) > f(x)$ for all x near x^*—that is, for all x such that $x^* - \varepsilon < x < x^* + \varepsilon$ for some $\varepsilon > 0$—then x^* is said to provide a *strict local maximum*. If $f(x^*) > f(x)$ for all x in the domain of f, then x^* provides a *strict global maximum*. Local and global minima are defined analogously.

Suppose $f(x)$ is twice continuously differentiable and attains its maximum at x^* on $a < x < b$. From the mean-value theorem (2.2),

$$f(x) - f(x^*) = f'(\bar{x})(x - x^*) \tag{1}$$

for some $\bar{x}$ between x and x^*. Since x^* maximizes f, the left side of (1) must be nonpositive and therefore the right side as well. Thus $f' \geq 0$ when $x < x^*$, and $f' \leq 0$ when $x > x^*$. Since f^* is continuous, we conclude that

$$f'(x^*) = 0. \tag{2}$$

Furthermore, from Taylor's theorem (2.4),

$$f(x) - f(x^*) - (x - x^*)f'(x^*) = (x - x^*)^2 f''(\bar{x})/2 \tag{3}$$

for some $\bar{x}$ between x and x^*. Since (2) holds and x^* is maximizing, the left side of (3) is nonpositive. This implies

$$f''(x^*) \leq 0. \tag{4}$$

Thus, *conditions (2) and (4) are necessary for a point x^* to maximize a twice continuously differentiable function $f(x)$ on the interior of its domain*. At a local maximum, the function is stationary (2) and locally concave (4).

Further, if x^* satisfies (2) and also

$$f''(x) < 0 \tag{5}$$

for all x near x^*, i.e., $x^* - \varepsilon < x < x^* + \varepsilon$, then it follows from (3) that $f(x^*) > f(x)$. Therefore, *(2) and (5) are sufficient conditions for a point x^* to provide a local maximum.*

Similar arguments show that necessary conditions for a local minimum are (2) and

$$f''(x^*) \geq 0. \tag{6}$$

Sufficient conditions for a local minimum are (2) and

$$f''(x) > 0, \tag{7}$$

for all x near x^*.

To find the maximum of a function of one variable, one compares the values of the function at each of the local maxima and at the boundary points of the domain, if any, and selects the largest. If the function is strictly concave over its entire domain (globally strictly concave), the maximizing point will be unique.

In Figure 4.1, the boundary point $x = g$ maximizes the function $f(x)$ over $a \leq x \leq g$. Points b and d are local maxima and satisfy (2) and (4). Points c and e are local minima, satisfying (2) and (6).

For a twice continuously differentiable function $f(x, y)$ of two variables, one can repeat the above arguments. Let x^*, y^* provide an interior maximum.

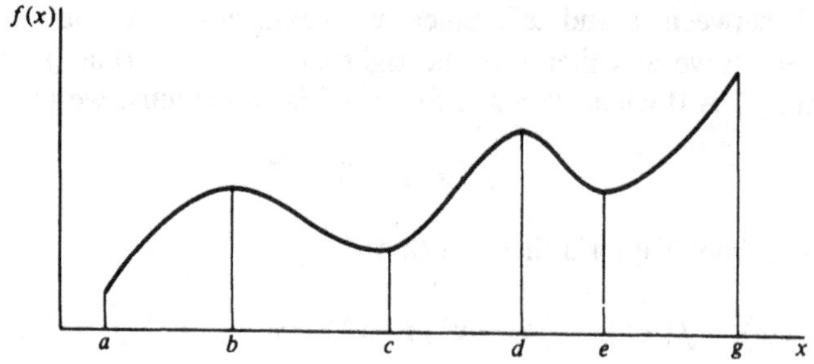

Figure 4.1

Mean-value theorem (2.8) gives

$$f(x, y) - f(x^*, y^*) = (x - x^*)f_x(\bar{x}, \bar{y}) + (y - y^*)f_y(\bar{x}, \bar{y}) \quad (8)$$

for some $\bar{x}$, $\bar{y}$ between x^*, y^* and x, y. Since x^*, y^* is maximizing, the left side is nonpositive for all x, y. Taking $y = y^*$, we find that $x - x^*$ and f_x must have opposite signs for any x, so that $f_x = 0$. Similarly, $f_y = 0$. Thus, it is necessary for f to be stationary at x^*, y^*:

$$f_x(x^*, y) = 0, \qquad f_y(x^*, y^*) = 0. \quad (9)$$

From the Taylor expansion (2.9),

$$f(x, y) - f(x^*, y^*) - hf_x(x^*, y^*) - kf_y(x^*, y^*)$$
$$= h^2 f_{xx}/2 + hkf_{xy} + k^2 f_{yy}/2, \quad (10)$$

where $h = x - x^*$; $k = y - y^*$, and the second partial derivatives on the right are evaluated at some point between x, y and x^*, y^*. But, since x^*, y^* is maximizing and since (9) holds, the left side of (10) must be nonpositive, so the right side must be as well. As shown in (3.6)–(3.10), at x^*, y^*

$$f_{xx} \le 0, \qquad \begin{vmatrix} f_{xx} & f_{xy} \\ f_{yx} & f_{yy} \end{vmatrix} = f_{xx}f_{yy} - f_{xy}^2 \ge 0. \quad (11)$$

Thus, (9) and (11), local stationarity and local concavity, are necessary for a local maximum at x^*, y^*. Similarly, one shows that local stationarity (9) and local convexity,

$$f_{xx} \ge 0, \qquad f_{xx}f_{yy} - f_{xy}^2 \ge 0, \quad (12)$$

are necessary for a local minimum at x^*, y^*. Sufficient conditions for a local optimum are (9) and strong inequalities in (11) (maximum) or (12) (minimum) for all x, y near x^*, y^*.

To find the minimum of a function $f(x, y)$, one compares the values of the function at each of the local maxima and along the boundary of the domain and selects the largest value. If $f(x, y)$ is strictly concave throughout its domain, then a local maximum will be the global maximum.

Example. Maximize $f(x, y) = xy + 9y - x^2 - y^3/12$.

Compute

$$f_x = y - 2x = 0, \qquad f_{xx} = -2, \qquad f_{xy} = 1,$$
$$f_y = x + 9 - y^2/4 = 0, \qquad f_{yy} = -y/2,$$
$$|H| = \begin{vmatrix} -2 & 1 \\ 1 & -y/2 \end{vmatrix} = y - 1.$$

The first order conditions are satisfied at $(1/2 + (37/4)^{1/2}, 1 + 37^{1/2})$ and at $(1/2 - (37/4)^{1/2}, 1 - 37^{1/2})$. At the first solution, $f_{xx} < 0$ and $|H| > 0$; thus it is a local maximum. At the second point, $|H| < 0$; thus it is neither a local maximum, nor a local minimum.

The way is clear to show that if $x^* = [x_1^*, \ldots, x_n^*]$ maximizes the twice continuously differentiable function $f(x_1, \ldots, x_n)$ on the interior of its domain, then

$$f_i(x^*) = 0, \qquad i = 1, \ldots, n. \tag{13}$$

Furthermore, let $x = [x_1, \ldots, x_n]$ and $h_i = x_i - x_i^*$. Then, by Taylor's theorem (2.11) we have

$$f(x) - f(x^*) - \sum_{i=1}^{n} h_i f_i(x^*) = (1/2) \sum_{i=1}^{n} \sum_{j=1}^{n} h_i h_j f_{ij}, \tag{14}$$

where the second partial derivatives on the right are evaluated at some point between x and x^*. Since x^* is maximizing and since (13) holds, the quadratic form on the right must be nonpositive, that is, negative semidefinite. Thus, the principal minors of the Hessian matrix of f must alternate in sign, starting with negative (review (3.15)). And if the quadratic form in (14) is negative definite at x^* and (13) holds, then x^* provides a local maximum.

Analogously, necessary conditions for a minimum of f are stationarity (13) and local convexity, that is, positive semidefiniteness of the Hessian. Positive definiteness of the Hessian and stationarity are sufficient conditions for a local minimum.

Section 5

Equality Constrained Optimization

A maximization problem subject to constraint is

$$\max \quad f(x, y) \tag{1}$$

$$\text{subject to} \qquad g(x, y) = b. \tag{2}$$

Find among the points x, y satisfying the constraint (2) the one giving the largest value to the objective function (1). One might solve (2) for y in terms of x, and then substitute into (1) to get an unconstrained problem in the single variable x. In principle, one can do this locally in the neighborhood of a point x_0, y_0 provided that $g_y(x_0, y_0) \neq 0$:

Implicit Function Theorem. *If $g(x, y)$ has continuous partial derivatives and if $g(x_0, y_0) = 0$ while $g_y(x_0, y_0) \neq 0$, then there is a rectangular region $x_1 \leq x \leq x_2$, $y_1 \leq y \leq y_2$ containing x_0, y_0 such that for every x in the interval, the equation $g(x, y) = 0$ determines exactly one value of $y = G(x)$. Furthermore, $g(x, G(x)) = 0$ in this region. Finally, $G(x)$ is continuous and differentiable, with*

$$dy/dx = G'(x) = -g_x/g_y. \tag{3}$$

The condition $g_y \neq 0$ prevents division by zero. Solve (2) for $y = G(x)$ and substitute into (1) to get

$$\max f(x, G(x)). \tag{4}$$

If x^* solves (4), then by the chain rule (1.5), we have at x^*

$$f_x + f_y G' = 0. \tag{5}$$

Substituting from (3) gives

$$f_x - f_y g_x / g_y = 0$$

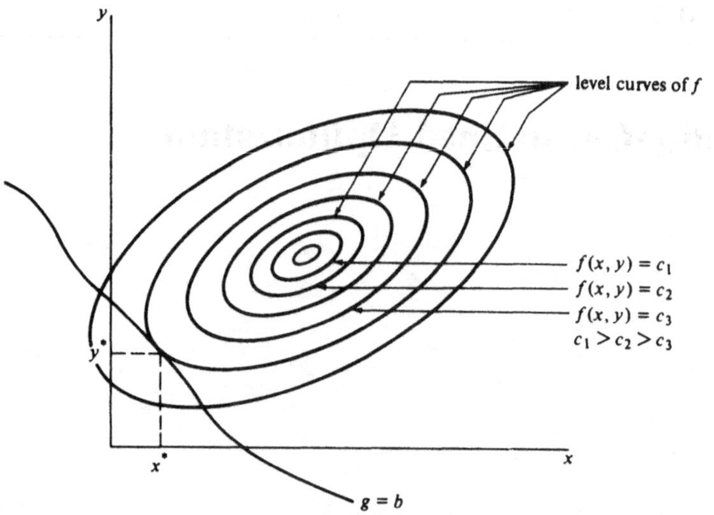

Figure 5.1

or equivalently, as long as $f_y \neq 0$,

$$f_x / f_y = g_x / g_y. \tag{6}$$

At the optimum, the maximum attainable level curve of f is just tangent to the constraining relation $g = b$ (see Figure 5.1).

Provided that $g_x \neq 0$, one can define

$$\lambda = f_x / g_x. \tag{7}$$

Rearranging (7) gives

$$f_x - \lambda g_x = 0 \tag{8}$$

and substituting from (7) into (6) gives

$$f_y - \lambda g_y = 0. \tag{9}$$

Conditions (8) and (9) are equivalent to (6). They can also be generated by forming the *Lagrangian* for (1) and (2), namely,

$$L(x, y, \lambda) = f(x, y) + \lambda[b - g(x, y)], \tag{10}$$

and setting the partial derivatives of (10) equal to zero

$$\begin{aligned} L_x &= f_x - \lambda g_x = 0, \\ L_y &= f_y - \lambda g_y = 0, \\ L_\lambda &= b - g(x, y) = 0. \end{aligned} \tag{11}$$

If it is not true that both $g_x = g_y = 0$, then the solution (x, y, λ) to (1) and (2) satisfies the three equations (11). Of course, (11) is also necessary for a minimum. There are two advantages to (10) and (11) over (6). First, (10) and (11) are readily generalized in a symmetric form to problems involving several variables and possibly several constraints. Second, the Lagrange multiplier has a useful interpretation.

To develop the interpretation, let $V(b)$ be the maximum value in (1). Since the constraint is satisfied at the optimum, we can write

$$V(b) = f(x, y) + \lambda[b - g(x, y)] \quad \text{at} \quad x = x^*(b), \quad y = y^*(b).$$

Differentiating with respect to b, and using the chain rule, gives

$$V'(b) = (f_x - \lambda g_x)\, dx/db + (f_y - \lambda g_y)\, dy/db + \lambda.$$

But, since x^*, y^* satisfy (11), this reduces to

$$V'(b) = \lambda. \tag{12}$$

Thus, the multiplier λ is the marginal valuation of the constraint. It tells the rate at which the maximum objective function value changes with a change in the right side of the constraint.

The second order condition for a solution to (1) and (2) can be developed from (4) and (5). The second derivative of (4) with respect to x will be nonpositive at a maximum. Hence,

$$d(f_x + f_y G')/dx = f_{xx} + 2f_{xy}G' + f_y G'' + f_{yy}(G')^2 \leq 0.$$

Substitute from (3) for G' and multiply through by g_y^2 to get

$$g_y^2 f_{xx} - 2f_{xy} g_x g_y + f_{yy} g_x^2 + g_y^2 f_y G'' \leq 0.$$

Substituting for

$$\begin{aligned} G''(x) &= dG'/dx = d(-g_x/g_y)/dx \\ &= -\{g_y[g_{xx} + g_{xy}G'(x)] - g_x(g_{yx} + g_{yy}G')\}/g_y^2 \end{aligned}$$

using $G' = -f_x/f_y$ finally gives

$$g_y^2 f_{xx} - 2f_{xy} g_x g_y + f_{yy} g_x^2 + 2f_y g_x g_{xy} - f_x g_x g_{yy} - f_y g_y g_{xx} \leq 0 \tag{13}$$

as a second order necessary condition satisfied by the solution (x^*, y^*) to (1) and (2). If (x^*, y^*) satisfies (6) and (13) with strong inequality, then (x^*, y^*) provides a local maximum. If g is linear, then (13) will be satisfied provided f is concave. The inequality of (13) is reversed for a minimum. Finally, it can be verified by evaluation of the determinant, using (7)–(10), that (13) is equivalent to

$$\begin{vmatrix} L_{xx} & L_{xy} & g_x \\ L_{yx} & L_{yy} & g_y \\ g_x & g_y & 0 \end{vmatrix} \geq 0. \tag{13'}$$

To solve a problem with n variables with m constraints,

$$\max \quad f(x_1, \ldots, x_n) \tag{14}$$

$$\text{subject to} \quad g_i(x_1, \ldots, x_n) = b_i, \quad i = 1, \ldots, m, \tag{15}$$

where $m < n$ and $f, g_1, \ldots, g_m$ are all twice continuously differentiable, one associates a Lagrange multiplier λ_i with the ith constraint of (15) and forms the Lagrangian

$$L(x_1, \ldots, x_n, \lambda_1, \ldots, \lambda_m) = f(x) + \sum_{i=1}^{m} \lambda_i [b_i - g_i(x)]. \tag{16}$$

Set the partial derivative of L with respect to each argument equal to zero

$$\partial L/\partial x_j = \partial f/\partial x_j - \sum_{i=1}^{m} \lambda_i \, \partial g_i/\partial x_j = 0, \qquad j = 1, \ldots, n, \tag{17}$$

$$\partial L/\partial \lambda_i = b_i - g_i = 0, \qquad i = 1, \ldots, m. \tag{18}$$

Finally, solve the $n + m$ equations (17) and (18) for $x_1, \ldots, x_n, \lambda_1, \ldots, \lambda_m$. The maximizing point will be found among the sets of solutions to (17) and (18).

Theorem 1. *Let $f, g_1, \ldots, g_m$ be twice continuously differentiable functions of $x = [x_1, \ldots, x_n]$. If there are vectors $x^* = [x_1^*, \ldots, x_n^*]$ and $\lambda^* = [\lambda_1^*, \ldots, \lambda_m^*]$ that satisfy* (17) *and* (18) *and if*

$$\sum_{j=1}^{n} \sum_{i=1}^{n} h_i h_j \, \partial^2 L(x^*, \lambda^*)/\partial x_i \, \partial x_j < 0 \tag{19}$$

for every nonzero vector $[h_1, \ldots, h_n]$ satisfying

$$\sum_{j=1}^{n} h_j \, \partial g_i(x^*)/\partial x_j = 0, \qquad i = 1, \ldots, m, \tag{20}$$

then f has a strict local maximum at x^* subject to (15). *If the inequality in* (19) *is reversed, then f has a strict local minimum.*

If f is strictly concave and each g_i is linear, then (19)–(20) will be satisfied by any point x^* satisfying (17) and (18). In this special case, a solution of (17) and (18) is a solution of (14) and (15).

For the more general case, the conditions for the Hessian of the Lagrangian to be negative definite (19), subject to the linear constraint (20), can be stated equivalently as follows. Suppose the *Jacobian matrix*

$$\left[\partial g_i/\partial x_j\right], \qquad i = 1, \ldots, m, \quad j = 1, \ldots, n \tag{21}$$

has rank m and suppose the variables are indexed so the submatrix of the first m rows of (21) has rank m. Then, conditions (19) and (20) can be replaced by the requirement that at x^* the determinants of the *bordered Hessian*

$$(-1)^k \begin{vmatrix} \partial^2 L/\partial x_1^2 & \cdots & \partial^2 L/\partial x_1 \partial x_k & \partial g_1/\partial x_1 & \cdots & \partial g_m/\partial x_1 \\ \vdots & & \vdots & \vdots & & \vdots \\ \partial^2 L/\partial x_k \partial x_1 & \cdots & \partial^2 L/\partial x_k^2 & \partial g_1/\partial x_k & \cdots & \partial g_m/\partial x_k \\ \partial g_1/\partial x_1 & \cdots & \partial g_1/\partial x_k & 0 & \cdots & 0 \\ \vdots & & \vdots & \vdots & & \vdots \\ \partial g_m/\partial x_1 & \cdots & \partial g_m/\partial x_k & 0 & \cdots & 0 \end{vmatrix} > 0 \tag{22}$$

for $k = m + 1, \ldots, n$. For a local minimum, $(-1)^k$ is replaced by $(-1)^m$.

The multipliers λ_j retain their interpretation as the marginal valuation of the associated constraint. Thus, defining V by

$$V(b_1, \ldots, b_m) = \max f(x_1, \ldots, x_n) \qquad \text{subject to } (15), \tag{23}$$

it can be shown, as before, that

$$\lambda_i = \partial V/\partial b_i, \qquad i = 1, \ldots, m. \tag{24}$$

Example.

$$\max \quad 2x + y$$
$$\text{subject to} \qquad x^2 + y^2 = 4.$$

The Lagrangian is

$$L = 2x + y + \lambda(4 - x^2 - y^2),$$

so

$$L_x = 2 - 2\lambda x = 0,$$
$$L_y = 1 - 2\lambda y = 0$$
$$L_\lambda = 4 - x^2 - y^2 = 0.$$

These equations are satisfied by

$$(x_1, y_1, \lambda_1) = (4/5^{1/2}, 2/5^{1/2}, 5^{1/2}/4)$$

and by

$$(x_2, y_2, \lambda_2) = -(x_1, y_1, \lambda_1).$$

For the second order condition, $m = 1$, $k = n = 2$, and the determinant of the bordered Hessian is

$$\begin{vmatrix} L_{xx} & L_{xy} & g_x \\ L_{yx} & L_{yy} & g_y \\ g_x & g_y & 0 \end{vmatrix} = \begin{vmatrix} -2\lambda & 0 & 2x \\ 0 & -2\lambda & 2y \\ 2x & 2y & 0 \end{vmatrix} = 8\lambda(x^2 + y^2),$$

which has the sign of λ. It is positive at the first stationary point and negative at the second, and therefore the first solution gives a local maximum and the second a local minimum.

Finally, we note a more general theorem that applies without the regularity assumptions that are typically made.

Theorem 2. *Let f and g_i $i = 1, \ldots, m$, be continuously differentiable functions of $x = (x_1, \ldots, x_n)$. If x^* is a local maximum or local minimum of f for all points x in a neighborhood of x^* satisfying* (15), *then there exist $m + 1$ real numbers $\lambda_0, \lambda_1, \ldots, \lambda_m$, not all zero, such that*

$$\lambda_0\, \partial f(x^*)/\partial x_j - \sum_{i=1}^{m} \lambda_i\, \partial g_i(x^*)/\partial x_j = 0, \qquad j = 1, \ldots, n. \quad (25)$$

If x^* is a local extremum, then the rank of the augmented Jacobian matrix, formed by appending the columns of $\partial f/\partial x_j$, $j = 1, \ldots, n$, to (21), is less than $m + 1$ at x^*. Furthermore, if that augmented matrix has the same rank as (21) at x^*, then $\lambda_0 \neq 0$ and one can set $\lambda_0 = 1$. If the augmented matrix has a larger rank than (21), then $\lambda_0 = 0$. This can happen if the feasible region is just one point. For instance,

$$\begin{aligned} &\max \quad x + y \\ &\text{subject to} \quad x^2 + y^2 = 0 \end{aligned}$$

has Lagrangian

$$L = \lambda_0(x + y) - \lambda_1(x^2 + y^2).$$

The equations

$$\begin{aligned} L_x &= \lambda_0 - 2\lambda_1 x = 0, \\ L_y &= \lambda_0 - 2\lambda_1 y = 0, \\ L_{\lambda_1} &= -(x^2 + y^2) = 0 \end{aligned}$$

have solution $\lambda_0 = x = y = 0$. They have no solution with $\lambda_0 = 1$.

FURTHER READING

Among the texts treating optimization subject to equality constraint are those by Avriel (Chapter 2), Hadley (Chapter 3) and Silberberg (1990) (Chapter 6).

Section 6

Inequality Constrained Optimization

At the solution to

$$\max \quad f(x, y) \tag{1}$$
$$\text{subject to} \quad g(x, y) \le b, \tag{2}$$

constraint (2) may not be tight; that is, it may hold as a strict inequality. Then, we expect condition (4.9) to hold. Alternatively, the constraint may be active, that is, hold as an equality at the maximum. In this case, one expects (5.6) or (5.11) to hold. These conditions can all be set forth as follows. Associate multiplier λ with (2), form the Lagrangian

$$L(x, y, \lambda) = f(x, y) + \lambda[b - g(x, y)], \tag{3}$$

and differentiate

$$L_x = f_x - \lambda g_x = 0, \tag{4}$$
$$L_y = f_y - \lambda g_y = 0, \tag{5}$$
$$L_\lambda = b - g \ge 0, \qquad \lambda \ge 0, \quad \lambda(b - g) = 0. \tag{6}$$

Conditions (4)–(6) are necessary for a solution to (1) and (2). The new conditions are (6). The first part restates (2). Nonnegativity of λ reflects the fact that increasing b enlarges the feasible region and therefore cannot reduce the maximum value attainable. The third part of (6) says that either the multiplier is zero or else the constraint is tight.

If the constraint is not tight, then $b > g$, so $\lambda = 0$ and (4) and (5) reduce to the familiar requirements of Section 4. The marginal valuation of the constraint is zero since it has no effect at the optimum. If the constraint is tight, then (4)–(6) replicate the requirements of Section 5.

Example.

$$\max \quad x^2 + y^2$$
$$\text{subject to} \quad 2x^2 + y^2 \le 4.$$

The Lagrangian is

$$L(x, y, \lambda) = x^2 + y^2 + \lambda(4 - 2x^2 - y^2).$$

The necessary conditions

$$2x - 4\lambda x = 0,$$
$$2y - 2\lambda y = 0,$$
$$2x^2 + y^2 \le 4, \qquad \lambda \ge 0, \quad \lambda(4 - 2x^2 - y^2) = 0,$$

are satisfied by the following points:

$$(x, y, \lambda): (0,0,0), (\pm 2^{1/2}, 0, 1/2), (0, \pm 2, 1).$$

The largest value of the objective function, namely 4, is attained at the last pair of points.

Nonnegativity requirements on variables induce modification of necessary conditions. If the solution x^* to

$$\max \quad f(x) \tag{7}$$
$$\text{subject to} \quad x \ge 0, \tag{8}$$

occurs at $x^* > 0$, then the conditions of Section 4 apply. If the maximum occurs at the boundary $x^* = 0$, however, then the function must be decreasing or stationay at $x = 0$. Thus, necessary conditions for solution (7) and (8) are (Figure 6.1.)

$$f'(x) \le 0, \qquad x \ge 0, \quad xf'(x) = 0. \tag{9}$$

These conditions could also be found by using (1)–(6) with $g = -x$, $b = 0$.

The addition of nonnegativity conditions

$$x \ge 0, \qquad y \ge 0 \tag{10}$$

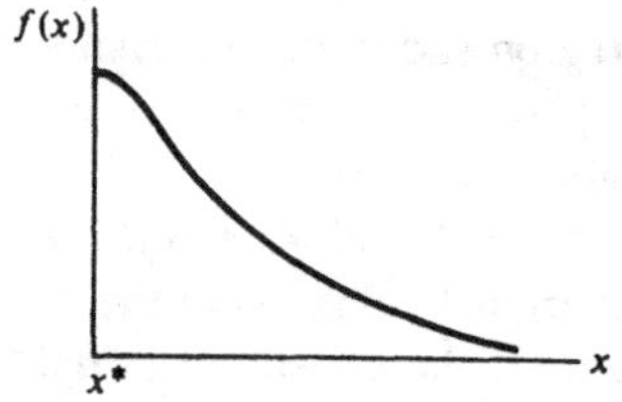

maximum on $x \ge 0$ at $x^* = 0$

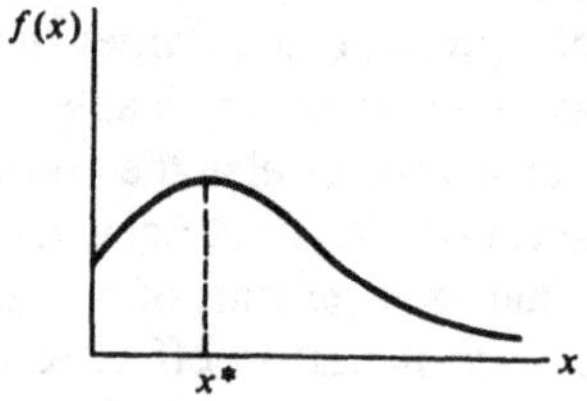

maximum on $x \ge 0$ at $x^* > 0$

Figure 6.1

to the problem (1) and (2) leads to the necessary conditions

$$\begin{aligned} f_x - \lambda g_x \le 0, \quad & x \ge 0, \quad x(f_x - \lambda g_x) = 0, \\ f_y - \lambda g_y \le 0, \quad & y \ge 0, \quad y(f_y - \lambda g_y) = 0, \\ b - g \ge 0, \quad & \lambda \ge 0, \quad \lambda(f - g) = 0. \end{aligned} \tag{11}$$

Example.

$$\begin{aligned} &\max \quad x^2 + y^2 \\ &\text{subject to} \quad 2x^2 + y^2 \le 4, \quad x \ge 0, \quad y \ge 0. \end{aligned}$$

The necessary conditions

$$\begin{aligned} 2x - 4\lambda x \le 0, \quad & x \ge 0, \quad x(2x - 4\lambda x) = 0, \\ 2y - 2\lambda y \le 0, \quad & y \ge 0, \quad y(2y - 2\lambda y) = 0, \\ 2x^2 + y^2 \le 4, \quad & \lambda \ge 0, \quad \lambda(4 - 2x^2 - y^2) = 0 \end{aligned}$$

are satisfied by (0, 0, 0) and by $(2^{1/2}, 0, 1/2)$, as well as by the optimal solution (0, 2, 1).

Necessary conditions for solution of

$$\max \quad f(x_1, \ldots, x_n) \tag{12}$$

$$\text{subject to} \quad g_i(x_1, \ldots, x_n) \le b_i, \quad i = 1, \ldots, m, \tag{13}$$

where $f, g_i, \ldots, g_m$ are continuously differentiable, are known as *Kuhn-Tucker* conditions. Under regularity conditions if $x^* = [x_1, \ldots, x_n]$ solves (12) and (13), then there is a vector $\lambda^* = [\lambda_1, \ldots, \lambda_m]$ such that letting

$$L(x, \lambda) = f + \sum_{i=1}^{m} \lambda_i (b_i - g_i), \tag{14}$$

x^*, λ^* satisfy

$$\begin{aligned} &\partial L/\partial x_j = \partial f/\partial x_j - \sum_{i=1}^{m} \lambda_i \, \partial g_i/\partial x_j = 0, \qquad j = 1, \ldots, n, \\ &\partial L/\partial \lambda_i = b_i - g_i \ge 0, \end{aligned} \tag{15}$$

$$\lambda_i \ge 0, \qquad \lambda_i(b_i - g_i) = 0, \qquad i = 1, \ldots, m. \tag{16}$$

If $x_j \ge 0$ is required, then the jth line of (15) is replaced by

$$\partial L/\partial x_j \le 0, \qquad x_j \ge 0, \quad x_j \, \partial L/\partial x_j = 0. \tag{15$'$}$$

If f is concave and each g_i is convex, then the necessary conditions (15) and (16) are also sufficient and a solution to (15) and (16) is a solution to (12) and (13).

The regularity condition mentioned above is also known as a *constraint qualification*. There are a number of formulations of the needed constraint qualification. Let $I(\boldsymbol{x}^*)$ be the set of indices of the constraints that are tight at $\boldsymbol{x}^*$:

$$I(\boldsymbol{x}^*) = \{i: g_i(\boldsymbol{x}^*) = b_i\}. \tag{17}$$

One constraint qualification is that the matrix of partial derivatives of active constraints $g_i(\boldsymbol{x}^*)$, $i \in I(\boldsymbol{x}^*)$, have full rank. A second is that all constraint functions g are convex and the constraint set has a nonempty interior. A third is developed as follows.

Inactive constraints will remain satisfied for a slight modification from x^* to $\boldsymbol{x}^* + \boldsymbol{h}$, where $\boldsymbol{h}$ is small in absolute value. If $\boldsymbol{x}^* + \boldsymbol{h}$ is to be feasible, we must have

$$g_i(\boldsymbol{x}^* + \boldsymbol{h}) \le b_i, \qquad i \in I(x^*).$$

Expand in a Taylor series around x^*:

$$g_i(\boldsymbol{x}^* + h) = g_i(\boldsymbol{x}^*) + \sum_{j=1}^{n} [\partial g_i(\boldsymbol{x}^*)/\partial x_j] h_j + R_2 \le b_i.$$

Neglecting the remainder term and recalling that $g_i(\boldsymbol{x}^*) = b_i$, this becomes

$$\sum_{j=1}^{n} [\partial g_i(\boldsymbol{x}^*)/\partial x_j] h_j \le 0, \qquad i \in I(\boldsymbol{x}^*), \tag{18}$$

for any feasible modification $\boldsymbol{h}$. It turns out that (18) may also be satisfied by an infeasible modification in case the regularity condition is not satisfied. For example, if the constraints are

$$\begin{aligned} g_1(x_1, x_2) &= -x_1 \le 0, \\ g_2(x_1, x_2) &= -x_2 \le 0, \\ g_3(x_1, x_2) &= -(1 - x_1)^3 + x_2 \le 0, \end{aligned}$$

then at $(1, 0)$, the second and third constraints are tight. Condition (18) becomes

$$\begin{aligned} 0h_1 - 1h_2 &\le 0 \\ 0h_1 + h_2 &\le 0, \end{aligned}$$

which is satisfied by $\boldsymbol{h} = (a, 0)$, where $a > 0$. But $\boldsymbol{x}^* + \boldsymbol{h} = (1 + a, 0)$ is not feasible for any $a > 0$ (see Figure 6.2). A constraint qualification rules out such a cusp. It indicates that there is no vector $\boldsymbol{h}$ satisfying (18) that improves the value of the objective function, that is, satisfying (18) and also

$$\sum_{j=1}^{n} h_j \, \partial f(\boldsymbol{x}^*)/\partial x_j > 0. \tag{19}$$

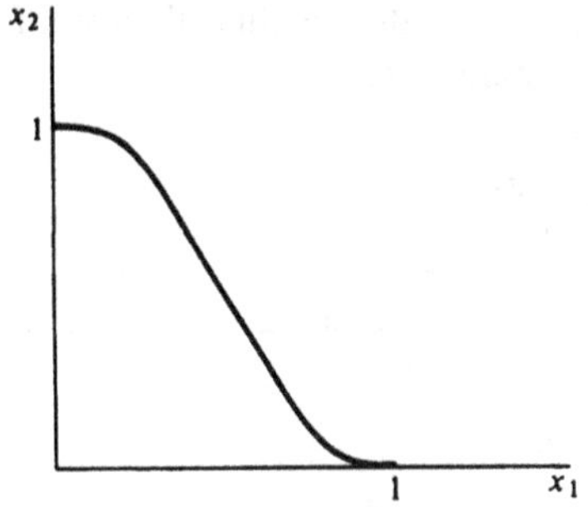

Figure 6.2

The nonexistence of an h that satisfies (18) and (19) is called the *constraint qualification* and is assured by either of the two other constraint qualifications mentioned earlier.

The Kuhn-Tucker theorem can be proved using Farkas' lemma.

Farkas' Lemma. *Let $\mathbf{q} = [q_1, \ldots, q_n]$ and $\mathbf{x} = [x_1, \ldots, x_n]$ be n-dimensional vectors and let A denote an $m \times n$ matrix*:

$$A = \begin{bmatrix} a_{11} & a_{12} & \cdots & a_{1n} \\ a_{21} & a_{22} & \cdots & a_{2n} \\ \vdots & \vdots & & \vdots \\ a_{m1} & a_{m2} & \cdots & a_{mn} \end{bmatrix}.$$

Then, the statement that

(i) $\mathbf{q} \cdot \mathbf{x} \le 0$ *for all x such that* $Ax \ge 0$

is equivalent to the statement that

(ii) *there exists an m-dimensional vector*

$$v = [v_1, \ldots, v_m] \ge 0 \qquad \text{such that} \quad vA + q = 0.$$

The two statements in Farkas' lemma can be written equivalently

(*i*) $\sum_{j=1}^{n} q_j x_j \le 0$ *for all* $[x_1, \ldots, x_n]$ *such that*

$$\sum_{j=1}^{n} a_{ij} x_j \ge 0, \qquad i = 1, \ldots, m,$$

and

(*ii*) *there exist* $v_i \ge 0$, $i = 1, \ldots, m$, *such that*

$$\sum_{i=1}^{m} v_i a_{ij} + q_j = 0, \qquad j = 1, \ldots, n.$$

One proof of this result uses the duality theory of linear programming. Consider the linear programming problem.

$$\max \sum_{j=1}^{n} q_j x_j$$

$$\text{subject to} \quad \sum_{j=1}^{n} a_{ij} x_j \geq 0, \quad i = 1, \ldots, m.$$

The dual linear program is

$$\min \sum_{i=1}^{m} 0 v_i$$

$$\text{subject to} \quad -\sum_{i=1}^{m} v_i a_{ij} = q_j, \quad j = 1, \ldots, n, \quad v_i \geq 0, \quad i = 1, \ldots, m.$$

If statement (i) holds, then the objective of the maximization problem is nonpositive for all feasible $[x_1, \ldots, x_n]$. The maximum that can be achieved is zero, and it certainly can be achieved with $x_j = 0, j = 1, \ldots, n$. Then, duality theory assures that the minimization problem has a solution as well, with value zero. Thus, the constraints of the minimization problem can be satisfied. That, in turn, means that statement (ii) is satisfied.

Conversely, suppose statement (ii) holds. Then the minimization problem has a feasible solution (with value zero). By duality, the maximization problem must also have a feasible solution with value zero. This, finally, implies that statement (i) holds. Now we have

Kuhn-Tucker Theorem. *If $f, g_1, \ldots, g_m$ are all differentiable and $\mathbf{x}^*$ is a local maximum, then there are multipliers $\boldsymbol{\lambda} = [\lambda_1, \ldots, \lambda_m]$, such that $\mathbf{x}^*, \boldsymbol{\lambda}$ satisfy* (15) *and* (16), *provided that there is no $\mathbf{h}$ satisfying* (18) *and* (19).

PROOF. Since there is no $\mathbf{h}$ satisfying (18) and (19), we must have

$$\sum_{j=1}^{n} h_j \, \partial f(x^*)/\partial x_j \leq 0$$

for all h such that

$$\sum_{j=1}^{n} h_j \, \partial g_i(x^*)/\partial x_j \leq 0, \qquad i \in I(x^*).$$

Identify h_j, $\partial f(x^*)/\partial x_j$, and $\partial g_i(x^*)/\partial x_j$ with x_j, q_j, and a_{ij}, respectively, in part (i) of Farkas' lemma. Then, identifying λ_i with v_i, part (ii) of Farkas' lemma says that there exist $\lambda_i \geq 0$, $i \in I(x^*)$, such that

$$\sum_{i \in I} \lambda_i \, \partial g_i(x^*)/\partial x_j + \partial f(x^*)/\partial x_j = 0, \qquad j = 1, \ldots, n.$$

Let $\lambda_i = 0$ for $i \notin I(x^*)$. Then (15) and (16) follow.

Without the regularity condition, we have the following result.

Fritz John Theorem. *Suppose $f, g_1, \ldots, g_m$ are continuously differentiable. If x^* is a solution to* (12) *and* (13), *there is a vector $\lambda^* = [\lambda_0^*, \lambda_1^*, \ldots, \lambda_m^*]$ such that x^*, λ^* satisfy*

$$\lambda_0 \, \partial f / \partial x_j - \sum_{i=1}^{m} \lambda_i \, \partial g_i / \partial x_j = 0,$$

$$\lambda \neq 0, \qquad \lambda \geq 0, \qquad \lambda_i (b - g_i) = 0, \qquad i = 1, \ldots, m.$$

FURTHER READING

Among the texts that discuss constrained optimization are those by Avriel (Chapter 3), Hadley (Chapter 6), Lasdon (Chapter 1), Zangwill (Chapter 2), and Silberberg (1990) (Chapter 14).

Section 7

Line Integrals and Green's Theorem

Let C denote a curve in the x-y plane; C is said to be *smooth* if it can be represented parametrically as

$$x = x(t), \qquad y = y(t), \qquad a \le t \le b$$

where x and y are continuously differentiable on $a \le t \le b$. Let A be the point $(x(a), y(a))$ and B be the point $(x(b), (b))$. The *orientation* of C is the direction from A toward B (see Figure 7.1). If A and B coincide but the curve does not otherwise intersect itself, then C is a simple closed curve.

Let $F(x, y)$ be a function defined for all (x, y) along C. Then, the line integral of F with respect to x along C is defined as

$$\int_C F(x, y)\, dx = \lim_{n\to\infty} \sum_{i=1}^{n} F(x(t_i), y(t_i))[x(t_i) - x(t_{i-1})],$$

provided the limit exists when $x(t-1) \to x(t)$, where $a = t_1 < t_2 < \cdots < t_{n-1} < t_n = b$, and can be evaluated by the ordinary integral

$$\int_C F(x, y)\, dx = \int_a^b F(x(t), y(t))\, x'(t)\, dt. \tag{1}$$

Similarly, the line integral of F with respect to y along C is defined as

$$\int_C F(x, y)\, dy = \lim_{n\to\infty} \sum_{i=1}^{n} F(x(t_i), y(t_i))[y(t_i) - y(t_{i-1})]$$

$$= \int_a^b F(x(t), y(t))\, y'(t)\, dt. \tag{2}$$

Reversing the orientation of C reverses the sign of the integral, since it amounts to reversing the limits of integration on the corresponding ordinary integral.

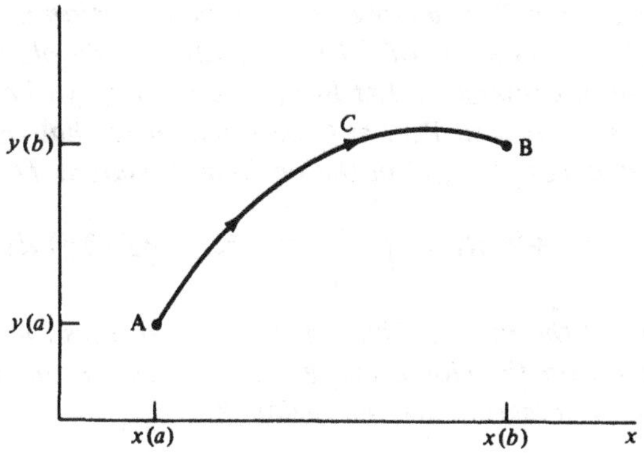

Figure 7.1

Example. Let C be the parabola represented parametrically by $x = 3t^2$, $y = 2t$, $0 \le t \le 1$, oriented in the direction of increasing t. Then, since $x'(t) = 6t$ and $y'(t) = 2$, we have the line integrals

$$\int_C x^2 y\,dx = \int_0^1 (3t^2)^2 2t6t\,dt = 108/7,$$

$$\int_C x^2 y\,dy = \int_0^1 (3t^2)^2 2t2\,dt = 6.$$

The expression

$$\int_C F(x, y)\,dx + G(x, y)\,dy$$

means

$$\int_C F(x, y)\,dx + \int_C G(x, y)\,dy.$$

Thus, for example,

$$\int_C x^2 y\,dx + x^2 y\,dy = 108/7 + 6.$$

In the special case that $y(t) = t$, we have $y'(t) = 1$, so that

$$\int_C G(x, t)\,dt = \int_a^b G(x(t), t)\,dt$$

and

$$\int_C F(x, t)\,dx + G(x, t)\,dt = \int_a^b \left[F(x, t)x' + G(x, t)\right]dt. \quad (3)$$

An important theorem, Green's theorem, explains how a line integral on a closed curve can be evaluated as an ordinary double integral.

Green's Theorem. *Let R be a closed and bounded region in the x-y plane, whose boundary C consists of a finite number of simple smooth closed curves that do not intersect. Let $F(x, y)$ and $G(x, y)$ have continuous first partial derivatives in R. Let C be oriented so that region is on the left as one advances along C in the positive direction. Then,*

$$\oint_C F\,dx + G\,dy = \int\int_R (\partial G/\partial x - \partial F/\partial y)\,dx\,dy. \tag{4}$$

The direction of the integral sign on the left indicates the orientation of the line integral on the closed curve. The expression on the right is the ordinary double integral over the region R.

Example. The area of a region R can be written as any of the following three line integrals.

$$\int_C -y\,dx = \int_C x\,dy = (1/2)\left[\int_C -y\,dx + x\,dy\right].$$

To check the third expression, for example, note that $F = -y, G = x$, so $\partial G/\partial x = 1$ and $\partial F/\partial y = -1$. Hence,

$$(1/2)\left[\int_C -y\,dx + x\,dy\right] = (1/2)\int\int_R [1-(-1)]\,dx\,dy = \int\int_R dx\,dy$$

as claimed.

APPENDIX B

DIFFERENTIAL EQUATIONS

Section 1

Introduction

The solution to an algebraic equation is a number or set of numbers that satisfy the equation. Solutions of the quadratic equation $t^2 - 8t + 15 = 0$ are the numbers *3* and *5* since they satisfy the equation.

Solutions of differential equations, on the other hand, are functions. Solving a differential equation means finding a function that, together with its derivatives, satisfies the specified equation. For example, a solution to the differential equation

$$y'(t) = dy/dt = 4 \tag{1}$$

is a function $y(t)$, whose derivative is equal to 4, namely, the linear functions $y(t) = a + 4t$, where a is any constant. Solutions of

$$y''(t) = d^2y/dt^2 = 2 \tag{2}$$

are functions $y(t)$ whose second derivative is 2, namely, functions of the form

$$y(t) = a + bt + t^2,$$

where a and b are constants. Solutions to

$$d^2y/dt^2 = dy/dt \qquad \text{or} \qquad y''(t) = y'(t) \tag{3}$$

are functions $y(t)$ whose first and second derivatives are equal, namely, functions of the firm

$$y(t) = be^t + a,$$

as may be verified by computing the derivatives.

Further examples of differential equations are

$$yy' = t, \tag{4}$$

$$(y'')^3 + y''(y')^4 - t^7 y = \sin t, \tag{5}$$

where $y = y(t)$, $y' = y'(t)$, and so on. A ***solution*** to a differential equation is a function $y(t)$ that, together with its derivatives, satisfies the differential equation. The ***general solution*** is the set of all solutions to the differential equation. A ***particular*** solution is obtained by specifying a value for the constant(s) of integration. Thus $y(t) = a + 2t$ is the general solution to (1), whereas $y(t) = 2t$ and $y(t) = -3 + 2t$ are particular solutions. Sometimes we seek the particular solution which passes through a given point.

Differential equations are classified in terms of the highest derivative appearing in them. This is referred to as their ***order***. Equations (1) and (4) are first order, whereas (2), (3), and (5) are of second order. A differential equation is said to be ***linear*** if the unknown function $y(t)$ and its derivatives each appear only to the first power. Otherwise, the equation is said to be ***nonlinear***. Thus, (1)–(3) are linear differential equations, whereas (4) and (5) are nonlinear.

A differential equation may be solved by the method known as *separation of variables* if it can be written as the equation of a term depending on y alone and a term depending on t alone. For example, the differential equation $g(y)y' = f(t)$ (where f depends only on t and g only on y) can be written

$$g(y)\,dy = f(t)\,dt$$

since $y' = dy/dt$. The variables are now separated and the solution is

$$\int g(y)\,dy = \int f(t)\,dt + c,$$

where c is an arbitrary constant.

Example 1. $y' = dy/dt = t^2$ may be formally rearranged to $dy = t^2\,dt$. Since the left side depends on y alone and the right on t alone, the variables have been separated. Integrating gives $y = t^3/3 + c$.

Example 2. $y' = aty$ can be rearranged to $dy/y = at\,dt$. The left side depends on y only and the right side depends only on t. Integrating gives $\ln y = at^2/2 + c$ or, equivalently, $y = e^{c + at^2/2}$ as the function sought.

A differential equation

$$f(t, y) + g(t, y)\,dy/dt = 0$$

or, equivalently,

$$f(t, y)\,dt + g(t, y)\,dy = 0$$

is said to be *exact* if there is some function $U(t, y)$ such that

$$dU \equiv U_t\,dt + U_y\,dy \equiv f\,dt + g\,dy.$$

Thus, the differential equation is exact if it is precisely the total differential of some function. The exact differential equation

$$dU = 0$$

has immediate solution

$$U(t, y) = c.$$

Example 3. The differential equation $t^2 y' + 2ty = 0$ may be written as $t^2\,dy + 2ty\,dt = 0$, which is equivalent to $d(t^2 y) = 0$. (Check this claim by differentiating: $d(t^2 y) = 2ty\,dt + t^2 dy$.) Integrating gives $t^2 y = c$, so

$$y = c/t^2.$$

Example 4. The differential equation $(1 + 2ty)y' + y^2 = 0$ may be written as $(1 + 2yt)\,dy + y^2\,dt = 0$, which is equivalent to $d(ty^2 + y) = 0$. The solution is $ty^2 + y = c$, which is an implicit equation for y as a function of t.

EXERCISE

Solve $yy' = t^2$. (Answer: $y^2 = 2t^3/3 + c$.)

FURTHER READING

There are many fine texts on differential equations, for example, those of Coddington and Levinson, Ince, Kaplan, Pontryagin, and Sanchez.

Section 2

Linear First Order Differential Equations

CONSTANT COEFFICIENTS

A first order linear differential equation with constant coefficients has the form

$$y'(t) + Py(t) = Q \tag{1}$$

where P and Q are given constants. To find a function $y(t)$ that satisfies (1), note that

$$d(e^{Pt}y(t))/dt = Pe^{Pt}y(t) + e^{Pt}y'(t) = e^{Pt}[y'(t) + Py(t)].$$

Thus, if equation (1) is multiplied by e^{Pt}, the left side will be an *exact differential*, that is, the total derivative with respect to t of some function. Formally multiplying through by dt then gives

$$d(e^{Pt}y) = Qe^{Pt}\,dt,$$

which may be integrated to

$$e^{Pt}y = e^{Pt}Q/P + c,$$

where c is a constant of integration. Multiplying by e^{-Pt} gives

$$y(t) = Q/P + ce^{-Pt} \tag{2}$$

as the family of functions that satisfy differential equation (1). It is called the *general solution* of (1).

To determine c, we need the value of the function y at one point. For example, if $y(0) = y_0$ were required, then, evaluating (2) at $t = 0$ gives $y_0 = Q/P + c$, and therefore $c = y_0 - Q/P$. The solution of (1) satisfying $y(0) = y_0$ is

$$y(t) = y_0e^{-Pt} + (1 - e^{-Pt})Q/P. \tag{3}$$

In sum, the procedure for solving a first order linear differential equation with constant coefficients (1) is to multiply through by e^{Pt}, called the *integrating factor*, and integrate. Use the value of the function at one point to evaluate the constant of integration. The point used is called an *initial condition* or *boundary condition*.

VARIABLE RIGHT SIDE

If the right side of (1) were not constant, but a known function of t, the procedure would be similar. For instance, given

$$y'(t) + Py(t) = ae^{bt}, \tag{4}$$

multiply by integrating factor e^{Pt} and separate variables:

$$d\left(e^{Pt}y(t)\right) = ae^{(b+P)t}\,dt.$$

Integrate:

$$e^{Pt}y(t) = ae^{(b+P)t}/(b+P) + c.$$

Therefore,

$$y(t) = ae^{bt}/(b+P) + ce^{-Pt} \tag{5}$$

is the family of functions satisfying (4). Again, c can be determined from a boundary condition.

VARIABLE COEFFICIENTS

The general form of a first order linear differential equation is

$$y'(t) + P(t)y(t) = Q(t), \tag{6}$$

where both $P(t)$ and $Q(t)$ are known functions and the function $y(t)$ is to be found. The integrating factor is $e^{\int P\,dt}$ since (recalling that $d(\int P\,dt)/dt = P$)

$$d\left(y(t)e^{\int P\,dt}\right)/dt = e^{\int P\,dt}\left[y'(t) + P(t)y(t)\right].$$

Therefore, multiplying (6) by the integrating factor $e^{\int P\,dt}$ and integrating gives

$$ye^{\int P\,dt} = \int Q(t)e^{\int P\,dt}\,dt + c$$

or finally

$$y = e^{-\int P\,dt}\int Q(t)e^{\int P\,dt}\,dt + ce^{-\int P\,dt}, \tag{7}$$

where c is the constant of integration. This is the general solution of (6). A particular solution is obtained by specifying the value of the constraint c.

If faced with a differential equation like (6), one should *not* try to use formula (7) directly. Rather, one should follow the same *procedure* as outlined above to obtain (7): multiply by $e^{\int P(t)\,dt}$ and integrate. We illustrate with

$$y'(t) + ay(t)/t = b. \tag{8}$$

The coefficient of y, namely a/t, is not constant. Following the outline above, the integrating factor will be

$$e^{\int (a/t)\,dt} = e^{a\ln t} = t^a.$$

Multiplying (8) through by t^a gives

$$t^a[y'(t) + ay(t)/t] = d(t^a y(t))/dt = bt^a.$$

Integrating

$$\begin{aligned} t^a y(t) &= bt^{a+1}/(a+1) + c && \text{if} \quad a \neq -1, \\ y(t)/t &= b \ln t + c && \text{if} \quad a = -1, \end{aligned}$$

so,

$$\begin{aligned} y(t) &= bt/(a+1) + ct^{-a}, && a \neq -1, \\ y(t) &= bt \ln t + ct, && a = -1, \end{aligned} \tag{9}$$

is the solution of (8).

If Q(t) $\equiv 0$ in (6), the resulting differential equation is said to be *homogeneous*; otherwise it is *nonhomogeneous*. One can check the samples given so far (as well as (6) and (7)) to confirm that the general solution of a first order linear differential equation consists of the sum of the general solution to the related homogeneous differential equation (obtained by setting $Q \equiv 0$ and solving) and a particular solution to the complete equation. For instance, the general solution to the homogeneous equation

$$y' + Py = 0$$

obtained from (1) is $y(t) = ce^{-Pt}$, while a particular solution to (1) is $y = Q/P$; the sum of these two functions is (2), the general solution to (1). While this observation is not needed for solving first order linear differential equations (since they can be solved by the procedure outlined above), the principle is a useful one for differential equations of higher order. It remains true that the general solution of a nonhomogeneous linear differential equation is the sum of the general solution to the related homogeneous equation and a particular solution of the complete equation.

In many applications of calculus of variations or optimal control theory, interest focuses on the behavior of the solution to a differential equation as the independent variable, usually time, increases indefinitely. The value that the solution approaches (if any) is referred to as the *steady state*, *stationary state*, or *equilibrium*. For example, from (2) it follows that if $P > 0$, then

$$\lim_{t\to\infty} y = \lim_{t\to\infty}(Q/P + ce^{-Pt}) = Q/P.$$

The particular solution $y = Q/P$ of (1) can also be interpreted as the stationary state. It could also be determined by asking the value of y that satisfies (6) with $y' = 0$.

EXERCISE

Solve $x'(t) + ax(t)/t = bt^2$.

Section 3

Linear Second Order Differential Equations

A second order linear differential equation can be put in the form

$$y''(t) + P(t)y'(t) + Q(t)y(t) = R(t), \tag{1}$$

where P, Q, and R are known functions and the function $y(t)$ is to be found. Equation (1) is called the complete equation. Associated with (1) is a differential equation obtained by replacing $R(t)$ by 0:

$$y''(t) + P(t)y'(t) + Q(t)y(t) = 0, \tag{2}$$

which is called the *reduced* equation. The complete equation is also said to be *nonhomogeneous* while the reduced equation is *homogeneous*. The reduced equation is of interest owing to

Theorem 1. *The general solution of the complete equation* (1) *is the sum of any particular solution of the complete equation and the general solution of the reduced equation* (2).

The general solution of the reduced equation is given in Theorem 2 below. Before citing it, we give a needed definition. The functions $y_1(t)$ and $y_2(t)$ are said to be linearly dependent on the interval $t_0 \le t \le t_1$ if there are constants c_1 and c_2, not both zero, such that

$$c_1 y_1(t) + c_2 y_2(t) \equiv 0, \qquad t_0 \le t \le t_1. \tag{3}$$

If no such identity holds, the functions are said to be *linearly independent*.

For example, the functions $y_1(t) = e^{2t}$ and $y_2(t) = 3e^{2t}$ are linearly dependent, since

$$3y_1(t) - y_2(t) = 0.$$

On the other hand, $y_1(t) = e^{2t}$ and $y_3(t) = e^{-2t}$ are linearly independent since

$$c_1 e^{2t} + c_2 e^{-2t} = e^{-2t}\left(c_1 e^{4t} + c_2\right)$$

can be zero on an interval only if $c_1 = c_2 = 0$.

Now we state

Theorem 2. *Any solution $y(t)$ of the reduced equation (2) on $t_0 \le t \le t_1$ can be expressed as a linear combination*

$$y(t) = c_1 y_1(t) + c_2 y_2(t), \qquad t_0 \le t \le t_1, \tag{4}$$

of any two solutions y_1 and y_2 that are linearly independent.

Since every solution of (2) is of form (4), (4) is called the *general solution* of (2). If we find any two linearly independent solutions of (2), then we can find all solutions of (2).

For example, both $y_1(t) = e^{2t}$ and $y_2(t) = e^{-2t}$ satisfy the differential equation

$$y''(t) - 4y(t) = 0. \tag{5}$$

Since these solutions are linearly independent, the general solution of (5) is

$$y(t) = c_1 e^{2t} + c_2 e^{-2t}.$$

Any solution of (5) must be of this form.

HOMOGENEOUS EQUATIONS WITH CONSTANT COEFFICIENTS

We consider differential equation (1) in which $P(t) = A$ and $Q(t) = B$ are constant. Suppose also that $R(t) = 0$. Since the general solution of $y' + Py = 0$ is $y = ce^{-Pt}$, one might guess that a solution of the second order differential equation

$$y''(t) + Ay'(t) + By(t) = 0 \tag{6}$$

would be of the form $y = ce^{rt}$ for appropriate choice of constants c and r. If $y(t) = ce^{rt}$ were a solution, it would satisfy (6). Compute $y' = rce^{rt}$, $y'' = r^2 ce^{rt}$ and substitute into (6):

$$ce^{rt}\left(r^2 + Ar + B\right) = 0.$$

Excluding the unhelpful case of $c = 0$, our trial solution satisfies (6) if and only if r is a solution to the quadratic equation

$$r^2 + Ar + B = 0.$$

Equation (7) is called the *characteristic equation* associated with (6). It has two roots, that may be found by the quadratic formula, namely

$$r_1, r_2 = -A/2 \pm \left(A^2 - 4B\right)^{1/2}/2. \tag{8}$$

There are three cases according to the sign of $A^2 - 4B$. Each is considered in turn.

Case 1. $A^2 > 4B$. In this case, the roots r_1 and r_2 are real and distinct. The general solution to (6) is therefore

$$y(t) = c_1 e^{r_1 t} + c_2 e^{r_2 t}, \tag{9}$$

where r_1 and r_2 are the roots (8) of the characteristic equation (7), while c_1 and c_2 are arbitrary constants.

Example. The characteristic equation associated with

$$y'' - 4y = 0$$

if $r^2 - 4 = 0$, which has real roots $r_1, r_2 = \pm 2$. Hence the general solution of the given differential equation is

$$y(t) = c_1 e^{2t} + c_2 e^{-2t}.$$

Case 2. $A^2 < 4B$. Here the roots r_1 and r_2 are complex conjugates:

$$r_1, r_2 = -A/2 \pm i(4B - A^2)^{1/2}/2 = a \pm bi \tag{10}$$

where $i \equiv (-1)^{1/2}$, and we define

$$a = -A/2, \qquad b = (4B - A^2)^{1/2}/2. \tag{11}$$

The general solution may be written

$$y(t) = e^{at}(c_1 e^{ibt} + c_2 e^{-ibt}). \tag{12}$$

Since it is true that

$$e^{\pm it} = \cos t \pm i \sin t, \tag{13}$$

(12) may be rewritten

$$y(t) = e^{at}(k_1 \cos bt + ik_2 \sin bt), \tag{14}$$

where $k_1 = c_1 + c_2$, $k_2 = c_1 - c_2$. We can also find real solutions, since if a complex function

$$y(t) = u(t) + iv(t) \tag{15}$$

is a solution to (6), then the real functions $u(t)$ and $v(t)$ are each solutions of (6) as well. To see this, use the assumption that (15) is a solution of (6):

$$u'' + Au' + Bu + i(v'' + Av' + Bv) = 0.$$

This equation holds only if the real part and the imaginary part are each zero:

$$u'' + AU' + Bu = 0,$$
$$v'' + Av' + Bv = 0.$$

This means that u and v are each solutions of (6), as claimed. Applying this result to (14), we conclude that the general real solution to (6) in Case (2) is

$$y(t) = e^{at}(k_1\cos bt + k_2\sin bt), \tag{16}$$

where a and b are defined in (11) and k_1 and k_2 are arbitrary constants.

Example. To solve the differential equation

$$y''(t) - 4y'(t) + 13y(t) = 0,$$

we write the characteristic equation

$$r^2 - 4r + 13 = 0$$

and solve

$$r_1, r_2 = 2 \pm 3i.$$

Hence, the general solution is

$$y(t) = e^{2t}(k_1\cos 3t + k_2\sin 3t).$$

Case 3. $A^2 = 4B$. The roots $r_1 = r_2 = -A/2$. Since the roots are repeated, we have only one solution $y(t) = c_1e^{-At/2}$, with one arbitrary constant. We need another. Try

$$y(t) = kte^{rt},$$

where the constants k and r are to be determined. For this trial solution, $y' = ke^{rt}(1 + rt)$, $y'' = rke^{rt}(2 + rt)$. Substituting into (6) gives

$$ke^{rt}\left[2r + A + t(r^2 + Ar + B)\right] = 0 \tag{17}$$

Equation (17) must hold for all t. This means that the coefficient of t must be zero, and therefore

$$r^2 + Ar + B = 0$$

or $r = -A/2$, since $A^2 = 4B$ in the present case. This choice of r also satisfies (17). Thus the general solution to (6) in this case is

$$y(t) = c_1e^{rt} + c_2te^{rt} = e^{rt}(c_1 + c_2t),$$

where $r = -A/2$.

Example. Associated with differential equation

$$y'' + 4y' + 4y = 0$$

is characteristic equation

$$r^2 + 4r + 4 = (r+2)^2 = 0$$

with repeated roots $r_1 = r_2 = -2$. The general solution to the differential equation is

$$y(t) = e^{-2t}(c_1 + c_2 t).$$

NONHOMOGENEOUS EQUATIONS WITH CONSTANT COEFFICIENTS

We can now find the general solution to any homogeneous second order linear differential equation with constant coefficients. To solve a nonhomogeneous differential equation, we need a particular solution to the complete equation. We focus on that. If the complete equation is of the form

$$y'' + Ay' + By = R$$

where B and R are constants, then a particular solution is the constant function $y = R/B$. More generally, if the coefficients A and B are constants, one can use the functional form R to suggest the functional form of the particular solution. This is called the *method of undetermined coefficients*. If R is a polynomial of degree n, one tries a general nth degree polynomial. For example, to find a particular solution of

$$y'' - y' + 2y = t^3 - t,$$

by

$$y(t) = at^3 + bt^2 + ct + d.$$

Differentiate and substitute into the differential equation. Collecting terms,

$$2at^3 + (2b - 3a)t^2 + 2(c - b + 3a)t + (2b - c + 2d) = t^3 - t.$$

For this to be satisfied for all t, the coefficients of like powers of t on each side of the equation must be equal, so

$$\begin{aligned} 2a &= 1, \\ 2b - 3a &= 0, \\ 2(c - b + 3a) &= -1, \\ 2b - c + 2d &= 0. \end{aligned}$$

Solving gives $a = 1/2$, $b = 3/4$, $c = -(5/4)$, and $d = -11/8$, so a particular solution is

$$y(t) = t^3/2 + 3t^2/4 - 5t/4 - 11/8.$$

If $R(t)$ contains trigonometric terms or exponential terms, the trial solution would contain such functional forms. For example, to find a particular solution to

$$y' - 3y = t^2 + e^t \sin 2t,$$

one would try a solution of the form

$$y = at^2 + bt + c + e^t(k_1 \sin 2t + k_2 \cos 2t).$$

Substituting into the differential equation and collecting terms gives

$$-3at^2 + (2a - 3b)t + (b - 3c) - 2(k_1 + k_2)e^t\sin 2t + 2(k_1 - k_2)e^t\cos 2t = t^2 + e^t\sin 2t.$$

Equating coefficients of like terms gives

$$-3a = 1, \qquad 2a - 3b = 0, \qquad b - 3c = 0,$$
$$-2(k_1 + k_2) = 1, \qquad 2(k_1 - k_2) = 0,$$

which have a solution

$$a = -1/3, \qquad b = -2/9, \qquad c = -2/27,$$
$$k_1 = k_2 = -1/4.$$

Therefore, a particular solution is $y(t) = -t^2/3 - 2t/9 - 2/27 - e^t(\sin 2t + \cos 2t)/4$.

A still more general technique for finding a particular solution is called the *variation of parameters* method. It may be applied even in cases in which the coefficients in the differential equation are not constant. It is developed as follows. Suppose that $y_1(t)$ and $y_2(t)$ are known linearly independent solutions of (2). Now consider the function

$$y(t) = c_1(t)y_1(t) + c_2(t)y_2(t), \tag{18}$$

where the functions c_1 and c_2 are to be chosen so that (18) is a particular solution to (1). Differentiating (18) gives

$$y' = c_1 y_1' + c_1' y_1 + c_2 y_2' + c_2' y_2. \tag{19}$$

Impose the restriction that

$$c_1' y_1 + c_2' y_2 = 0. \tag{20}$$

Differentiate (19), using (20):

$$y'' = c_1 y_1'' + c_2 y_2'' + c_1' y_1' + c_2' y_2'. \tag{21}$$

Substitute (18)–(21) into (1) and collect terms:

$$c_1(y_1'' + Py_1' + Qy_1) + c_2(y_2'' + Py_2' + Qy_2) + c_1' y_1' + c_2' y_2' = R. \tag{22}$$

But, since y_1 and y_2 are solutions to (2), the expression in parentheses in (22) are zero, and (22) reduces to

$$c_1' y_1' + c_2' y_2' = R. \tag{23}$$

Thus the functions $c_t'(t)$ and $c_2'(t)$ satisfy the pair of linear equations (20) and (23), which can be solved algebraically for c_1' and c_2'. Integrate the results to find c_1 and c_2 and hence a particular solution (18) of (10). Then, the general solution to (1) is

$$y(t) = a_1 y_1(t) + a_2 y_2(t) + c_1(t) y_1(t) + c_2(t) y_2(t). \tag{24}$$

where a_1 and a_2 are arbitrary constants.

Example. Solve

$$y'' + 2y' + y = e^{-t}/t^2. \tag{25}$$

The characteristic equation of the homogeneous equation is

$$r^2 + 2r + 1 = 0$$

with roots $r_1 = r_2 = -1$. Thus, two linearly independent solutions of the homogeneous equation are

$$y_1(t) = e^{-t} \quad \text{and} \quad y_2(t) = te^{-t}. \tag{26}$$

A particular solution of the complete equation is

$$\begin{aligned} y(t) &= c_1(t) y_1(t) + c_2(t) y_2(t) \\ &= c_1(t) e^{-t} + c_2(t) te^{-t}, \end{aligned} \tag{27}$$

where $c_1'(t)$ and $c_2'(t)$ satisfy (20) and (23), which are in this case

$$c_1'(t) e^{-t} + c_2'(t) te^{-t} = 0, \tag{28}$$

$$-c_1'(t) e^{-t} + c_2'(t)(e^{-t} - te^{-t}) = e^{-t}/t^2. \tag{29}$$

Solving algebraically, we find that

$$c_1'(t) = -1/t \quad \text{and} \quad c_2'(t) = 1/t^2. \tag{30}$$

Integrating gives

$$c_1(t) = -\ln|t| \quad \text{and} \quad c_2(t) = -1/t. \tag{31}$$

Therefore, substituting from (31) into (27) gives the particular solution

$$y(t) = -\ln|t| e^{-t} - e^{-t}. \tag{32}$$

Combining (26) and (32) and collecting terms gives the general solution to (25):

$$y(t) = e^{-t}(a_1 + a_2 t - \ln|t|). \tag{33}$$

Section 4

Linear *n*th Order Differential Equations

SOLUTIONS TO HOMOGENEOUS EQUATIONS

Let $y^{(n)}(t) = d^n y/dt^n$ denote the *n*th derivative of $y(t)$. Then, an *n*th order homogeneous linear differential equation with constant coefficients has the form

$$y^{(n)}(t) + p_1 y^{(n-1)}(t) + p_2 y^{(n-2)}(t) + \cdots + p_{n-1} y^{(1)}(t) + p_n y(t) = 0 \tag{1}$$

where the p_i are given constants.

Extending the definition given earlier, we say that the functions $y_1(t)$, $y_2(t), \ldots, y_n(t)$ are *linearly dependent* on $t_0 \le t \le t_1$ if there exist constants $c_1, c_2, \ldots, c_n$, not all zero, such that

$$c_1 y_1(t) + c_2 y_2(t) + \cdots + c_n y_n(t) \equiv 0, \qquad t_0 \le t \le t_1.$$

If no such set of n constants exists, the functions $y_1, \ldots, y_n$ are *linearly independent* on $t_0 \le t \le t_1$. A set of n linearly independent solutions of (1) always exists. If $y_1(t), \ldots, y_n(t)$ is a set of linearly independent solutions of (1), then the general solution is

$$y(t) = c_1 y_1(t) + c_2 y_2(t) + \cdots + c_n y_n(t). \tag{2}$$

Following the lead suggested by the cases of $n = 1$ and $n = 2$, we try $y = e^{rt}$ as a solution. Putting the trial solution in (1) (after computing the relevant derivatives) yields

$$e^{rt}(r^n + p_1 r^{n-1} + p_2 r^{n-2} + \cdots + p_{n-1} r + p_n) = 0. \tag{3}$$

Our proposed solution satisfies (1) if (3) is satisfied. Thus, r must satisfy the *characteristic equation*

$$r^n + p_1 r^{n-1} + \cdots p_{n-1} r + p_n = 0 \tag{4}$$

associated with (1). Polynomial equation (4) has, by the fundamental theorem of algebra, exactly n roots. They need not be distinct or real.

Corresponding to each real nonrepeated root r of (4), there is a solution of (1) of the form

$$y(t) = e^{rt}.$$

Corresponding to a real root r that has multiplicity m—that is, repeated m times—we have the m linearly independent solutions

$$y_j(t) = t^{j-1}e^{rt}, \qquad j = 1, \ldots, m.$$

Complex roots come in pairs, whose elements are complex conjugates. As before, corresponding to any pair of complex conjugate roots $a \pm bi$ are the two real linearly independent solutions

$$e^{at}\cos bt \qquad \text{and} \qquad e^{at}\sin bt.$$

If complex roots are repeated, say of multiplicity q, we have the following $2q$ linearly independent solutions:

$$t^{j-1}e^{at}\cos bt \qquad \text{and} \qquad t^{j-1}e^{at}\sin bt, \qquad j = 1, \ldots, q.$$

Thus, in sum, the differential equation (1) has general solution (2), where $y_1(t), \ldots, y_n(t)$ are n linearly independent solutions of (1) corresponding to the n roots $r_1, \ldots, r_n$ of the characteristic equation (4).

Example. The differential equation

$$y^{(4)} + 4y^{(3)} + 13y^{(2)} + 36y^{(1)} + 36y = 0$$

has characteristic equation

$$r^4 + 4r^3 + 13r^2 + 36r + 36 = 0,$$

or

$$(r+2)^2(r^2+9) = 0,$$

with the four roots $-2, -2, 3i, -3i$. The general solution of the differential equation is therefore

$$y(t) = e^{-2t}(c_1 + c_2 t) + c_3\cos 3t + c_4\sin 3t.$$

NONHOMOGENEOUS EQUATIONS

A nonhomogeneous nth order linear differential equation with constant coefficients has the form

$$y^{(n)} + p_1 y^{(n-1)} + p_2 y^{(n-2)} + \cdots + p_{n-1}y^{(1)} + p_n y = R(t). \qquad (5)$$

If $y^*(t)$ is a particular solution to (5) and $c_1 y_1(t) + \cdots + c_n y_n(t)$ is the general solution to the associated homogeneous system (1), then

$$y(t) = y^*(t) + c_1 y_1(t) + c_2 y_2(t) + \cdots + c_n y_n(t) \tag{6}$$

is the general solution of (5). Thus, the solution to (5) is composed of two parts, namely the general solution to the corresponding homogeneous equation plus a particular solution to the nonhomogeneous equation. Since the former has been discussed, we focus on the latter.

A particular solution to (5) can be found by the method of undetermined coefficients suggested in the last section. One tries a particular solution of the same general functional form as (5) and then seeks coefficient values so the trial solution will satisfy the differential equation. For instance, to find a particular solution to

$$y^{(3)} - y^{(2)} + y = t^2,$$

we try

$$y = at^2 + bt + c.$$

Substituting into the differential equation gives

$$at^2 + bt + c - 2a = t^2.$$

Equating coefficients of like powers of t gives the particular solution

$$y = t^2 + 2.$$

Since the terms of y and y' are missing from

$$y^{(3)} - y^{(2)} = t^2,$$

our trial solution for this differential equation will be a fourth degree polynomial

$$y(t) = at^4 + bt^3 + ct^2 + dt + e.$$

Substituting, collecting terms, and equating coefficients of like powers of t gives the particular solution

$$y(t) = -t^4/12 - t^3/3 - 2t^2.$$

A general method of finding a particular solution is the method of variation of parameters, developed as in Section 3 of this appendix. It applies, even if the coefficients p_i are not constant, but requires knowing n linearly independent solutions of the associated homogeneous equation.

Let $y_1(t), \ldots, y_n(t)$ be linearly independent solutions of (1). Then, we form

$$y(t) = c_1(t) y_1(t) + \cdots + c_n(t) y_n(t) \tag{7}$$

where the functions $c_1(t), \ldots, c_n(t)$ are to be found. Impose the restrictions

$$\sum_{i=1}^{n} c_i' y_i = 0, \sum_{i=1}^{n} c_i' y_i' = 0, \ldots, \sum_{i=1}^{n} c_i' y_i^{(n-2)} = 0. \tag{8}$$

Substitute (7) and its derivatives into (5), taking into account restrictions (8). This gives

$$\sum_{i=1}^{n} c_i' y_i^{(n-1)} = R(t). \tag{9}$$

Now the n linear equations (8) and (9) can be solved algebraically for $c_1', \ldots, c_n'$. Integrate the results to find $c_1, \ldots, c_n$ and thereby, through (7), a particular solution to (5).

Example. Solve

$$y''' - y'' = t^2.$$

The characteristic equation associated with the homogeneous part is $r^3 - r^2 = 0$, which has roots

$$r_1 = 0, \qquad r_2 = 0, \qquad r_3 = 1.$$

Thus, three linearly independent solutions to the homogeneous equation are

$$y_1(t) = 1, \qquad y_2(t) = t, \qquad y_3(t) = e^t.$$

The trial solution is

$$y(t) = c_1(t) + tc_2(t) + e^t c_3(t),$$

where the coefficients satisfy (8) and (9):

$$\begin{aligned} c_1'(t) + c_2'(t) + c_3'(t)e^t &= 0, \\ c_2'(t) + c_3'(t)e^t &= 0, \\ c_3'(t)e^t &= t^2. \end{aligned}$$

Solving and integrating:

$$\begin{aligned} &c_1' = t^3 - t^2, \qquad c_2' = -t^2, \qquad c_3' = t^2 e^{-t}; \\ &c_1 = t^4/4 - t^3/3, \qquad c_2 = -t^3/3, \qquad c_3 = -e^{-t}(t^2 + 2t + 2), \end{aligned}$$

so a particular solution is

$$y(t) = -t^4/12 - t^3/3 - t^2 - 2t - 2.$$

EQUIVALENT SYSTEMS

An nth order linear differential equation is equivalent to a certain system of n first order linear differential equations. For example, given

$$y'' + Ay' + By = R, \tag{7'}$$

one can define x by

$$y' = x. \tag{8'}$$

Then $y'' = x'$. Substituting into (7′) gives

$$x' + Ax + By = R. \tag{9'}$$

Thus, (8′) and (9′) constitute a pair of first order linear differential equations equivalent to the single second order differential equation (7′). (Equivalence may be verified by starting from (8′) and (9′), differentiating (8′) with respect to t, and substituting from (9′) into the result to get (7′).)

More generally, given (5), one can define successively the functions $z_1, \ldots, z_{n-1}$, by

$$\begin{aligned} y' &= z_1, \\ (y'' =)\, z_1' &= z_2, \\ (y''' =)\, z_2' &= z_3, \\ &\vdots \\ (y^{(n-1)} =)\, z_{n-2}' &= z_{n-1}. \end{aligned} \tag{10}$$

Substituting from (10) into (5), one gets

$$z_{n-1}' + p_1 z_{n-1} + p_2 z_{n-2} + \cdots + p_{n-1} z_1 = R. \tag{11}$$

Now (10) and (11) constitute a system of n first order linear differential equations in the n functions $y, z_1, \ldots, z_{n-1}$. Equivalence between (10) and (11) and (5) is verified by substituting the equivalent expressions in terms of y and its derivatives from (10) to (11) to get (5). Since a system of n first order linear differential equations is equivalent to a single nth order linear differential equation, results and methods developed for one imply corresponding results and methods for the other.

Section 5

A Pair of Linear Equations

We consider the system of two linear first order differential equations

$$dx/dt = a_1 x(t) + b_1 y(t) + p(t),$$
$$dy/dt = a_2 x(t) + b_2 y(t) + g(t), \tag{1}$$

where a_1, a_2, b_1, b_2 are given constants, $p(t)$ and $g(t)$ are given functions, and $x(t)$ and $y(t)$ are to be found. The solution will be two functions $x(t)$, $y(t)$ that satisfy both equations. Recall that the solution of a single linear differential equation is the general solution to the related homogeneous equation plus a particular solution to the nonhomogeneous equation. A similar theorem applies here.

The homogeneous system corresponding to (1) is

$$dx/dt = a_1 x + b_1 y,$$
$$dy/dt = a_2 x + b_2 y. \tag{2}$$

One method for solving the system (2) is to reduce the pair to a single second order linear differential equation and apply the method of Section 3. Differentiating the first equation of (2) totally with respect to t yields

$$x'' = a_1 x' + b_1 y'.$$

Substitute for x', y', y from (2) and collect terms to get

$$x'' - (a_1 + b_2)x' + (a_1 b_2 - b_1 a_2)x = 0. \tag{3}$$

Hence, if the roots of the characteristic equation

$$r^2 - (a_1 + b_2)r + a_1 b_2 - b_1 a_2 = 0 \tag{4}$$

associated with (3) are real and distinct, then the solution of (3) is

$$x(t) = c_1 e^{r_1 t} + c_2 e^{r_2 t}, \tag{5}$$

where r_1 and r_2 are roots of (4).

Rearranging the second equation of (2), we get

$$y = (x' - a_1 x)/b_1. \tag{6}$$

Substituting the known first for (5) for x (and hence x') gives

$$y(t) = \left[(r_1 - a_1)c_1 e^{r_1 t} + (r_2 - a_1)c_2 e^{r_2 t}\right]/b_1. \tag{7}$$

Thus, the general solution to (2) is (5) and (7) if the roots of (4) are real and distinct. The method of finding the general solution in the case of repeated or complex roots is similar.

A second method of solving (2) directly is available. Our discussion of the solution to the single equation further suggests that we try $x(t) = Ae^{rt}$, $y(t) = Be^{rt}$ as particular solutions to (2). Substitution of these proposed solutions into (2) yields

$$\begin{aligned} rAe^{rt} &= a_1 Ae^{rt} + b_1 Be^{rt}, \\ rBe^{rt} &= a_2 Ae^{rt} + b_2 Be^{rt}. \end{aligned} \tag{8}$$

Divide by e^{rt} throughout (8), collect terms, and rewrite in matrix notation:

$$\begin{bmatrix} a_1 - r & b_1 \\ a_2 & b_2 - r \end{bmatrix} \begin{bmatrix} A \\ B \end{bmatrix} = \begin{bmatrix} 0 \\ 0 \end{bmatrix} \tag{9}$$

In order that the solution to (9) be other than $A = B = 0$, the coefficient matrix in (9) must be singular, or equivalently, its determinant must be zero.

$$\begin{vmatrix} a_1 - r & b_1 \\ a_2 & b_2 - r \end{vmatrix} = 0 \tag{10}$$

Expanding (10) yields a quadratic equation in r:

$$r^2 - r(a_1 + b_2) + a_1 b_2 - a_2 b_1 = 0, \tag{11}$$

referred to as the *characteristic equation* of (2). Note that this is exactly the characteristic equation (4) found above! It has roots

$$r = (a_1 + b_2)/2 \pm \left[(a_1 + b_2)^2 - 4(a_1 b_2 - a_2 b_1)\right]^{1/2}/2. \tag{12}$$

Note for future reference that (12) implies that the two roots r_1 and r_2 satisfy

$$\begin{aligned} r_1 r_2 &= a_1 b_2 - a_2 b_1, \\ r_1 + r_2 &= a_1 + b_2. \end{aligned} \tag{13}$$

If $r_1 \neq r_2$, the general solution to (2) is

$$x(t) = A_1 e^{r_1 t} + A_2 e^{r_2 t},$$

$$y(t) = B_1 e^{r_1 t} + B_2 e^{r_2 t}, \tag{14}$$

where A_1 and A_2 are determined from initial conditions, (12) defines r_1 and r_2, and B_1 and B_2 re determined from (8):

$$B_1 = (r_1 - a_1) A_1 / b_1, \qquad B_2 = (r_2 - a_1) A_2 / b_1. \tag{15}$$

This agrees with the solution (4) and (7) found above.

If the two roots in (12) are identical, then as before we try solutions of the form $x(t) = (A_1 + A_2 t)e^{rt}$, $y(t) = (B_1 + B_2 t)e^{rt}$. The result is

$$x(t) = (A_1 + A_2 t)e^{rt}$$

$$y(t) = [(r - a_1)(A_1 + A_2 t) + A_2]e^{rt}/b_1, \tag{16}$$

where the coefficients B_i have been determined in terms of A_1, A_2, and the parameters of (2). If the roots are complex conjugates, the solution can be stated in terms of real trigonometric functions in a fashion similar to that employed earlier.

We have now indicated how to find the general solution to the homogeneous system (2). There is a theorem that asserts that the general solution to the nonhomogeneous system (1) is the sum of the general solution to (2) and a particular solution to (1). A particular solution to (1) can be obtained by, for instance, the method of variation of parameters. In case p and g are constants, then a particular constant solution x_s, y_s can be found by solving the system of equations

$$a_1 x_s + b_1 y_s + p = 0,$$

$$a_2 x_s + b_2 y_s + g = 0. \tag{17}$$

Extension to systems of several linear first order differential equations is straightforward.

EQUILIBRIUM BEHAVIOR

A point x_s, y_s at which $x' = y' = 0$ is called an *equilibrium point*, *steady state*, or *stationary point*. An equilibrium is *stable* if $\lim_{t\to\infty} x(t) = x_s$ and $\lim_{t\to\infty} y(t) = y_s$. For system (1) with p and g *constant*, there are several cases.

Case IA. The roots of (12) are real and distinct, with $r_2 < r_1 < 0$. From (14), the equilibrium is stable in this case. It is called a *stable mode*. From

(13), the *Routh-Hurwitz* conditions

$$a_1 b_2 - a_2 b_1 > 0, \qquad a_1 + b_2 < 0 \tag{18}$$

hold. These conditions are necessary and sufficient for stability of (1).

Case IB. The roots are real and distinct, with $r_1 > r_2 > 0$. Since both roots are positive, the solution $x(t)$, $y(t)$ in (14) grows without bound as t increases. The equilibrium x_s, y_s is an *unstable mode.*

Case IC. The roots are real and distinct, with $r_1 > 0 > r_2$. If $A_1 \neq 0$, then the term in (14) with positive root will dominate the solution and both $x(t)$ and $y(t)$ grow without bound. However, if $A_1 = 0$ and $A_2 \neq 0$, then the solution (14) will converge to the equilibrium (x_s, y_s) as t increases. This equilibrium is called a *saddlepoint*.

Case ID. The roots are real and distinct, with $r_1 = 0 > r_2$. From (13), $a_1 / b_1 = a_2 / b_2$ in this case, so that equations (17) are either inconsistent or redundant. In the former case there is no equilibrium. In the latter, any point satisfying $a_1 x_s + b_1 y_s + p = 0$ is an equilibrium. The initial condition determines which of these equilibria will be approached as t increases.

Case IE. The roots are real and distinct, with $r_2 = 0 < r_1$. This is similar to Case ID, except that the solution $x(t)$, $y(t)$ moves away from the equilibria (unless $x(0) = x_s$, $y(0) = y_s$).

Case II. The roots are complex conjugates:

$$r_1, r_2 = a \pm bi,$$

where

$$a = (a_1 + b_2)/2, \qquad b = \left[4(a_1 b_2 - a_2 b_1) - (a_1 + b_2)^2\right]^{1/2}/2. \tag{19}$$

The solution is of the form

$$\begin{aligned} x(t) &= e^{at}(k_1 \cos bt + k_2 \sin bt) + x_s, \\ y(t) &= e^{at}(c_1 \cos bt + c_2 \sin bt) + y_s. \end{aligned} \tag{20}$$

There are several possibilities.

Case IIA. The roots are pure imaginaries, with $a = 0$. The solution $x(t)$, $y(t)$ oscillates within fixed bounds. In the x, y space, the trajectories are ellipses about x_s, y_s.

Case IIB. The roots are complex, with negative real parts, and $a < 0$. The solution (20) oscillates and tends toward x_s, y_s. The equilibrium is called a *stable focus.*

Case IIC. The roots are complex, with positive real parts, and $a > 0$. The solution (20) oscillates and moves away from x_s, y_s (unless $x(0) = x_s$, $y(0) = y_s$). The equilibrium is an *unstable focus.*

Case III. The roots are real and equal, $r_1 = r_2 \neq 0$. The solution is (16). The equilibrium is stable if $r < 0$ and unstable if $r > 0$.

One can sketch the behavior of the solution in the (x, y) plane where a point represents $(x(t), y(t))$ and movement in the direction of the arrow represents movement with advancing time.

In nonlinear systems, there is an additional sort of equilibrium, called a *limit cycle.* This is a closed curve in the x, y plane. A stable *limit cycle* is one toward which trajectories converge. An *unstable limit cycle* is one from which trajectories diverge.

NONLINEAR SYSTEMS

Some insight in the nature of an equilibrium of a nonlinear differential equation system can be obtained by study of the approximating linear differential equation system. For example, suppose the system

$$x' = f(x, y), \qquad y' = g(x, y) \tag{21}$$

has an isolated equilibrium at (x_s, y_s). That is, (x_s, y_s) satisfies

$$f(x_s, y_s) = 0, \qquad g(x_s, y_s) = 0, \tag{22}$$

and there is a neighborhood of (x_s, y_s) containing no other equilibria of (21). (Thus limit cycles are not under consideration here.) The approximating linear differential equation system in the neighborhood of (x_s, y_s) is found by expanding the right side of (21) around (x_s, y_s) by Taylor's theorem, retaining only linear terms. Thus,

$$x' = f(x_s, y_s) + f_x(x_s, y_s)(x - x_s) + f_y(x_s, y_s)(y - y_s)$$
$$y' = g(x_s, y_s) + g_x(x_s, y_s)(x - x_s) + g_y(x_s, y_s)(y - y_s).$$

But, in view of (22), this reduces to

$$x' = a_1(x - x_s) + b_1(y - y_s)$$
$$y' = a_2(x - x_s) + b_2(y - y_s), \tag{23}$$

where the constants are

$$a_1 = f_x(x_s, y_s), \qquad a_2 = g_x(x_s, y_s)$$
$$b_1 = f_y(x_s, y_s), \qquad b_2 = g_y(x_s, y_s). \tag{24}$$

In the neighborhood of (x_s, y_s), the solution to (21) behaves like that of (23) in the same neighborhood, and the behavior of (23) can be determined following the analysis for (1).

EXERCISE

Show the solution to the system

$$\begin{aligned} x' &= a_1 x + b_1 y + c_1 z, \\ y' &= a_2 x + b_2 y + c_2 z, \\ z' &= a_3 x + b_3 y + c_3 z \end{aligned}$$

is

$$\begin{aligned} x(t) &= A_1 e^{r_1 t} + A_2 e^{r_2 t} + A_3 e^{r_3 t}, \\ y(t) &= B_1 e^{r_1 t} + B_2 e^{r_2 t} + B_3 e^{r_3 t}, \\ z(t) &= C_1 e^{r_1 t} + C_2 e^{r_2 t} + C_3 e^{r_3 t}, \end{aligned}$$

where r_1, r_2 and r_3 are roots of the determinant equation

$$\begin{vmatrix} a_1 - r & b_1 & c_1 \\ a_2 & b_2 - r & c_2 \\ a_3 & b_3 & c_3 - r \end{vmatrix} = 0$$

provided the roots are distinct.

Section 6

Existence and Uniqueness of Solutions

One way to verify that a solution to a differential equation exists is to produce it. However, if the equation is not readily solved, or if it contains functions that are not fully specified, then it is desirable to have a theorem assuring the existence of a solution.

Theorem 1. *Suppose $f(t, y)$ is a single valued continuous function with a continuous partial derivative $f_y(t, y)$ on a rectangular domain D*

$$|t - t_0| \leq a, \quad |y - y_0| \leq b$$

around the point t_0, y_0. Let M be the upper bound of $|f(t, y)|$ in D and let h be the smaller of a and b/M. Then there exists a unique continuous function of t defined for all t such that $|t - t_0| < h$ that satisfies the differential equation $dy/dt = f(t, y)$ and is equal to y_0 at t_0.

If the differential equation involves a parameter p,

$$dy/dt = f(t, y; p),$$

where $f(t, y; p)$ is single valued, continuous, and continuously differentiable with respect to its arguments in D for $p_1 \leq p \leq p_2$, then the solution depends continuously upon p and is differentiable with respect to p when $|t - t_0| < h$.

Remark 1. *Existence and uniqueness of the solution are assured only in a neighborhood of (t_0, y_0). For example, the solution of $dy/dt = y^2$ with $(t_0, y_0) = (1, -1)$ is $y = -1/t$, but this is not defined at $t = 0$.*

The results of Theorem 1 may be extended to a system of first order differential equations:

$$dy_i/dt = f_i(t, y_1, y_2, \ldots, y_n), \quad i = 1, \ldots, n. \tag{1}$$

Theorem 2. *Suppose the functions $f_i(t, y_1, y_2, \ldots, y_n)$, $i = 1, \ldots, n$, are single valued, continuous, and continuously differentiable in their last n arguments on the domain D defined by $|t - t_0| \le a$ and $|y_i - y_i^0| \le b_i$, $i = 1, \ldots, n$. Let M be the greatest of the upper bounds of $f_1, f_2, \ldots, f_n$ in this domain. Let h be the smallest of $a, b_1/M, \ldots, b_n/M$. Then there is a unique set of continuous solutions of the system of Equations* (1) *for t such that $|t - t_0| < h$ and that the solutions assume the given values $y_1^0, \ldots, y_n^0$ at $t = t_0$.*

Remark 2. *Since the differential equation of order n,*

$$d^n yx/dt^n = f(t, y, y^1, \ldots, y^{(n-1)}) \tag{2}$$

is equivalent to the set of first order equations

$$dy/dt = y_1, \qquad dy_1/dt = y_2, \ldots,$$

$$dy_{n-2}/dt = y_{n-1}, \qquad dy_{n-1}/dt = f(t, y, y_1, \ldots, y_{n-1}),$$

it follows that if f is continuous and is continuously differentiable in its last n arguments, then (2) *has a unique continuous solution that, together with its first $n - 1$ derivatives (which are also continuous), will assume an arbitrary set of initial conditions for $t = t_0$.*

If the functions f_i in (1) are linear in $y_1, \ldots, y_n$, then the existence of solutions on a larger domain can be established.

Theorem 3. *If in* (1), *we have*

$$dy_i/dt = f_i = \sum_{j=1}^{n} a_{ij}(t) y_j + b_i(t), \qquad i = 1, \ldots, n, \tag{3}$$

and if the coefficients $a_{ij}(t)$, $i = 1, \ldots, n$, $j = 1, \ldots, n$, and $b_i(t)$, $i = 1, \ldots, n$, are continuous on $t_0 \le t \le t_1$, then there is a unique set of continuous functions $y_t(t), \ldots, y_n(t)$ having continuous derivatives on $t_0 \le t \le t_1$ that satisfy (3) *and have prescribed values*

$$y_1(t^*) = k_1, \ldots, y_n(t^*) = k_n$$

at a point t^* of $t_0 \le t \le t_1$.

Remark 3. The problems just given are all initial-value problems. The given data all pertain to a single point, say, $y_i(t_0)$, $i = 1, \ldots, n$, for problem (1), or $(y(t_0), y'(t_0), \ldots, y^{(n-1)}(t_0))$ for problem (2). In differential equations arising from calculus of variations or optimal control problems, the boundary conditions typically are divided between specifications at two points in time. Thus, for a second order equation, instead of $y(t_0)$, $y'(t_0)$ we may be given $y(t_0)$ and $y(t_1)$. These situations are called boundary value problems. Conditions under which the general boundary value problem has a solution are more difficult.

References

Aarrestad, Jostein, "Optimal Savings and Exhaustible Resource Extraction in an Open Economy," *Journal of Economic Theory*, 19 (October 1978), 163–179.

Allen, R. G. D., *Mathematical Analysis for Economists*, New York: St. Martin's, 1938.

Amit, R., "Petroleum Reservoir Exploitation: Switching from Primary to Secondary Recovery," *Operations Research*, 34 (July–August 1986), 534–549.

Anderson, W. H. L., "Production Scheduling, Intermediate Goods, and Labor Productivity," *American Economic Review*, 60 (March 1970), 153–162.

Arnold, Ludwig, *Stochastic Differential Equations: Theory and Applications*, New York: Wiley, 1974.

Arrow, K. J., "Optimal Capital Policy, the Cost of Capital, and Myopic Decision Rules," *Annals of the Institute of Statistical Mathematics*, 16 (1964), 21–30.

Arrow, K. J., "Optimal Capital Policy with Irreversible Investment," in *Value, Capital and Growth* (J. N. Wolfe, ed.), Edinborough: Edinborough University Press, 1968, pp. 1–20.

Arrow, K. J. and A. C. Enthoven, "Quasi-Concave Programming," *Econometrica*, 29 (October 1961), 779–800.

Arrow, K. J. and M. Kurz, *Public Investment, the Rate of Return, and Optimal Fiscal Policy*, Baltimore: Johns Hopkins University Press, 1970.

Arvan, L. and L. N. Moses, "Inventory Investment and the Theory of the Firm," *American Economic Review*, 72 (March 1982), 186–193.

Avriel, Mordecai, *Nonlinear Programming*, Englewood Cliffs, N.J.: Prentice-Hall, 1976.

Bakke, V. L., "A Maximum Principle for an Optimal Control Problem with Integral Constraints," *Journal of Optimization Theory and Applications*, 14 (1974), 32–55.

Basar, T. and G. J. Olsder, *Dynamic Noncooperative Game Theory*, London: Academic Press, 1982.

Baum, R. F., "Existence Theorems for Lagrange Control Problems with Unbounded Time Domain," *Journal of Optimization Theory and Applications*, 19 (May 1976), 89–116.

Beckmann, Martin J., *Dynamic Programming of Economic Decisions*, New York: Springer-Verlag, 1968.

Bellman, R., *Dynamic Programming*, Princeton, N.J.: Princeton University Press, 1957.

Bellman, R. and Stuart Dreyfus, *Applied Dynamic Programming*, Princeton, N.J.: Princeton University Press, 1962.

Ben-Porath, Y., "The Production of Human Capital and the Life Cycle of Earnings," *Journal of Political Economy*, 75 (August 1967), 352–365.

Bensoussan, A., E. G. Hurst, and B. Naslund, *Management Applications of Modern Control Theory*, Amsterdam: North-Holland, 1974.

Bensoussan, A., P. R. Kleindorfer, and C. S. Tapiero, eds., *Applied Optimal Control*, TIMS Studies in Management Sciences, Vol. 9, Amsterdam: North-Holland, 1978.

Benveniste, L. M. and J. A. Scheinkman, "On the Differentiability of the Value Function in Dynamic Models of Economics," *Econometrica*, 47 (May 1979), 727–732.

Benveniste, L. M., "Duality Theory for Dynamic Optimization Models of Economics," *Journal of Economic Theory*, 1980.

Berkovitz, Leonard D., *Optimal Control Theory*, New York: Springer-Verlag, 1974.

Berkovitz, Leonard D., "Optimal Control Theory," *American Mathematical Monthly*, 83 (April 1976), 225–239.

Blinder, A. S. and Y. Weiss, "Human Capital and Labor Supply: A Synthesis," *Journal of Political Economy*, 84 (June 1976), 449–472.

Bliss, Gilbert Ames, *Lectures on the Calculus of Variations*, Chicago: University of Chicago Press, 1946.

Boorse, H. A., L. Motz, and J. H. Weaver, *The Atomic Scientists*, New York: Wiley, 1989.

Brito, D. L. and W. H. Oakland, "Some Properties of the Optimal Income Tax," *International Economic Review*, 18 (June 1977), 407–423.

Brock, W. A., *Introduction to Stochastic Calculus: A User's Manual*, unpublished, 1976.

Brock, W. A., "The Global Asymptotic Stability of Optimal Control: A Survey of Recent Results," in *Frontiers of Quantitative Economics* (Michael Intriligator, ed.), Vol. III, Amsterdam: North-Holland, 1977, pp. 207–237.

Brock, W. A. and A. Haurie, "On Existence of Overtaking Optimal Trajectories over an Infinite Time Horizon," *Mathematics of Operations Research*, 1 (November 1976), 337–346.

Brock, W. A. and J. A. Scheinkman, "Global Asymptotic Stability of Optimal Control Systems with Applications to the Theory of Economic Growth," *Journal of Economic Theory*, 12 (February 1976), 164–190.

Brock, W. A., "The Global Asymptotic Stability of Optimal Control with Applications to Dynamic Economic Theory," in *Applications of Control Theory to Economic*

Analysis, (J. Pitchford and S. Turnovsky, eds.), New York: North-Holland, 1977, pp. 173–208.

Brock, W. A., "Differential Games with Active and Passive Players," in *Mathematical Economics and Game Theory*, (R. Henn and O. Moeschlin, eds.), Berlin: Springer-Verlag, 1977.

Brown, Gardner, "An Optimal Program for Managing Common Property Resources with Congestion Externalities," *Journal of Political Economy*, 82 (January/February 1974), 163–173.

Bryson, Arthur E., Jr. and Yu-Chi Ho, *Applied Optimal Control*, Washington, D.C.: Hemisphere Publishing Corp., 1975.

Budelis, J. J. and A. R. Bryson, Jr., "Some Optimal Control Results for Differential-Difference Systems," *IEEE Transactions on Automatic Control*, 15 (April 1970), 237–241.

Calvo, G. A., "Efficient and Optimal Utilization of Capital Services," *American Economic Review*, 65 (March 1975), 181–186.

Caputo, M. R., "Comparative Dynamics via Envelope Methods in Variational Calculus," *Review of Economic Studies*, 57 (1990), 689–697.

Caputo, M. R., "How to do Comparative Dynamics on the Back of an Envelope in Optimal Control Theory," *Journal of Economic Dynamics and Control*, 14 (1990), 655–683.

Caputo, M. R., "New Qualitative Properties in the Competitive Nonrenewable Resource Extracting Model of the Firm," *International Economic Review*, 31 (November 1990), 829–839.

Case, J., *Economics and the Competitive Process*, New York: New York University Press, 1979.

Cass, David and Karl Shell, "The Structure and Stability of Competitive Dynamical Systems," *Journal of Economic Theory*, 12 (February 1976), 31–70.

Cesari, L., "Existence Theorems for Optimal Solutions in Pontryagin and Lagrange Problems," *SIAM Journal of Control A*, 3 (1966), 475–498.

Clark, Colin W., *Mathematical Bioeconomics: The Optimal Management of Renewable Resources*, New York: Wiley, 1976.

Clark, Colin W. and Gordon R. Munro, "The Economics of Fishing and Modern Capital Theory: A Simplified Approach," *Journal of Environmental Economics and Management*, 2 (1975), 92–106.

Clark, Colin W., Frank H. Clarke, and Gordon R. Munro, "The Optimal Exploitation of Renewable Resource Stocks: Problems of Irreversible Investment," *Econometrica*, 47 (January 1979), 25–47.

Clemhout, S. and H. Y. Wan, Jr., "A Class of Trilinear Differential Games," *Journal of Optimization Theory and Applications*, 14 (1974), 419–424.

Clemhout, S. and H. Y. Wan, Jr., "Interactive Economic Dynamics and Differential Games," *Journal of Optimization Theory and Applications*, 27 (1979), 7–29.

Clemhout, S. and H. Y. Wan, Jr., "On Games of Cake Eating," in *Dynamic Policy Games in Economics* (F. van der Ploeg and A. J. de Zeeniv, eds.), New York: Elsevier, 1989.

Coddington, Earl A. and Norman Levinson, *Theory of Ordinary Differential Equations*, New York: McGraw-Hill, 1955.

Connors, Michael M. and Daniel Teichroew, *Optimal Control of Dynamic Operations Research Models*, Scranton, Pa.: International Textbook Co., 1967.

Constantinides, George M. and Scott F. Richard, "Existence of Optimal Simple Policies for Discounted Cost Inventory and Cash Management in Continuous Time," *Operations Research*, 26 (July/August 1978), 620–636.

Cooter, Robert, "Optimal Tax Schedules and Rates: Mirrlees and Ramsey," *American Economic Review*, 68 (December 1978), 756–768.

Cropper, M. L., "Regulating Activities with Catastrophic Environmental Effects," *Journal of Environmental Economics and Management*, 3 (June 1976), 1–15.

Cropper, M. L., "Health, Investment in Health, and Occupational Choice," *Journal of Political Economy*, 85 (December 1977), 1273–1294.

Cullingford, G. and J. D. C. A. Prideaux, "A Variational Study of Optimal Resource Profiles," *Management Science*, 19 (May 1973), 1067–1081.

Cummings, Ronald G. and William G. Schulze, "Optimal Investment Strategy for Boomtowns," *American Economic Review*, 68 (June 1978), 374–385.

Dasgupta, P. and G. Heal, "The Optimal Depletion of Exhaustible Resources," *The Review of Economic Studies*, 1974 Symposium, pp. 3–29.

Davis, B. E., "Investment and Rate of Return for the Regulated Firm," *Bell Journal of Economics and Management Science*, 1 (Fall 1970), 245–270.

Davis, B. E. and D. J. Elzinga, "The Solution of an Optimal Control Problem in Financial Modelling," *Operations Research*, 19 (October 1971), 1419–1433.

Davis, H. T., *Introduction to Nonlinear Differential and Integral Equations*, New York: Dover, 1962.

De Bondt, R. R., "Limit Pricing, Uncertain Entry, and the Entry Lag," *Econometrica*, 44 (September 1976), 939–946.

Dockner, E., "On the Relation Between Dynamic Oligopolistic Competition and Long-Run Competitive Equilibrium," *European Journal of Political Economy*, 4 (1988), 47–64.

Dockner, E., G. Feichtinger, and S. Jorgensen, "Tractable Classes of Non-Zerosum Open-Loop Differential Games," *Journal of Optimization Theory and Applications*, 45 (1985), 179–197.

Dockner, E., G. Feichtinger, and A. Mehlmann, "Noncooperative Solutions for a Differential Game Model of a Fishery," *Journal of Economic Dynamics and Control*, 13 (January 1989), 1–20.

Dorfman, Robert, "An Economic Interpretation of Optimal Control Theory," *American Economic Review*, 59 (December 1969), 817–831.

Dreyfus, Stuart E., *Dynamic Programming and the Calculus of Variations*, New York: Academic Press, 1965.

Driskill, R. H. and S., McCafferty, "Dynamic Duopoly with Adjustment Costs: A Differential Game Approach," *Journal of Economic Theory*, 49 (December 1989), 324–338.

El-Hodiri, M. A., E. Loehman, and A. Whinston, "Optimal Investment with Time Lags," *Econometrica*, 40 (November 1972), 1137–1146.

Elsgolc, L. E., *Calculus of Variations*, Reading, Mass.: Addison-Wesley, 1962.

Epstein, L. G., "The Le Chatelier Principle in Optimal Control Problems." *Journal of Economic Theory*, 19 (October 1978), 103–122.

Evans, G. C., "The Dynamics of Monopoly," *American Mathematics Monthly*, 31 (February 1924), 77–83.

Evans, G. C., *Mathematical Introduction to Economics*, New York: McGraw-Hill, 1930.

Feichtinger, G., "The Nash Solution to a Maintenance Production Differential Game," *European Journal of Operational Research*, 10 (1982), 165–172.

Feichtinger, G., "Optimal Policies for Two Firms in a Noncooperative R&D Race," in *Optimal Control Theory and Economic Analysis*, (G. Feichtinger, ed.), Amsterdam: North Holland, 1982, 373–397.

Feichtinger, G., "A Differential Game Solution to a Model of Competition Between a Thief and the Police," *Management Science*, 29 (1983), 686–699.

Feichtinger, G. and S. Jorgensen, "Differential Game Models in Management Science," *European Journal of Operational Research*, 14 (1983), 137–155.

Feichtinger, G. and R. F. Hartl, *Optimalle Kontrolle Okonomischer Prozesse*, Berlin: de Gruyter, 1986.

Fershtman, C., "Identification of Classes of Differential Games for Which the Open-Loop is a Degenerated Feedback Nash Equilibrium," *Journal of Optimization Theory and Applications*, 55 (1987), 217–231.

Fershtman, C. and M. I. Kamien, "Dynamic Duopolistic Competition with Sticky Prices," *Econometrica*, 55 (September 1987), 1151–1164.

Fershtman, C. and M. I. Kamien, "Turnpike Properties in a Finite-Horizon Differential Game: Dynamic Duopoly with Sticky Prices," *International Economic Review*, 31 (February 1990), 49–60.

Fershtman, C. and E. Muller, "Capital Accumulation Games of Infinite Duration," *Journal of Economic Theory*, 33 (1984), 322–339.

Fershtman, C. and E. Muller, "Turnpike Properties of Capital Accumulation Games," *Journal of Economic Theory*, 38 (1986), 167–177.

Fershtman, C. and E. Muller, "Capital Investment and Price Agreements in Semicollusive Markets," *The Rand Journal of Economics*, 17 (1986), 214–226.

Feynman, R. P., R. B. Leighton, and M. Sands, *The Feynman Lectures on Physics*, (Vol. II), Reading, Mass.: Addison-Wesley, 1964.

Fischer, Stanley, "The Demand for Index Bonds," *Journal of Political Economy*, 83 (June 1975), 509–534.

Fleming, Wendell H. and Raymond W. Rishel, *Deterministic and Stochastic Optimal Control*, New York: Springer-Verlag, 1975.

Forster, Bruce A., "Optimal Consumption Planning in a Polluted Environment," *Economic Record*, 49 (December 1973), 534–545.

Forster, Bruce A., "On a One State Variable Optimal Control Problem: Consumption-Pollution Trade-Offs," in *Applications of Control Theory to Economic Analysis* (J. D. Pitchford and S. J. Turnovsky, eds.), New York: North-Holland, 1977.

Friedman, A., *Differential Games*, New York: Wiley, 1971.

Gaines, R. E., "Existence of Solutions to Hamiltonian Dynamical Systems of Optimal Growth," *Journal of Economic Theory*, 12 (February 1976), 114–130.

Gaines, R. E., "Existence and the Cass-Shell Stability Condition for Hamiltonian Systems of Optimal Growth," *Journal of Economic Theory*, 15 (June 1977), 16–25.

Gelfand, I. M. and S. V. Fomin, *Calculus of Variations*, Englewood Cliffs, N.J.: Prentice-Hall, 1963.

Goldstein, H., *Classical Mechanics*, Reading, Mass.: Addison-Wesley, 1950.

Goldstine, H. H., *A History of the Calculus Variations*, New York: Springer-Verlag, 1980.

Gonedes, N. J. and Z. Lieber, "Production Planning for a Stochastic Demand Process," *Operations Research*, 22 (July/August 1974), 771–787.

Gould, J. P., "Diffusion Processes and Optimal Advertising Policy," in *Microeconomic Foundations of Employment and Inflation Theory*, (E. S. Phelps, ed.), New York: Norton, pp. 338–368.

Grossman, Michael, "On the Concept of Health Capital and the Demand for Health," *Journal of Political Economy*, 80 (March, 1972), 223–255.

Hadley, G., *Nonlinear and Dynamic Programming*, Reading, Mass.: Addison-Wesley, 1964.

Hadley, G. and M. C. Kemp, *Variational Methods in Economics*, New York: North-Holland, 1971.

Haley, W. J., "Human Capital: The Choice Between Investment and Income," *American Economic Review*, 63 (December 1973), 929–944.

Halkin, Hubert, "Necessary Conditions for Optimal Control Problems with Infinite Horizons," *Econometrica*, 42 (March 1974), 267–272.

Harris, Milton, "Optimal Planning under Transaction Costs: The Demand for Money and Other Assets," *Journal of Economic Theory*, 12 (April 1976), 298–314.

Hartl, R. F. and S. P. Sethi, "A Note on the Free Terminal Time Transversality Condition," *Zeitschrift fur Operations Research*, 27 (1983), 203–208.

Hartl, R. F., "Arrow-type Sufficient Optimality Conditions for Nondifferentiable Optimal Control Problems with State Constraints," *Applied Mathematics and Optimization*, 14 (1986), 229–247.

Hartl, R. F., "A Survey of the Optimality Conditions for Optimal Control Problems with State Variable Inequality Constraints," in *Operational Research*, (T. Brams, ed.), New York: Elsevier, 1984, 423–433.

Haurie, A., "Optimal Control on an Infinite Time Horizon: The Turnpike Approach," *Journal of Mathematical Economics*, 3 (March 1976), 81–102.

Heckman, James, "A Life Cycle Model of Earnings, Learning, and Consumption," *Journal of Political Economy*, 84, Part II (August 1976), 511–544.

Henderson, Dale W. and Stephen J. Turnovsky, "Optimal Macroeconomic Policy Adjustment under Conditions of Risk," *Journal of Economic Theory*, 4 (February 1972), 58–71.

Hestenes, Magnus R., *Calculus of Variations and Optimal Control Theory*, New York: Wiley, 1966.

Hochman, E. and O. Hochman, "On the Relations Between Demand Creation and Growth in a Monopolistic Firm," *European Economic Review*, 6 (January 1975), 17–38.

Hotelling, H., "The Economics of Exhaustible Resources," *Journal of Political Economy*, (April 1931), 137–175.

Howard, Ronald, "Dynamic Programming," *Management Science*, 12 (January 1966), 317–348.

Ijiri, Y. and G. L. Thompson, "Applications of Mathematical Control Theory to Accounting and Budgeting (The Continuous Wheat Trading Model)," *The Accounting Review*, 45 (April 1970), 246–258.

Ince, E. L., *Ordinary Differential Equations*, New York: Dover, 1956.

Intriligator, Michael D., *Mathematical Optimization and Economic Theory*, Englewood Cliffs, N.J.: Prentice-Hall, 1971.

Intriligator, Michael D. (ed.), *Frontiers of Quantitative Economics*, Vol. III-B, Amsterdam: North-Holland, 1977.

Isaacs, R., *Differential Games*, New York: Wiley, 1965.

Jacobson, D. H., M. M. Lele, and J. L. Speyer, "New Necessary Conditions of Optimality for Control Problems with State Variable Inequality Constraints," *Journal of Mathematical Analysis and Applications*, 35 (1971), 255–284.

Jacquemin, A. P. and J. Thisse, "Strategy of the Firm and Market Structure: An Application of Optimal Control Theory," in *Market Structure and Corporate Behavior* (K. Cowling, ed.), London: Gray-Mills, 1972, 61–84.

Jen, F. C. and L. Southwick, "Implications of Dynamic Monopoly Behavior," *American Economic Review*, 59 (March 1969), 149–158.

Jorgenson, D. W., "Technology and Decision Rules in the Theory of Investment Behavior," *Quarterly Journal of Economics*, 87 (November 1973), 523–542.

Kamien, M. I. and E. Muller, "Optimal Control with Integral State Equations," *Review of Economic Studies*, 43 (October 1976), 469–473.

Kamien, M. I. and N. L. Schwartz, "Expenditure Patterns for Risky R&D Projects," *Journal of Applied Probability*, 8 (March 1971a), 60–73.

Kamien, M. I. and N. L. Schwartz, "Optimal Maintenance and Sale Age for a Machine Subject to Failure," *Management Science*, 17 (April 1971b), B495–504.

Kamien, M. I. and N. L. Schwartz, "Limit Pricing with Uncertain Entry," *Econometrica*, 39 (May 1971c), 441–454.

Kamien, M. I. and N. L. Schwartz, "Sufficient Conditions in Optimal Control Theory," *Journal of Economic Theory*, 3 (June 1971d), 207–214.

Kamien, M. I. and N. L. Schwartz, "Risky R&D with Rivalry," *Annals of Economic and Social Measurement*, 3 (January 1974a), 267–277.

Kamien, M. I. and N. L. Schwartz, "Patent Life and R&D Rivalry," *American Economic Review*, 64 (March 1974b), 183–187.

Kamien, M. I. and N. L. Schwartz, "Product Durability under Monopoly and Competition," *Econometrica*, 42 (March 1974c), 289–301.

Kamien, M. I. and N. L. Schwartz, "Disaggregated Intertemporal Models with an Exhaustible Resource and Technical Advance," *Journal of Environmental Economics and Management*, 4 (1977a), 271–288.

Kamien, M. I. and N. L. Schwartz, "Optimal Capital Accumulation and Durable Goods Production," *Zeitschrift für Nationalokonomie*, 37 (1977b), 25–43.

Kamien, M. I. and N. L. Schwartz, "Optimal Exhaustible Resources Depletion with Endogenous Technical Change," *Review of Economic Studies*, 45 (February 1978a) 179-196.

Kamien, M. I. and N. L. Schwartz, "Self-Financing of an R&D Project," *American Economic Review*, 68 (June 1978b), 252-261.

Kamien, M. I. and N. L. Schwartz, "Potential Rivalry, Monopoly Profits, and the Pace of Inventive Activity," *Review of Economic Studies*, 45 (October 1978c), 547-557.

Kamien, M. I. and N. L. Schwartz, "Role of Common Property Resources in Optimal Planning Models with Exhaustible Resources," in *Restructuring the Economics of Natural Resources* (V. Kerry Smith and John V. Krutilla, eds.), Resources for the Future, 1981a.

Kamien, M. I. and N. L. Schwartz, "Technical Change Inclinations of a Resource Monopolist," in *Essays in Contemporary Fields of Economics*, (George Horwich and James P. Quirk, eds.), Purdue Research Foundation, 1981b.

Kamien, M. I. and N. L. Schwartz, *Market Structure and Innovation*, Cambridge: Cambridge University Press, 1982.

Kaplan, Wilfred, *Ordinary Differential Equations*, Reading, Mass.: Addison-Wesley, 1958.

Keller, Joseph B., "Optimal Velocity in a Race," *American Mathematical Monthly*, 81 (May 1974a), 474-480.

Keller, Joseph B., "Optimum Checking Schedules for Systems Subject to Random Failure," *Management Science*, 21 (November 1974b), 256-260.

Kemp, M. C. and N. V. Long, "Optimal Control Problems with Integrands Discontinuous with Respect to Time," *Economic Record*, 53 (September 1977), 405-420.

Koo, Dalia, *Elements of Optimization with Applications in Economics and Business*, New York: Springer-Verlag, 1977.

Kotowitz, Y. and F. Matthewson, "Informative Advertising and Welfare," *American Economic Review*, 69 (June 1979), 284-294.

Kupperman, Robert H. and Harvey A. Smith, "Optimal Sale of a Commodity Stockpile," *Management Science*, 16 (July 1970), 751-758.

Kurz, Mordecai, "The General Instability of a Class of Competitive Growth Processes," *Review of Economic Studies*, 35 (April 1968a), 155-174.

Kurz, Mordecai, "Optimal Economic Growth and Wealth Effects," *International Economic Review*, 9 (October 1968b), 348-357.

Kydland, F., "Noncooperative and Dominant Player Solutions in Discrete Dynamic Games," *International Economic Review*, 16 (June 1975), 321-335.

Lasdon, Leon, *Optimization Theory for Large Systems*, New York: Macmillan, 1970.

Lee, E. B. and L. Markus, *Foundations of Optimal Control Theory*, New York: Wiley, 1967.

Legey, L. M., Ripper, and P. Varaiya, "Effects of Congestion on the Shape of the City," *Journal of Economic Theory*, 6 (April 1973), 162-179.

Leland, Hayne E., "The Dynamics of a Revenue Maximizing Firm," *International Economic Review*, 13 (June 1972), 376-385.

Leung, Siu Fai, "Transversality Conditions and Optimality in a Class of Infinite Horizion Continuous Time Economic Models," *Journal of Economic Theory,* 54 (June 1991), 224–233.

Levhari, D. and N. Liviatan, "On Stability in the Saddlepoint Sense," *Journal of Economic Theory*, 4 (February 1972), 88–93.

Levhari, D. and L. J. Mirman, "The Great Fish War: An Example Using a Dynamic Cournot-Nash Solution," *Bell Journal of Economics*, 11 (1980), 322–334.

Levhari, D. and E. Sheshinksi, "Lifetime Excess Burden of a Tax," *Journal of Political Economy*, 80 (January/February 1972), 139–147.

Lewis, Tracy R. and Richard Schmalensee, "Nonconvexity and Optimal Exhaustion of Renewable Resources," *International Economic Review*, 18 (October 1977), 535–552.

Livesey, D. A., "Optimum City Size: A Minimum Congestion Cost Approach," *Journal of Economic Theory*, 6 (April 1973), 144–161.

Long, Ngo Van, "International Borrowing for Resource Extraction," *International Economic Review*, 15 (February 1974), 168–183.

Long, Ngo Van, "Resource Extraction Under the Uncertainty About Possible Nationalization," *Journal of Economic Theory*, 10 (February 1975), 42–53.

Long, Ngo Van and N. Vousden, "Optimal Control Theorems," in *Applications of Control Theory to Economic Analysis* (J. D. Pitchford and S. J. Turnovsky, eds.), New York: North-Holland, 1977.

Loury, Glenn C., "The Minimum Border Length Hypothesis Does Not Explain the Shape of Black Ghettos," *Journal of Urban Economics*, 5 (April 1978), 147–153.

Lucas, Robert E., Jr., "Optimal Investment Policy and the Flexible Accelerator," *International Economic Review*, 8 (February 1967), 78–85.

Luce, R. D. and H. Raiffa, *Games and Decisions*, New York: Wiley, 1957.

Lusky, Rafael, "Consumers' Preferences and Ecological Consciousness," *International Economic Review*, 16 (February 1975), 188–200.

Maccini, Louis, "On Optimal Delivery Lags," *Journal of Economic Theory*, 6 (April 1973), 107–125.

Magill, Michael J. P., "Some New Results on the Local Stability of the Process of Capital Accumulation," *Journal of Economic Theory*, 15 (June 1977a), 174–210.

Magill, Michael J. P., "The Origin of Cycling in Dynamic Economic Models Arising from Maximizing Behaviour," Center for Mathematical Studies in Economics and Management Science, Discussion Paper 296, Northwestern University, 1977b.

Malcomson, James M., "Replacement and the Rental Value of Capital Equipment Subject to Obsolescence," *Journal of Economic Theory*, 10 (February 1975), 24–41.

Malliaris, A. G. and W. A. Brock, *Stochastic Methods in Economics and Finance*, New York: North Holland, 1982.

Mangasarian, O. L., "Sufficient Conditions for the Optimal Control of Nonlinear Systems," *SIAM Journal Control*, 4 (February 1966), 139–152.

Mann, D. H., "Optimal Theoretical Advertising Stock Models: A Generalization Incorporating the Effects of Delayed Response from Promotional Expenditure," *Management Science*, 21 (March 1975), 823–832.

McIntrye, J. and B. Paieworsky, "On Optimal Control with Bounded State Variables," in *Advances in Control Systems* (C. T. Leondes, ed.), New York: Academic Press, 1967.

Mehlmann, A., *Applied Differential Games*, New York: Plenum, 1988.

Merton, Robert C., "Lifetime Portfolio Selection under Uncertainty: The Continuous-Time Case," *Review of Economics and Statistics*, 51 (August 1969), 247–257.

Merton, Robert C., "Optimal Consumption and Portfolio Rules in a Continuous Time Model," *Journal of Economic Theory*, 31 (December 1971), 373–413.

Michel, P., "On the Transversality Conditions in Infinite Horizon Optimal Control Problems," *Econometrica*, 50 (1982), 975–985.

Miller, R. E., *Dynamic Optimization and Economic Applications*, New York: McGraw-Hill, 1979.

Mills, E. S. and D. M. De Ferranti, "Market Choices and Optimum City Size," *American Economic Review*, 61 (May 1971), 340–345.

Mirrlees, J. A., "Optimum Growth When Technology is Changing," *Review of Economic Studies*, 34 (January 1967), 95–124.

Mirrlees, J. A., "An Exploration in the Theory of Optimum Income Taxation," *Review of Economic Studies*, 38 (April 1971), 175–208.

Mirrlees, J. A., "The Optimum Town," *Swedish Journal of Economics*, 74 (March 1972), 114–135.

Muller, E. and Y. C. Peles, "Optimal Dynamic Durability," *Journal of Economic Dynamics and Control*, 14 (1990), 709–719.

Mussa, Michael and Sherwin Rosen, "Monopoly and Product Quality," *Journal of Economic Theory*, 18 (August 1978), 301–317.

Myerson, R. B., *Game Theory: Analysis of Conflicts*, Cambridge: Harvard University Press, 1991.

Nash, J. F., "Equilibrium Points in n-person Games," *Proceedings of the National Academy of Sciences*, 36 (1950), 48–49.

Nemhauser, George L., *Introduction to Dynamic Programming*, New York: Wiley, 1966.

Nerlove, Marc and Kenneth J. Arrow, "Optimal Advertising Policy under Dynamic Conditions," *Economica*, 29 (May 1962), 129–142.

Nickell, Stephen, "On the Role of Expectations in the Pure Theory of Investment," *Review of Economic Studies*, 41 (January 1974), 1–19.

Ohm, J. Y., *Differential Game Approach to the Complementary Monopoly Problem* (unpublished Ph.D. dissertation), Evanston: Northwestern University, 1986.

Oniki, H., "Comparative Dynamics (Sensitivity Analysis) in Optimal Control Theory," *Journal of Economic Theory*, 6 (June 1973), 265–283.

Pekelman, Dov, "Simultaneous Price-Production Decisions," *Operations Research*, 22 (July/August 1974), 788–794.

Pekelman, Dov, "Production Smoothing with Fluctuating Price," *Management Science*, 21 (January 1975), 576–590.

Pekelman, Dov, "On Optimal Utilization of Production Processes," *Operations Research*, (March/April 1979), 260–278.

Peterson, D. W., "Economic Significance of Auxiliary Functions in Optimal Control," *International Economic Review*, 14 (February 1973), 234-252.

Phelps, E., *Golden Rules of Economic Growth*, New York: Norton, 1966.

Phelps, E., "Taxation of Wage Income for Economic Justice," *Quarterly Journal of Economics*, 87 (August 1973), 331-354.

Plourde, C. and D. Yeung, "Harvesting of a Transboundary Replenishable Fish Stock: A Noncooperative Game Solution," *Marine Resource Economics*, 6 (1989), 57-71.

Pontryagin, L. S., *Ordinary Differential Equations*, Reading, Mass.: Addison-Wesley, 1962.

Pontryagin, L. S., V. G. Boltyanksii, R. V. Gamkrelidze, and E. F. Mischchenko, *The Mathematical Theory of Optimal Processes*, New York: Wiley, 1962.

Ramsey, F., "A Mathematical Theory of Savings," *Economic Journal* (1928); reprinted in AEA, *Readings on Welfare Economics* (K. J. Arrow and T. Scitovsky, eds.), Homewood, Ill.: Richard D. Irwin, 1969.

Raviv, Artur, "The Design of an Optimal Insurance Policy," *American Economic Review*, 69 (March 1979), 84-96.

Reinganum, J. F., "Dynamic Games of Innovation," *Journal of Economic Theory*, 25 (1981), 21-41.

Reinganum, J. F., "A Dynamic Game of R&D: Patent Protection and Competitive Behavior," *Econometrica*, 50 (1982), 671-688.

Reinganum, J. F., "A Class of Differential Games for which the Open-loop and Closed-loop Equilibria Coincide," *Journal of Optimization Theory and Applications*, 36 (1982), 646-650.

Reinganum, J. F. and N. L. Stokey, "Oligopoly Extraction of a Common Property Resource: The Importance of the Period of Commitment in Dynamic Games," *International Economic Review*, 26 (1985), 161-173.

Reynolds, S., "Capacity Investment, Preemption and Commitment in an Infinite Horizon Model," *International Economic Review*, 28 (1987), 67-88.

Roberts, A. W. and D. E. Varberg, *Convex Functions*, New York: Academic Press, 1973.

Robinson, Bruce and Chet Lakhani, "Dynamic Price Models for New-Product Planning," *Management Science*, 21 (June 1975), 1113-1122.

Robson, A. J., "Sequential Exploitation of Uncertain Deposits of a Depletable Natural Resource," *Journal of Economic Theory*, 21 (August 1979), 88-110.

Robson, A. J., "Sufficiency of the Pontryagin Conditions for Optimal Control when the Time Horizon is Free," *Journal of Economic Theory*, 24 (June 1981), 437-445.

Rockafellar, R. Tyrrell, "Saddlepoints of Hamiltonian Systems in Convex Lagrange Problems Having a Nonzero Discount Rate," *Journal of Economic Theory*, 12 (February 1976), 71-113.

Roos, C. F., "A Mathematical Theory of Competition, " *American Journal of Mathematics*, 46 (1925), 163-175.

Roos, C. F., "A Dynamic Theory of Economics," *Journal of Political Economy*, 35 (1927), 632-656.

Ryder, Harl E., Jr. and Geoffrey M. Heal, "Optimal Growth with Intertemporally Dependent Preferences," *Review of Economic Studies*, 40 (January 1973), 1-31.

Ryder, Harl E., Jr., F. P. Stafford and P. E. Stephan, "Labor, Leisure and Training over the Life Cycle," *International Economic Review*, 17 (October 1976), 651-674.

Salop, S. C., "Wage Differentials in a Dynamic Theory of the Firm," *Journal of Economic Theory*, 6 (August 1973), 321-344.

Sampson, A. A., "A Model of Optimal Depletion of Renewable Resources," *Journal of Economic Theory*, 12 (April 1976), 315-324.

Samuelson, P. A., "Efficient Paths of Capital Accumulation in Terms of the Calculus of Variations," in *Mathematical Methods in the Social Sciences, 1959* (Kenneth J. Arrow, Samuel Karlin, and Patrick Suppes, eds.), Stanford: Stanford University Press, 1960, pp. 77-88.

Samuelson, P. A., "A Catenary Turnpike Theorem Involving Consumption and the Golden Rule," *American Economic Review*, 55 (1965), 486-496.

Samuelson, P. A., "The General Saddlepoint Property of Optimal-Control Motions," *Journal of Economic Theory*, 5 (August 1972), 102-120.

Samuelson, P. A. and R. M. Solow, "A Complete Capital Model Involving Heterogeneous Capital Goods," *Quarterly Journal of Economics*, 70 (November 1956), 537-562.

Sanchez, David A., *Ordinary Differential Equations and Stability Theory: An Introduction*, San Francisco: Freeman, 1968.

Sasieni, M. W., "Optimal Advertising Expenditure," *Management Science*, 18, Part II (December 1971), 64-71.

Scheinkman, Jose Alexandre, "Stability of Separable Hamiltonians and Investment Theory," *Review of Economic Studies*, 45 (October 1978), 559-570.

Schmalensee, Richard, "Resource Exploitation Theory and the Behavior of the Oil Cartel," *European Economic Review*, 7 (April 1976), 257-279.

Seater, John J., "Job Security and Vacancy Contacts," *American Economic Review*, 69 (June 1979), 411-419.

Seierstad, Atle, and Knut Sydsaeter, "Sufficient Conditions in Optimal Control Theory," *International Economic Review*, 18 (June 1977), 367-391.

Seierstad, Atle, and Knut Sydsaeter, *Optimal Control Theory with Economic Applications*, New York: North Holland, 1987.

Selten, R., "Reexamination of the Perfectness Concept for Equilibrium Points in Extensive Games," *International Journal of Game Theory*, 4 (1975), 25-55.

Sethi, Suresh P., "Optimal Control of the Vidale-Wolfe Advertising Model," *Operations Research*, 21 (July/August 1973), 998-1013.

Sethi, Suresh P., "Dynamic Optimal Control Models in Advertising: A Survey," *SIAM Review*, 19 (1977a), 685-725.

Sethi, Suresh P., "Nearest Feasible Paths in Optimal Control Problems: Theory, Examples, and Counterexamples," *Journal of Optimization Theory and Applications*, 23 (December 1977b), 563-579.

Sethi, Suresh P., "A Survey of Management Science Applications of the Deterministic Maximum Principle," *TIMS Studies in the Management Sciences*, 9 (1978), 33-67.

Sethi, Suresh P. and T. W. McGuire, "Optimal Skill Mix: An Application of the Maximum Principle for Systems and Retarded Controls," *Journal of Optimization Theory and Applications*, 23 (October 1977), 245–275.

Sethi, Suresh P. and Gerald L. Thompson, *Optimal Control Theory and Applications to Management Science*, Boston: Martinus Nijhoff, 1981.

Shell, Karl (ed.), *Essays on the Theory of Optimal Dynamic Growth*, Cambridge, Mass.: MIT Press, 1967.

Silberberg, E., "A Revision of Comparative Statistics Methodology in Economics, or How to do Comparative Statics on the Back of an Envelope," *Journal of Economic Theory*, 7 (1974), 159–172.

Silberberg, E., *The Structure of Economics: A Mathematical Analysis,* (2nd edition), New York: McGraw-Hill, 1990.

Simaan, M. and T. Takayama, "Game Theory Applied to Dynamic Duopoly Problems with Production Constraints," *Automatica*, 14 (1978), 161–166.

Smith, D. R., *Variational Methods in Optimization*, Englewood Cliffs, N.J.: Prentice-Hall, 1974.

Smith, Vernon L., "Economics of Production from Natural Resources," *American Economic Review*, 58 (June 1968), 409–431.

Smith, Vernon L., "An Optimistic Theory of Exhaustible Resources," *Journal of Economic Theory*, 9 (December 1974), 384–396.

Smith, Vernon L., "The Primitive Hunter Culture, Pleistocene Extinction, and the Rise of Agriculture," *Journal of Political Economy*, 83 (August 1975), 727–755.

Smith, Vernon L., "Control Theory Applied to Natural and Environmental Resources," *Journal of Environmental Economics and Management*, 4 (1977), 1–24.

Solow, R. M. and W. S. Vickery, "Land Use in a Long Narrow City," *Journal of Economic Theory*, 3 (December 1971), 430–447.

Southwick, L. and S. Zionts, "An Optimal-Control-Theory Approach to the Education-Investment Decision," *Operations Research*, 22 (November/December 1974), 1156–1174.

Spence, Michael and David Starrett, "Most Rapid Approach Paths in Accumulation Problems," *International Economic Review*, 16 (June 1975), 388–403.

Starr, A. W. and Y. C. Ho, "Nonzero-Sum Differential Games," *Journal of Optimization Theory and Applications*, 3 (1969), 184–208.

Steinberg, A. M. and H. C. Stalford, "On Existence of Optimal Controls," *Journal of Optimization Theory and Applications*, 11 (1973), 266–273.

Stokey, N. and R. E. Lucas, Jr., *Recursive Methods in Economics*, Cambridge, Mass.: Harvard University Press, 1989.

Strotz, R. H., "Myopia and Inconsistency in Dynamic Utility Maximization," *Review of Economic Studies*, 23 (1955/1956), 165–180.

Sweeney, J. L., "Housing Unit Maintenance and the Mode of Tenure," *Journal of Economic Theory*, 8 (June 1974), 111–138.

Takayama, Akira, *Mathematical Economics* (2nd Edition), Cambridge: Cambridge University Press, 1985.

Tapiero, Charles, "Optimum Advertising and Goodwill Under Uncertainty," *Operations Research*, 26 (May/June 1978), 450–463.

Taylor, James G., "Comments on a Multiplier Condition for Problems with State Variable Inequality Constraints," *IEEE Transactions on Automatic Control*, AC-17 (October 1972), 743–744.

Taylor, James G., "On Boundary Conditions for Adjoint Variables in Problems with State Variable Inequality Constraints," *IEEE Transactions on Automatic Control*, AC-19 (August 1974), 450–452.

Telser, L. G. and R. L. Graves, "Continuous and Discrete Time Approaches to a Maximization Problem," *Review of Economic Studies*, 35 (July 1968), 307–325.

Thompson, Gerald L., "Optimal Maintenance Policy and Sale Date of Machine," *Management Science*, 14 (May 1968), 543–550.

Tomiyama, K., "Two Stage Optimal Control Problems with Optimality Conditions," *Journal of Economic Dynamics and Control*, 9 (1985), 317–337.

Tomiyama, K. and R. J. Rossana, "Two Stage Optimal Control Problems with an Explicity Switch Point Dependence," *Journal of Economic Dynamics and Control*, 13 (1989), 319–338.

Treadway, A. B., "On Rational Entrepreneurial Behavior and the Demand for Investment," *Review of Economic Studies*, 36 (April 1969), 227–239.

Treadway, A. B., "Adjustment Costs and Variable Inputs in the Theory of the Competitive Firm," *Journal of Economic Theory*, 2 (December 1970), 329–347.

Tsutsui, S. and K. Mino, "Nonlinear Strategies in Dynamic Duopolistic Competition with Sticky Prices," *Journal of Economic Theory*, 52 (1990), 136–161.

Tu, Pierre N. V., *Introductory Optimization Dynamics*, New York: Springer-Verlag, 1984.

Uzawa, H., "Optimal Growth in a Two-Sector Model of Capital Accumulation," *Review of Economic Studies*, 31 (January 1964), 1–24.

Varaiya, P. P., *Notes on Optimization*, New York: Van Nostrand Reinhold, 1972.

Vind, Karl, "Control Systems with Jumps in the State Variables," *Econometrica*, 35 (April 1967), 273–277.

von Neumann, J. and O. Morgenstern, *Theory of Games and Economic Behavior*, Princeton: Princeton University Press, 1944.

von Rabenau, B., "Optimal Growth of a Factory Town," *Journal of Urban Economics*, 3 (April 1976), 97–112.

Vousden, Neil, "Basic Theoretical Issues of Resource Depletion," *Journal of Economic Theory*, 6 (April 1973), 126–143.

Vousden, Neil, "International Trade and Exhaustible Resources: A Theoretical Model," *International Economic Review*, 15 (February 1974), 149–167.

Yaari, M. E., "Uncertain Lifetime, Life Insurance, and the Theory of the Consumer," *Review of Economic Studies*, 32 (April 1965), 137–150.

Zangwill, Willard, *Nonlinear Programming*, Englewood Cliffs, N.J.: Prentice-Hall, 1969.

Author Index

R

S

T

V

W

Y

Z

Subject Index

D

A CATALOG OF SELECTED

DOVER BOOKS

IN SCIENCE AND MATHEMATICS

Mathematics–Algebra and Calculus

VECTOR CALCULUS, Peter Baxandall and Hans Liebeck. This introductory text offers a rigorous, comprehensive treatment. Classical theorems of vector calculus are amply illustrated with figures, worked examples, physical applications, and exercises with hints and answers. 1986 edition. 560pp. 5 3/8 x 8 1/2. 0-486-46620-5

ADVANCED CALCULUS: An Introduction to Classical Analysis, Louis Brand. A course in analysis that focuses on the functions of a real variable, this text introduces the basic concepts in their simplest setting and illustrates its teachings with numerous examples, theorems, and proofs. 1955 edition. 592pp. 5 3/8 x 8 1/2. 0-486-44548-8

ADVANCED CALCULUS, Avner Friedman. Intended for students who have already completed a one-year course in elementary calculus, this two-part treatment advances from functions of one variable to those of several variables. Solutions. 1971 edition. 432pp. 5 3/8 x 8 1/2. 0-486-45795-8

METHODS OF MATHEMATICS APPLIED TO CALCULUS, PROBABILITY, AND STATISTICS, Richard W. Hamming. This 4-part treatment begins with algebra and analytic geometry and proceeds to an exploration of the calculus of algebraic functions and transcendental functions and applications. 1985 edition. Includes 310 figures and 18 tables. 880pp. 6 1/2 x 9 1/4. 0-486-43945-3

BASIC ALGEBRA I: Second Edition, Nathan Jacobson. A classic text and standard reference for a generation, this volume covers all undergraduate algebra topics, including groups, rings, modules, Galois theory, polynomials, linear algebra, and associative algebra. 1985 edition. 528pp. 6 1/8 x 9 1/4. 0-486-47189-6

BASIC ALGEBRA II: Second Edition, Nathan Jacobson. This classic text and standard reference comprises all subjects of a first-year graduate-level course, including in-depth coverage of groups and polynomials and extensive use of categories and functors. 1989 edition. 704pp. 6 1/8 x 9 1/4. 0-486-47187-X

CALCULUS: An Intuitive and Physical Approach (Second Edition), Morris Kline. Application-oriented introduction relates the subject as closely as possible to science with explorations of the derivative; differentiation and integration of the powers of x; theorems on differentiation, antidifferentiation; the chain rule; trigonometric functions; more. Examples. 1967 edition. 960pp. 6 1/2 x 9 1/4. 0-486-40453-6

ABSTRACT ALGEBRA AND SOLUTION BY RADICALS, John E. Maxfield and Margaret W. Maxfield. Accessible advanced undergraduate-level text starts with groups, rings, fields, and polynomials and advances to Galois theory, radicals and roots of unity, and solution by radicals. Numerous examples, illustrations, exercises, appendixes. 1971 edition. 224pp. 6 1/8 x 9 1/4. 0-486-47723-1

AN INTRODUCTION TO THE THEORY OF LINEAR SPACES, Georgi E. Shilov. Translated by Richard A. Silverman. Introductory treatment offers a clear exposition of algebra, geometry, and analysis as parts of an integrated whole rather than separate subjects. Numerous examples illustrate many different fields, and problems include hints or answers. 1961 edition. 320pp. 5 3/8 x 8 1/2. 0-486-63070-6

LINEAR ALGEBRA, Georgi E. Shilov. Covers determinants, linear spaces, systems of linear equations, linear functions of a vector argument, coordinate transformations, the canonical form of the matrix of a linear operator, bilinear and quadratic forms, and more. 387pp. 5 3/8 x 8 1/2. 0-486-63518-X

Browse over 9,000 books at www.doverpublications.com